Stats with Cats

Stats with Cats

The Domesticated Guide to Statistics, Models, Graphs, and Other Breeds of Data Analysis

Charles Kufs

Stats with Cats: The Domesticated Guide to Statistics, Models, Graphs, and Other Breeds of Data Analysis

Published by Wheatmark®
610 East Delano Street, Suite 104
Tucson, Arizona 85705 U.S.A.
www.wheatmark.com

ISBN: 978-1-60494-472-3
LCCN: 2010928284

To my parents,
without whom I would not exist
in this or any other timeline.

Contents

Part 4: Statistical Foreplay

Part 5: A Model for Modeling

Part 6: Saving the World One Analysis at a Time

Acknowledgments

Thousands of candles can be lighted from a single candle, and the life of the candle will not be shortened. Happiness never decreases by being shared.

Buddha

I would like to thank the numerous friends, classmates, teachers, coworkers, bosses, clients, and professional buddies I've encountered over the years for the support and wisdom they have given to me. I would also like to acknowledge the cynical statistiphobes, the knuckleheaded number-junkers, the shoot-from-the-lip story sellers, the egotistical autocrats, the tyrannical reviewers, and the sanctimonious mudslingers who are also a part of a professional statistician's education and who, I gotta admit despite their being nagging pains in the posterior, provided the best stories.

I want to acknowledge my sister, Maryanne, and my brother, Bill, who have always been there for me when I needed support. I would especially like to thank all my children, both the ones with and the ones without fur, claws, and tails. They provided me indispensable support and enough pictures to fill the book. I would especially like to thank my English-speaking, can-opener-operating children—Lisa, Peter, Krista and Rob, and Miranda—for knowingly and unknowingly feeding me ideas, examples, and perspectives on statistics that have benefitted both the book and myself as a professional. I couldn't have survived without all you guys. And to my grandson Xander, may you find your way to all the people, places, and things that bring you happiness.

Finally, I would like to thank all the folks at Wheatmark, especially Grael Norton, for helping me get in the door, and Kat Gautreaux for helping me get *Stats with Cats* out the door. Let's do it again someday.

Onyx

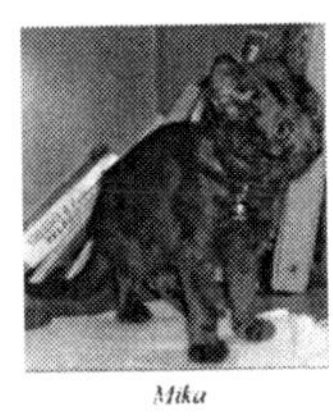

Mika

Tiger

Critter

Obsidian (Sid)

Kerpow

Preface
Welcome to Your Nightmare

There is terror in numbers. [...] Perhaps we suffer from a trauma induced by grade-school arithmetic.
Darrell Huff, *How to Lie with Statistics*, p. 60

You close your eyes, and suddenly you're back in school sitting through an interminably mind-numbing statistics class. You're not sure what language you're hearing, but it's not English. "Summadasquares." "Degreaserfreedom." You hear "Sigma, Alpha, Mu," but this is no frat party. Professor Zahlmeister drones on, "You must minimizen ze deviations und achieven Normality." You want to scream, but you cannot wake up. After all, it is Statistics 101.

Statistics is a branch of mathematics ... no, wait. Statistics is a branch of ... a branch of everything. Statistics examines properties of groups of numbers, whether they involve political polls and other opinion surveys, pharmaceutical effectiveness, sales and marketing, species counts, stock market trends, chemical reaction kinetics, batting averages, environmental contamination maps, elementary particle properties, consumer credit, and the list goes on and on and on.

Statistics is common to almost all fields of inquiry—social and natural sciences, sports, business, education, library and information science, and even music and art. Its popularity is attributable at least in part to its applicability to any type of data. If you can measure it, you can analyze it with statistics. If you're creative enough, you might even be able to analyze things you can't measure very well. Statistical methods can be used for analyzing data whether they are based on theories or natural laws or nothing in particular. Statistics are an integral part of everyday life in America. Without statistics, there would be no U.S. Census, IRS audits, Nielsen ratings of TV shows, or IQ and SAT tests. Baseball announcers would have nothing to talk about between pitches.

Ya Gotta Do What Ya Gotta Do

By three methods we may learn wisdom: First, by reflection, which is noblest; second, by imitation, which is easiest; and third by experience, which is the bitterest.
Confucius

Let's not kid ourselves. You're not reading this book because football season is over for the year and you have nothing better to do. You no doubt have a dataset that defies all the forms of analysis you're familiar with. You need a new tool for your data interpretation toolbox. You figure that statistics may be the magic bullet you need. But you're not a statistician. You barely made it through the one statistics class you took a decade ago. You're not going to sit through any more courses. Besides, your deadline was last week.

Then again, maybe you're a stressed-out graduate student faced with a pile of dissertation data. Your advisor, for whom none of your efforts are either too burdensome or good enough, wants you to make your analysis *more quantitative*. You need some advice, some *real* advice.

Or maybe you're a statistician looking for a few new tricks. Maybe you've been out of school for only a few years and are sometimes unsure of yourself—not the statistics part—all the other stuff like organizing projects, dealing with clients, and writing reports. You're not here to relearn statistics. You just want to see how you can use what you already know more effectively. And you need to have a way to decide when it's reasonable to ask for help and how to do it without looking inadequate.

The book needs more pictures of me.

So, whatever your reason for reading this book is, *ya gotta do what you gotta do*. If you already have a dataset burning out your brain cells faster than watching reruns of Paris Hilton on *The Simple Life*, start by skimming all the chapters. Here's what you'll find:

PART I. The Lost Treasures of Statistics 101

Part I is a focused review of some of the jargon and concepts you'll need to know to do your own analysis. Chapter 1, Reality Statistics, starts by explaining why number crunching is becoming so prevalent in our everyday lives. Chapter 2, Data Speak, describes the basic jargon of statistics, including data, samples, and variables. You'll hear a lot about these throughout the book. Chapter 3, Designer Datasets, explains matrices and file formats. Chapter 4, Hellbent on Measurement, describes the variety of measurement scales that can be used to characterize phenomena. Chapter 5, Catch an Error by the Tail, describes variance and why it is such a fundamental concept in statistics. Chapter 6, The Zen of Modeling, explains what models are and how statistics uses models to create models. Chapter 7, Assuming the Worst, describes five fundamental assumptions inherent in statistical inference and what happens when the assumptions are violated. Finally, Chapter 8, Perspectives on Objectives, describes the five broad goals a statistical analysis may have. Whether you completed your statistics education recently or years ago, Part I will refresh your memory about the basics you'll need to analyze data, and perhaps introduce a few concepts that may have eluded you in the past.

PART II. Frisky Business

No matter if you're going to build a condo for your cat or analyze some data, you'll need to have the skills, the tools, the materials, the plans, and the availability of expert help if you need it. That's what Part II is all about. Chapter 9, The Statistical Do-It-Yourselfer, helps you decide if you should do the data analysis yourself or get someone to do it for you. Chapter 10, Manage to Get It Right, describes how to set up a data analysis project so that it gets done right the first time. Chapter 11, Weapons of Math Production, is about the software and information sources you'll need to be aware of. Chapter 12, Tales of the Unprojected, describes problems that you may encounter on projects involving data. If you are a recent graduate involved in data analysis, you'll find Part II a bit depressing, but very useful.

PART III. Is That a Dataset in Your Pocket?

When you took Statistics 101, you worked with flawless little datasets cheerily provided by your instructor. Now you have to set up that data file for yourself. Part III will show you how. Chapter 13, In Search of ... Variables, provides a strategy for deciding what variables to measure and how to measure them. Chapter 14, Not-So-Simple Samples, explains how to select samples, including guidance on how to decide how many you'll need. Chapter 15, The Heart and Soul of Variance Control, describes how to recognize sources of variability and control them to improve your analysis. Finally, Chapter 16, Functional File Formats, provides a template for putting your samples and variables into a format that statistical software can analyze.

PART IV. Statistical Foreplay

You've heard the admonition *garbage in, garbage out.* Well before you actually start analyzing your data, you have to be sure your dataset isn't garbage. Chapter 17, Getting the Numbers Right, describes common types of data errors and how to find them. Chapter 18, Getting the Right Numbers, describes what you can do about duplicate data, missing data, censored data, and outliers. Chapter 19, Kicking the Data Tires, describes what to calculate, what to plot, and what to look for when you first explore your data. Chapter 20, Teaching Old Data New Tricks, describes six ways you can augment your dataset to make your analysis more thorough. You'll find you really need to do all these things before you try more complicated data analyses.

PART V. A Model for Modeling

Advanced data analyses involve using, building, and evaluating models. Chapter 21, Modelus Operandi, describes the process you go through to create a statistical model. Chapter 22, The Land beyond Statistics 101, describes advanced statistical analysis techniques you probably didn't hear much about in school. Chapter 23, Models and Sausages, discusses why even the most credible models can fail and what you might do about it.

PART VI. Saving the World One Analysis at a Time

Outside the utopia of college textbooks, you'll find that data analysis can be a very messy undertaking. Chapter 24, Grasping at Flaws, provides some ideas for how you can comment on a statistical analysis even if you don't know a lot about statistics. And finally, Chapter 25, The TerraByte Zone, provides some suggestions for how you can put the things you've learned to good use. There's also a glossary at the end so that you can look up any arcane terminology I might inadvertently drop.

Keep going. There's even better stuff on the inside.

So, whether you're a business person or a researcher or other professional who needs to conduct some statistical analyses, or supervise someone else who is conducting a statistical analysis, or review a statistical analysis done by someone else, this is a book you'll *want* to read. If you're a college student considering going into any of the subdisciplines of statistics as a career, or a graduate student who has to do a statistical analysis as part of a thesis or dissertation, this is a book you'll *need* to read. And if you're a veteran data analyst, well, this is a book that will look *really good* when displayed prominently on your desk. It will, without doubt, impress all your bosses and coworkers.

Why Cats?

If I had my choice of matter
I would rather be with cats
from "I'm a Man," by Jimmy Miller and Steve Winwood

I didn't just sit down and write this book. I've been writing it in bits and pieces for over a decade. I wanted to bridge the gap between the scores of introductory textbooks on statistics and the real world of solving problems with messy data. I wanted to provide an uber-practical guide for the hesitant do-it-yourselfers and the untested statistician-wannabes. For that reason, I ended up throwing out the equivalent of two books to get to this book. My cats were thrilled to help me with the shredding part. The material I scrapped wasn't bad, just... conventional. I decided I didn't want to trudge down that too-well-beaten path.

No. No. No. This all has to be rewritten.

Academic books on data analysis are often like cement—heavy, hard, dense, impenetrable. The contents of this book are unlike what you'll find in introductory textbooks on statistics. This book doesn't explain probability, the central limit theorem, hypothesis testing, and other fundamental statistical topics. You won't find a lot of equations or descriptions of esoteric statistical tests you have to program or calculate by hand. You can find that information in a hundred other books. What you will find are topics, like data scrubbing, minimizing variance, model building, and critiquing statistical reports, which you'll need to complete your own statistical analyses. This book will complement any other textbook on statistics you choose to read. Think of *Stats with Cats* as a textbook for Statistics 101.5. Read it while you're taking Statistics 101 or after you've completed the course, and it will help you place pieces of the puzzle that may have eluded you. It will take you from statistics in the classroom to statistics on the job.

With a few exceptions, I've avoided referencing scholarly works because they only confuse and frustrate most novices. If you want the actual articles, you can find them in most college libraries. You may also be able to purchase them on the Internet. A few clicks on a search engine ought to deliver. I *have* cited a few of the more reader-friendly books and websites that I like, but don't limit your searches to these. Your tastes may be different from mine.

Following the words of Samuel Johnson, I tried "*to make new things familiar and familiar things new.*" I tried to facilitate comprehension by using graphics, examples, stories, quotes, and perhaps way too many analogies. I tried to put in a little bit of humor. Hopefully, you'll recognize it when you see it. If you like reading between the lines, there are a few cultural references for you to find. Some are a bit obscure, so it'll be a good opportunity to learn some things besides statistics. There are also a few songs. Sing these out loud, especially if you are reading the book in a library or on public transportation. Feel free to write your own lyrics, too. Finally, I included pictures of my cats. Publishers claim that sales of books with pictures of cats are substantially higher than sales of books without pictures of cats. I thought it was a hypothesis worth testing, hence, *Stats with Cats.*

So I hope you will be able to find some educational and entertainment value in this book, or failing that, be content with how *really good* it looks on your bookshelf.

Charlie Kufs
AKA TerraByte
Willow Grove, Pennsylvania

PART 1

The Lost Treasures of Statistics 101

If you took an introductory course in statistics, you probably learned about descriptive statistics, graphing, theoretical frequency distributions, like the Normal distribution, correlation and regression, and hypothesis testing. Part I of *Stats with Cats* focuses on information you'll need to put your Stats 101 knowledge to work. The eight chapters in Part I are:

Chapter 1—Reality Statistics, why you should consider analyzing data on your job (and in your life).

Chapter 2—Data Speak, some of the basic jargon of applied statistics

Chapter 3—Designer Datasets, explains matrices and file formats.

Chapter 4—Hellbent on Measurement, describes the variety of measurement scales that can be used to characterize phenomena.

Chapter 5—Catch an Error by the Tail, describes variance and why it is such a fundamental concept in statistics.

Chapter 6—The Zen of Modeling, explains what models are and how statistics uses models to create models.

Chapter 7—Assuming the Worst, describes five fundamental assumptions inherent in statistical inference and what happens when the assumptions are violated.

Chapter 8—Perspectives on Objectives, describes the five broad goals a statistical analysis may have.

Whether you received your statistics education recently or years ago, Part I will refresh your memory about the basics you'll need to analyze data, and perhaps introduce a few concepts that may have eluded you in the past.

1 Reality Statistics

Everybody who has completed high school has learned some forms of statistics. Your class grades were *averages* of scores you received for tests and other efforts. If you were graded on a curve, you were exposed to the Normal distribution, standard deviations, and confidence limits. Scores on standardized tests, like the SAT, were presented in percentiles. You probably learned how to create scatter plots, pie and bar charts, and maybe other ways to display data. If you took some math, you might even have learned about equations for a straight line and some elementary curves. So, you've been exposed to at least enough statistics to read *USA Today* or the *Wall Street Journal*, and that's the most that many people would do with statistics were it not for the evolution of the computer.

Times Change; Statistics Transform

When I took office {1993}, only high energy physicists had ever heard of what is called the World Wide Web.... Now {1996} even my cat has its own web page.
Bill Clinton, former United States president

During the 1970s, statistical analyses were done on mainframe computers that were as big as refrigerators and even cars. They were sequestered in their own climate-controlled quarters, waited on command and reboot by a priesthood of system operators. In contrast, personal computers (PCs) were like mammals during the Jurassic period, hiding in protected niches while the mainframe dinosaurs ruled. Statisticians wrote their own programs, either in a standalone programming language like FORTRAN or COBOL, or in the language of one of the few commercially available statistical software packages, like SAS or SPSS. There were no GUIs (Graphical User Interfaces). The statistical packages were easier to use than the programming languages, but they were still complicated and expensive mainframe programs. Only the government, universities, and major corporations could afford their annual licenses.

To conduct a statistical analysis, first you had to write a program. Then you had to wait in line for an available keypunch machine so that you could transfer your program code and all your data onto 3¼ by 7⅜ inch computer punch cards. After that, you waited so that you could feed the cards through the mechanical card reader. Finally, you waited for the mainframe to run your program and the printer to output your results. When you picked up your output from the priesthood who tended the sacred processing units, sometimes all you got was a page of error codes. You had to decide what to do next and start the process all over again. Life wasn't slower back then, it just required more waiting.

A lot has changed since then. Punch cards and their supporting machinery are extinct. Mainframes are an endangered species, having been exiled to specialty niches by PCs. Inexpensive statistical packages that run on PCs, on the other hand, have multiplied like rabbits. All of these packages have GUIs. Even the venerable ancients, SAS and SPSS, have evolved point-and-click faces (although you can still write code if you want). Now you can run even the most complex statistical analysis in less time than it takes to drink a cup of coffee.

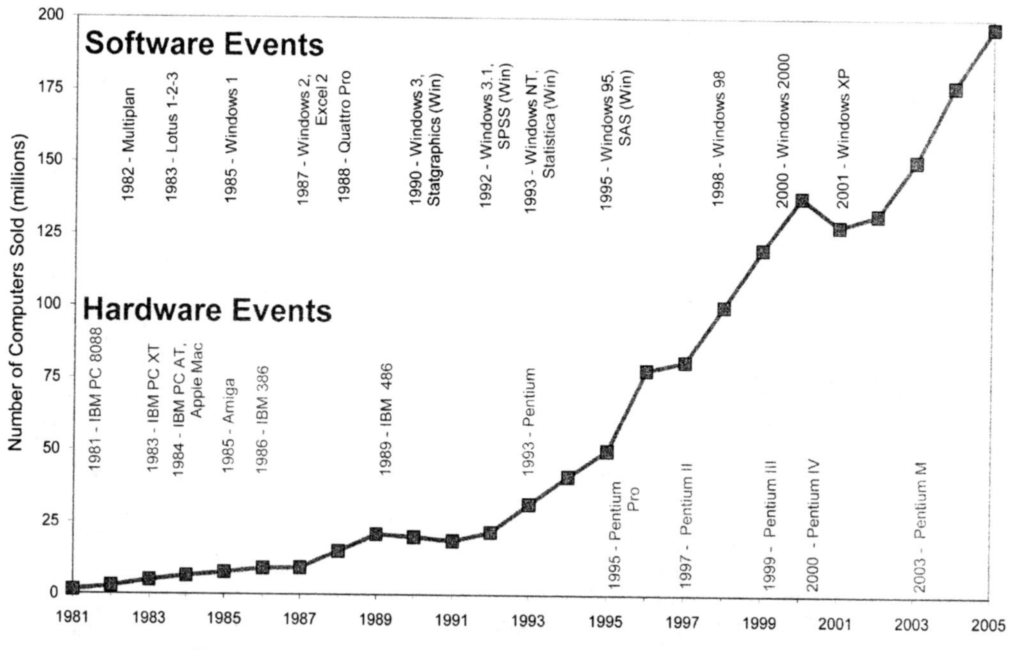

Figure 1. Numbers of Personal Computers Sold from 1981 to 2005.

So what could have spawned all these changes? It was the PC, the rodent of the Jurassic 1970s. Before 1974, most PCs were built by hobbyists from kits. The MITS Altair is generally acknowledged as the first personal computer, although there are more than a few other claimants.[1] In 1975, MITS sold about 6,000 Altairs. In 1980, almost a million PCs were sold. Then in 1981, IBM introduced their 8088 PC. The number of PCs

1 As with biological organisms, it's sometimes difficult to say when a new technological species originated. What differentiates a calculator from a microcomputer from a personal computer? Digital electronics was developed in the 1930s and 1940s. By the 1950s, plans and kits for microcomputers, some analog rather than digital, were available from several companies. The MITS Altair was probably the first PC to be produced in quantity. It used a new version of BASIC from the startup company Micro-Soft. And though MITS and the Altair are gone, Microsoft has survived and it's the survivors who write the history books. See www.computerhope.com/issues/ch000984.htm and www.blinkenlights.com/pc.shtml for an interesting look back.

sold in a year (Figure 1)[2] increased from about 2 million in 1981 to almost 200 million in 2005. From the early 1990s, sales of PCs have been fueled by Pentium-speed, GUIs, the Internet, and affordable, user-friendly software, including spreadsheets with statistical functions.

The maturation of the Internet, in particular, has led to unlimited opportunities. You no longer have to have access to a huge library of books to do a statistical analysis. There are thousands of websites with reference materials for statistics. Instead of purchasing one expensive reference, you can now consult a dozen different discussions on the same topic, free. If you find a book you want to keep as a handy reference, you can buy electronic access to it. No dead trees will clutter your office. If you can't find a reference book with what you need, there are discussion groups where you can post your questions. Perhaps most importantly, though, data that would have been difficult or impossible to obtain a decade ago are now just a few mouse clicks away. It's almost as if some great intelligence designed things to happen this way.

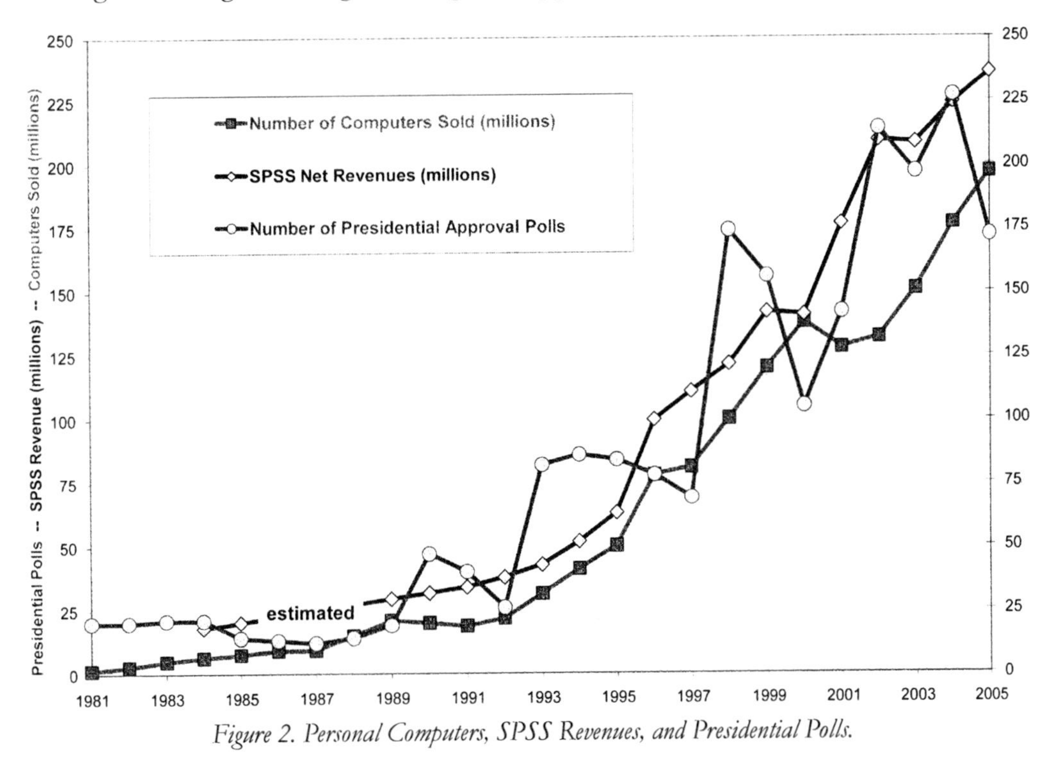

Figure 2. Personal Computers, SPSS Revenues, and Presidential Polls.

So, with computer sales skyrocketing and the Internet becoming as addictive as crack, it's not surprising that the use of statistics might also be on the increase. Consider the trends shown in Figure 2.[3] As in Figure 1, the squares represent the number of computers sold from 1981 to 2005. The diamonds, which follow a trend similar to computer sales, represent revenues for SPSS, Inc., the makers of the software formerly known as

2 Data for Figures 1 and 2 came from http://arstechnica.com/old/content/2005/12/total-share.ars.

3 Additional data for Figure 2 came from 137.99.36.203/CFIDE/roper/presidential/webroot/presidential_rating.cfm and spss.com.

Statistical Package for the Social Sciences. So, sales of at least one of the major pieces of statistical software have also grown substantially over the past decade. They probably all have.

With the availability of more computers and more statistical software, you might expect that there may be more statistical analyses being done. That's a tough trend to quantify, but consider this. The circles in Figure 2, representing the number of polls conducted on presidential approval, follow a trend similar to the trends for computer sales and SPSS revenues.

Table 1. Number of Presidential Polls and Pollsters from January 1969 to April 2007

President		Weeks of Polls	Number of Polls	Polls per Month	Number of Polling Groups	Most Nationwide Presidential Polls Conducted
G. W. Bush	Second Term *	116	408	14.0	33	Gallup/CNN/USA (14%) Fox/OpinDynamics (10%)
	First Term	208	782	15.0	41	Gallup/CNN/USA (18%) Fox/OpinDynamics (10%)
Clinton	Second Term	207	506	9.8	27	Gallup/CNN/USA (20%) CBS/NYT (10%)
	First Term	209	331	6.3	5	Gallup/CNN/USA (33%) CBS/NYT (22%)
G. H. W. Bush		207	135	2.6	1	Gallup (100%)
Reagan	Second Term	205	54	1.0	1	Gallup (100%)
	First Term	207	83	1.6	1	Gallup (100%)
Carter		202	92	1.8	1	Gallup (100%)
Ford		122	37	1.2	1	Gallup (100%)
Nixon *		289	97	1.3	1	Gallup (100%)

* Partial term in office

Data Source: 137.99.36.203/CFIDE/roper/presidential/webroot/presidential_rating.cfm

Table 1 summarizes increases in the numbers of polls and pollsters since the 1960s. Before 1988, there were on average only one or two presidential approval polls conducted per month. Within a decade, that number had increased to more than a dozen. Perhaps even more revealing is the increase in the number of pollsters. Before 1990, the Gallup Organization was pretty much the only organization conducting presidential approval polls. Now, there are several dozen. These pollsters don't just ask about presidential approval, either. There are a plethora of polls for every issue of real importance and most of the issues of contrived importance to society. Many of these polls are repeated to look for changes in opinions over time, between locations, and for different demographics. And that's just political polls. There has been the same increase in polling for marketing, product development, and other business applications. Even without including nonprofessional polls conducted on the Internet, the growth of polling has been exponential. So, there should be no doubt that there are many more statistical analyses being done today than even a decade ago.

Times change. There are no more elevator operators because untrained riders can just press a button and the doors close automatically. Gas station attendants, store cashiers, and bank tellers are being replaced by self-service mechanisms. Professional television actors and writers are being replaced by wannabes on reality TV shows like *American Idol* and *Survivor*. So it should come as no surprise that some people who can't program the clock on their microwave will think nothing of doing all kinds of data analyses. Statistics are no longer the exclusive domain of professionals. If there can be reality TV, why not reality statistics too?

Safety in Numbers

I am not going to give you a number for it because it's not my business to do intelligent work.

Donald Rumsfeld, two-time former secretary of defense

While campaigning for the Republican nomination for president in 1976, Ronald Reagan told a story about a "welfare queen" who used eighty names, thirty addresses, and twelve Social Security numbers to fraudulently collect over $150,000 of government assistance. The story was apocryphal but resonated with the press and Americans eager for stereotypes and sound bites.

Statistics are antidotes to anecdotes. Politicians, preachers, and parents can get away with telling tales to illustrate points they want to make because their followers trust them and want to believe them whether they are telling the truth or not. Other professionals, though, can't rely on their audience having such unquestioning faith. Scientists need hard data to dispel their disbelief, not just whimsical stories. Businessmen want to see the numbers before they spend their money.

Just about every profession there is relies on statistics to some extent.[4] Sure, professionals tell their fair share of anecdotes, but if they're honest, only to back up their

4 OK, maybe not the oldest one.

numbers. And for most professions, if they're going to use numbers, they're pretty much stuck with statistics. There are few other quantitative ways to analyze data in business and the social sciences. Natural scientists and engineers, on the other hand, have lots of alternatives, like deterministic models.[5]

Deterministic models are based on established principles and are assumed to produce answers that are correct with little or no variation. In a deterministic approach, established theories are used to construct a general model. Data are then collected for the model on a specific instance to which the model is being applied. Then the model's predictions are compared to observations of actual conditions. Take for example, a groundwater flow model. Groundwater elevations and aquifer permeabilities are measured in the field and used in a specific equation[6] (i.e., model) to calculate average rates of groundwater movement. If the flow rates aren't reasonable based on some in-field observations, some data inputs are adjusted to produce results that do seem reasonable. This is called model calibration. The assumption is that the model is correct while the data inputs to the model may be somewhat in error.

But there are situations in which you may choose not to use a deterministic model. For one, there may not be any model (i.e., equation) available to guide your analysis of the data. This is a big issue with the social and behavioral sciences. Or there may be too many intervening factors to enable the successful application of an established model. Biochemical models might be effective in a petri dish, for example, but fall apart when applied to lab rats. Finally, you may not have the expertise or resources (e.g., software) to use a particular deterministic model.

In contrast, statistical approaches do not start with specific mathematical models. Data are collected on what are thought to be important characteristics of a phenomenon, and then a model is developed *from* the data. The assumption is that the data are pretty much *correct* (more on what this means later) and the model may be somewhat in error. If the model's predictions are reasonable based on some independent observations, the model is considered *valid* (more on what this means later too).

Why Statistics?

It is a mistake to think you can solve any major problems just with potatoes.
Douglas Adams, *Life, the Universe, and Everything*, 1982

If you were required to take statistics in school, you probably asked your advisor, "Why do I have to take statistics? Why would I ever need to use statistics?" Here are a few of the reasons.

---

5 A model is just a representation of an object or phenomenon with a surrogate that can be manipulated more easily. A model can either be a physical representation, like model planes used in wind tunnels, or a mathematical representation, an equation. It is unlikely that *America's Next Top Model* will be an equation.

6 Average groundwater flow velocity equals hydraulic gradient times hydraulic conductivity divided by effective porosity. Feel free to forget I mentioned this.

- **Statistics provide a starting point and a course of action**—If you're in the natural sciences, you'll probably have some basic principles, laws, or at least theories to start with in analyzing data. Even some of those were discovered or verified by statistical observation. If you're in the social sciences, business, economics, or most other fields, though, you're got little to go on besides statistics. Anecdotes aren't worth much. Statistics gives you a place to start by having you focus on the population, so you know what to sample, and the phenomenon, so you know what to measure and how to measure it. Once you have laid this groundwork, statistics provides a variety of methods to analyze the data.
- **Statistics give you more ways to analyze data**— Statistics is a colossal workshop with more tools than you could ever use in a career. Statistics allows you to describe, correlate, detect differences, group, separate, reorganize, identify, predict, smooth, and model. And it's not just the variety of tools for doing different things; there are also many tools for doing the same thing in different ways. Want to find the center of a data distribution? You can use the arithmetic mean, the geometric mean, the harmonic mean, trimmed and winsorized means, weighted means, the median, the trimean, or the mode. Each has its own special use, like the variety of types of screwdrivers used by a mechanic. With a statistician's toolbox, you can gain far more insight from your data than you might from any other type of analysis.
- **Statistics examine both accuracy and precision**—Any marksman will tell you that it's not enough to be able to hit a target. You have to be able to hit it where you aim and do it consistency. That's accuracy and precision. Many data analysis techniques focus on accuracy and forget all about precision. But variability, uncertainty, and risk don't go away by just ignoring them. Statistics is all about understanding variability.
- **Statistics examine both trends and anomalies**—Most forms of analysis focus on finding similarities and patterns in data. Statistics, in particular, can be used to find linear and nonlinear trends, cycles, steps, shocks, clusters, and many other types of patterns. What's more, statistics can be used to identify and explore divergent or anomalous cases, which don't fit general patterns. Sometimes it is these outliers rather than the trends that reveal the information most crucial in an analysis.
- **Statistics tell you how much information you need**—In data analysis, more is not always better. It's not unusual to have too much data to make sense of using only graphs and tables. Statistics provides a variety of ways to help you decide about how many samples you need to achieve a certain objective. Statistics provides ways to judge the quality of the data and compensate for misleading variability. Statistics can also tell you if your data are redundant, and if so, provide ways to reassemble the data more efficiently.
- **Statistics provide standardization**—You can usually convince people who are reviewing your work that your data analysis is legitimate because it uses well-

known, professionally accepted, statistical procedures. Likewise, it's easier to use statistics as the basis for any standardized procedures you specify that others use because most people know some statistics. For example, Government regulations frequently require the use of statistics to report and analyze datasets, such as crime rates, pharmaceutical effectiveness, environmental impact, occupational safety, public health, and educational testing.

The catvalry arrives in time to save the project.

So you see, statistics has a lot to offer you, whether there is a strong theoretical basis to your field of practice or not. That's why your advisors want you to learn about it.

But as you can imagine, there's a catch with using a statistical approach. There are two, in fact. First, you have to have real data, usually a lot of them, to develop and apply a statistical model. You can't take a few reference values from the literature or make up estimates based on your experience the way you can with some deterministic models. Second, even after putting great effort into collecting the right data for a statistical model, you may find that the model doesn't work very well. It's a valid answer; it means you have to explore other types of measurements or other approaches to modeling the phenomenon (if it can be modeled at all). Unfortunately, most bosses, clients, academic advisors, and journal editors don't appreciate the value of such findings.

2 Data Speak

Whether you plan to do a statistical analysis yourself or contract with a professional, you'll need to know a few basic terms and concepts. That's the purpose of this chapter. Once you learn about data, samples, variables, metadata, matrices, and data files, you'll be able to *technobabble* with the best of 'em.

Data, Information, Knowledge, and Wisdom[1]

Information is not knowledge. Knowledge is not wisdom. Wisdom is not truth. Truth is not beauty. Beauty is not love. Love is not music. Music is the best.

From "Packard Goose" by Frank Zappa, 1979

To most laymen, *data*[2] refer to *information*. Many people use the two terms synonymously. Consult a dictionary, and you might find that the word *data* means "factual information," or sometimes "factual information derived from research," and occasionally "factual information derived from research that has been prepared for analysis." But really, data are just facts, unadorned numbers or text. They are like atoms. There are many types, and they exist alone or in combinations with others, but they are only building blocks.

Information is data in context. Information has two parts—what the data are and what the data are about. Facts about data, such as units of measurement, are called metadata. Information is the combination of data and the associated metadata. If data are like atoms, information is like chemical compounds, composed of many atoms yet still considered a building block for something more complex.

Data analysis, actually information analysis, is the process of examining data and metadata to identify patterns and relationships. Through this process, information becomes knowledge. Knowledge is an appreciation of what a collection of information

1 There are several notions of the progression from data to understanding or wisdom. Search the Internet for these terms to learn more.

2 A single piece of information is a datum. Saying data with a plural verb may sound wrong, but it's not. If the phrase "what do the data mean" sounds dissonant to you, just think "data points" whenever you see the word "data." I've been trying to get this right my entire professional life to no avail.

means, in other words, *how* the facts are interrelated. If you combine information in just the right way, you can create something greater, like chemical compounds put together in the right way can create a cell with a nucleus in which chromosomes and mitochondria correspond to pieces of related knowledge.

As a data analyst, your mission is to convert data into knowledge. Then, combined with other knowledge, concepts of *why* data display patterns and relationships can be synthesized. Understanding the fundamental principles governing knowledge is wisdom. Wisdom is like an organism composed of cells (knowledges) made from chemical compounds (information) comprised of atoms (data).

Wisdom is what you need to make informed decisions from data. Decisions made with the knowledge of an analysis but without the experience and synthesis that makes wisdom will be less informed. Likewise, making a decision based on even an overload of information will be less informed than a decision based on the knowledge that comes from a sound analysis of the information. If you are a decision maker, you need to understand the distinction between information, knowledge, and wisdom so that you can also account for uncertainties.

Wisdom is what you need to make informed decisions from data. Decisions made with the knowledge of an analysis but without the experience and synthesis that makes wisdom will be less informed. Likewise, making a decision based on even an overload of information will be less informed than a decision based on the knowledge that comes from a sound analysis of the information. If you are a decision maker, you need to understand the distinction between information, knowledge, and wisdom so that you can also account for uncertainties. Wisdom is what you need to make informed decisions from data. Decisions made with the knowledge of an analysis but without the experience and synthesis that makes wisdom will be less informed. Likewise, making a decision based on even an overload of information will be less informed than a decision based on the knowledge that comes from a sound analysis of the information. If you are a decision maker, you need to understand the distinction between information, knowledge, and wisdom so that you can also account for uncertainties.

Samples

> *The proof of the pudding is in the eating. By a small sample we may judge the whole piece.*
>
> *Don Quixote de la Mancha,* Part I, Book I, ch. 4
> by Miguel de Cervantes Saavedra

A sample is a part of a bigger entity. A *good* sample is a part of a bigger entity and has all the same properties as the bigger entity (at least the properties you consider to be important). Such a sample is also called a representative sample. Think of going to a food market that offers free samples to encourage you to buy their products. Say the market was offering a sample of pizza. The sample should include crust, sauce, and toppings such as cheese and mushrooms. If you were given only a piece of the crust, you wouldn't have a very good idea of what the entire pizza tastes like. That sample (i.e., the crust) wouldn't have the same properties (i.e., sauce, toppings) as the bigger entity (i.e., the pizza), so it would not be representative of the pizza. In other words, it wouldn't be a good sample. Good samples are representative of the larger entity from which they are drawn. This is perhaps the most underemphasized concept in statistics. You can't make an inference from a small group of samples to the population from which the samples were taken if the samples are not representative of the population (at least on the important properties).

Can I have tuna on my pizza?

Now for the really confusing part. The term *sample* has more than one meaning in statistics. A sample can be a single entity on which information is collected. Such an individual sample is also called an observation, subject, record, case, individual, or what the entity is, such as patient or student. Think of this meaning of *sample* as *example*. In contrast, a collection of such individual samples is also called a sample because it is part of a larger collection of entities called a population. Whether the term *sample* refers to an individual entity or a collection of entities is usually taken from the context in which it is used.

Representativeness is a fundamental characteristic of a sample. An individual sample must fairly represent the attributes of the population it is part of or else it is considered a statistical anomaly. A sample-of-a-population must fairly represent the larger population or else statistics calculated from the sample will not fairly characterize the population. So a sample can be an individual item in a larger sample, which in turn, is part of a larger population of individuals.

Say you were to participate in a statistical study of an herbal supplement. The objective of the study might be to determine if the supplement provided some benefit for weight control. In the study, you would be considered to be a sample or a patient, selected from a larger statistical population, namely, the population of persons who might take the supplement. For you to be considered a good sample of the population, you would probably have to be overweight, but it might not matter what your eye color, tax bracket, or religion were.

Confused? Don't panic. If you can handle the distinctions between to, too, tutu, and two, you'll be able to handle the distinction between sample and sample.

Variables

> Spock: *I've had to program some of the variables from memory.*
> Kirk: *What are some of the variables?*
> Spock: *Availability of fuel components, mass of a vessel through a time continuum, and probable location of humpback whales, in this case, the Pacific Basin.*
>
> *Star Trek IV: The Voyage Home*

Variables contain the pieces of information you collect from or about each of your samples. Variable values change from sample to sample. That's why they're called variables rather than constants. If the information weren't different, there would be no variability and no need for statistics.

Variables used in statistical analyses, sometimes called attributes, must have some bearing on the problem. For example, your SAT score might be a piece of data unique to you that could be analyzed, but it would probably have little relevance to the hypothetical study on herbal supplement mentioned previously. Instead, the experimenter might record information about you, such as your age, weight, and sex. These pieces of information would be considered to be variables or attributes. Then, after taking the supplement for some period of time, you might have your weight measured again, your blood pressure might be tested, a sample of your blood might be taken for analysis, and you might be asked if you feel any benefit from the supplement. These pieces of information would also be considered to be variables.

So, variables are types of information, samples are sources of information, and data are the actual pieces of information. Why is this important, you ask? Here's why. All statistical calculations begin with a matrix. A matrix is nothing more than a rectangular array of numbers arranged into rows and columns. To do a statistical analysis, you must arrange your matrix so that each row represents a different sample, each column represents a different variable, and the elements of the matrix are data. If you don't understand the difference between samples, variables, and data, you won't be able to set up your matrix properly for statistical analysis. Chapter 3 explains this in more detail.

Metadata

To a collector of curios, the dust is metadata.

David Weinberger in *Everything Is Miscellaneous*

You can't have good data unless you have good metadata. Metadata are data about data. Metadata consist of information about the identity of a sample, or its relationship to the population, or how a variable value was generated, or how reliable the data are. For example, metadata may include:

- **Identification of Sample**—A unique identifier such as a sample number, social security number, credit card number, or other designation that can be used to track data back to their source. Location (e.g., address, coordinates) and time (e.g., date of sample collection) can also be used for identifying samples.
- **Relation of Sample to Population**—Information about sample characteristics that are thought to be important attributes of the population, such as a patient's sex or a consumer's economic class. This type of metadata is sometimes included in statistical analyses.
- **How Data Were Generated**—Information about the instrument and process that were used to create the data, such as the sampler (i.e., the person doing the

data collection). Sometimes this information refers to a source of data like the U.S. Census or the General Social Survey (GSS).

- **Quality of Data**—Information on issues that might affect the quality of the data, such as meter calibration, data units, data validation flags, and general comments.

Metadata are sometimes stored alongside data in a dataset, especially if the metadata are different for different samples. If the information is the same for all samples, metadata are usually provided separately from the data.

It's All Greek

When I use a word, it means just what I choose it to mean, neither more nor less.
Humpty Dumpty, in *Through the Looking Glass* by Lewis Carroll

With the exception of a few terms introduced in later chapters, you now know the fundamental jargon you'll need to start your analysis. But, there's still a superabundance of esoteric statistical slang that you might hear from a statistician you hire to help you do your analysis. There's an even greater chance you'll run into *statspeak* when you start reading websites, books, and worst of all, journals on topics in statistics. You'll see what I mean if you read some of the article titles in the Journal of the American Statistical Association (at www.amstat.org/).

To simplify statistical jargon, think of three distinctions:

- **Named Things**—Statistical procedures, especially statistical tests, are often modified to accommodate some special circumstance or to have some desirable property. When this occurs, the new procedure is commonly named after the originators. Thus, there are statistical tests named after Dixon, Tukey, Wilcoxon, Scheffe, Mann-Whitney, Shapiro-Wilk, Kolmogorov-Smirnov, Fisher, Kruskal-Wallis, Levene, Durbin-Watson, Hotelling, Dunnett, and Bonferroni. And those are just some well-known ones off the top of my head. Dig into the literature, and you'll find hundreds more. If someone mentions such a named procedure, don't panic. Nobody knows everything. Just ask what the test is supposed to do. If you took an introductory course in statistics and know about the t-test, the F-test, and the χ^2-test, you're in great shape for understanding most of the tests you might run into.

- **Created Words**—Some statistical jargon might just as well be a foreign language because the words have no common meaning in the English language outside of statistics. Examples of such words include: *kurtosis, leptokurtic, platykurtic, skewness, covariance, autoregressive, logit, probit, eigenvalue, median, outlier, stationarity, winsorizing, communality, multicollinearity,* and my personal favorite, *homoscedasticity.* If you're at a bar and you hear any of these words being bandied around, slip quietly out the door and run for your life. Any statistician who uses these words with innocent civilians without explanation is a sadist.

Table 2. Examples of English Words That Have a Different Meaning in Statistics

Word	Meaning to a Statistician	Meaning to a Nonstatistician
bagging	A method for combining predictions from many data mining models	What the cashier does with your groceries when you're done paying
blocking	A technique for controlling variation in ANOVA	What the offensive line does during football season
brushing	Interactively selecting data points on an on-screen graph to access other information associated with the point	What you do with your toothpaste and toothbrush
breakdown	Splitting data into groups to calculate descriptive statistics and correlations	What happens to your car when you're in a hurry to get somewhere
censoring	Data with a real but undetermined value, usually less than or greater than all other values in a dataset.	Restricting free speech; removing material considered to be offensive from books or other media
confidence	Absence of type I errors	Ego stability
discriminate	Classify observations by a statistical model; a good thing.	To make distinctions based on race, creed, ethnicity, age or other category without regard to individual merit; a bad thing
errors	Differences between observed values and values predicted from a statistical model; residuals	Mistakes
mode	The most frequently appearing number in a set of numbers	A manner of acting, such as being in "relaxation mode."
Monte Carlo	A simulation procedure for evaluating the properties or performance of a statistic	The quarter of Monaco known for its resorts and casinos; a hotel in Las Vegas
Normal	Follows a Gaussian (bell-shaped) distribution	Typical, routine, sane
residual	Differences between observed values and values predicted from a statistical model; errors	Money made by musicians and actors when their works are replayed.
sample	An individual observation or multiple observations that are part of a population	A piece, a bit, a taste.

- **Alternative Meanings**—The most confusing statistical jargon just might be words in most people's everyday vocabulary that have a very different statistical meaning. For example, when you hear the word *mean*, your mind has to sort out the word's connotation. It can signify to intend, as in *say what you mean*. It can be used to associate, as in *spring means flowers*. It can refer to resources or methods, as in *by any means*. It can indicate character, as in *she has a mean streak*. It can imply exceptional skill, as in *he has a mean fastball*. And of course, in statistics, *mean* means *average*." Table 2 has a few more examples. If you don't realize that some words in English have different meanings in statistics, you can get confused very

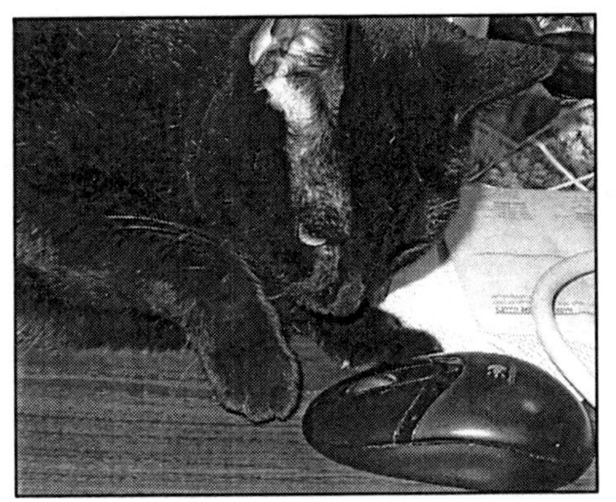

Doesn't look like a mouse to me.

quickly. I've had well-meaning report editors change *median* to *medium*, *nonparametric* to *not parametric* and *nonsignificant* to *insignificant*.

Don't feel that you're alone in the quagmire of statistical jargon. Like dialects of the English language, different statistical specialties have their own jargon and ways of expressing ideas. Data mining, time-series forecasting, quality control, nonlinear modeling, biometrics, econometrics, and geostatistics are all examples of statistical specialties that use terms not used in the other specialties. Imagine a Cajun talking to a Pennsylvania Dutch. They both speak dialects of English, but it might as well be Greek.

3 Designer Datasets

You probably didn't think much about datasets in your Statistics 101 course. The data file was packaged ready-to-go by your instructor and you did the assigned calculations using a calculator or spreadsheet software. Now that you're on your own, you're going to have to build those datasets yourself. Many datasets you create will probably be small and uncomplicated. Occasionally, though, you'll have to deal with putting hundreds of variables and thousands of samples into a dataset for analysis. So you'll have to know quite a bit about designing datasets for a statistical analysis. And there's no better place to begin than learning about the template from which all statistical analyses begin—the matrix.

The Matrix

Unfortunately, no one can be told what the Matrix is. You have to see it for yourself.
Morpheus, *The Matrix*

Matrices are convenient ways to assemble data so that computers can perform mathematical calculations. If you are familiar with spreadsheets, they are a kind of matrix. In a spreadsheet you have rows and columns that define rectangular areas, called cells. In statistics, the rows of the spreadsheet represent individual samples, cases, records, observations, entities that you're making measurements on, sample collection points, survey respondents, organisms, or any other point or object on which information is collected. The columns represent variables, the measurements or the conditions or the types of information you're recording. The columns can correspond to instrument readings, survey responses, biological parameters, meteorological data, economic or business measures, or any other types of information. You usually have several sets of variables for a given set of samples. Together, the rows and the columns of the spreadsheet define the cells, which is where the data is stored. Samples (rows), variables (columns), and data (cells) are the matrix that goes into a statistical analysis.

For example, say you had conducted a ten-question opinion survey involving five hundred respondents. You might set up a data matrix consisting of ten columns (variables) and five hundred rows (samples). Actually, you'll also need an additional column (and a few more as discussed later) to uniquely identify each sample. Straightforward, right?

Consider this example from environmental statistics. Say you are monitoring groundwater at a landfill where you have 15 wells. You collect water samples from each well and send the 15 samples to a lab where they are analyzed for 100 chemicals that might be polluting the groundwater. But which results came from which well? No problem, just add a column for a unique designation for each well. Then you would set up a data matrix consisting of 101 columns for the well identification and the chemical concentrations (actually, a lot more as discussed later) and 15 rows for the 15 wells. OK so far?

Now say you return to the landfill every quarter for a year to resample those 15 wells. There are a couple of ways you might add this data to your spreadsheet at the end of the year:

- Option 1: You could define new columns for the 100 chemicals each quarter, to create a matrix of 15 rows for each well and 400 columns for the 100 chemicals in each of the four quarters.
- Option 2: Keep the 100 columns for the chemical concentrations and add 15 rows for each new quarterly sample, so you would have 101 columns and 60 (15 wells sampled in four quarters) rows. But there's another problem. You have a unique identification for each well, but you won't be able to tell when the four samples from a well were collected. You need to add another column to uniquely identify the time the well was sampled. So you would end up with 102 columns and 60 rows.
- Option 3: Set up separate spreadsheets in the same format for each of the four sampling rounds. So you would have four spread sheets, each with 101 columns and 15 rows.

Table 3 summarizes the three options for formatting this dataset.

Table 3. Options for Formatting the Example Dataset

	Number of Rows	Number of Columns for:		Number of Spreadsheets	Total Number of Data Elements
		Identification	Data		
Option 1	15	1	400	1	6,000
Option 2	60	2	100	1	12,000
Option 3	15	1	100	4	6,000

Now, the dataset design is a bit more complicated. Each of the three options is a legitimate way to arrange the data, but each design has some limitations.

Option 1 is impractical for a couple of reasons. If you create different variables containing the same type of information, you won't be able to combine them easily to evaluate all the data from one sample source. More importantly, if you are creating the dataset with spreadsheet software, you won't be able to define that many columns. Most spreadsheet programs have limits of a few hundred columns.

Option 3 would be fine for calculating statistics for each sampling round. Sometimes, the best way to organize a dataset is to split it into several sheets or matrices in this way. Sheets may be split based on observations (rows) or variables (columns). For example, a set of sheets may share the same variables but include different samples based on study area (location), sampling round (time), or other sample attributes. In the same way, a set of sheets may share the same samples but include different groups of variables, such as analyte concentrations, field measurements, or other types of information. Several similarly formatted matrices comprise a data cube.

But what if you wanted to combine or compare rounds? Then option 3 might not be practical.[1] In this case, the best way to set up the data would be option 2, in which all the sampling rounds are on one spreadsheet and one additional column is created to describe the sampling round or a date of sample collection. As it turns out, this format is probably preferable for most of the types of statistical analysis you might conduct because it allows more flexibility.

The point to remember is this: *the format in which you save your data will dictate how you can analyze them.* So if you want to conduct a particular type of statistical analysis on a dataset, be sure you know what software you plan to use and what format the software and the procedure require the data to follow.

File Formats

> *Today there is a severe risk of [incompatible file formats] reaching a point where historians and archivists would actually be unable to read the processed documents and the threads by which decisions were made.*
>
> Simon Phipps, chief open source officer for Sun Microsystems

A spreadsheet is a type of data structure called a flat file. Think of a flat file as data written on a single sheet of paper. Flat files are two-dimensional and simple to conceptualize and implement. Unfortunately, they're not particularly efficient for data storage and management. As the number of data increases, the size of the sheet of paper becomes too big to handle.

Relational databases address this inefficiency. Think of relational databases as notebooks containing many pieces of paper. They are not only conceptually simple, but like a notebook, they can be indexed so that they are also very efficient to store, query, and manage. A relational database is in essence a set of matrices from which the duplicated data and empty space have been removed.[2] Each individual matrix, called a table, is re-

1 Most spreadsheet software allows formulas to cross over several sheets. Most statistical software, however, uses a single sheet, that is, a matrix.

2 Except for identification metadata, called keys, which allow the matrices to be linked together.

lated to the other tables in the database. In this way, a relational database can be designed to hold data without much duplication, segregate the data into appropriate collections (tables), and relate all the tables to each other to facilitate queries. Large collections of data are usually managed in relational databases.

Besides the duplication of data within a file, another fundamental design difference between a relational database and a matrix is how information types (i.e., the variables) are arranged. In a matrix, variables are arrayed as columns. When there are a large number of information types, the matrix becomes very wide and difficult to work with. Furthermore, many brands of spreadsheet software also limit the number of rows. Excel 2003, for example, is limited to 256 columns and 65,536 rows. This is a problem for many types of data sources, such as long opinion surveys, economic indicators, business metrics, and analytical chemistry data from environmental samples. For example, the Environmental Protection Agency's standard method for detecting organic compounds in water requires at least a result (i.e., the concentration of the analyte in the sample), a data quality flag,[3] and a detection limit.[4] Thus, you need three columns to represent each analytical parameter, and the method can detect over a hundred compounds. Thus, it would be impossible to fit the data into a single Excel 2003 spreadsheet because there would be far more than the 256 columns the software allows.[5]

Nope, it's not a beanstalk.

One way this problem is addressed in a relational database is to array both the samples and the variables as rows. For example, one column might contain the variable names. Instead of being represented by one line, each sample would be represented by one line for each variable. This arrangement, sometimes called a beanstalk, works great for getting a lot of information into a database while still making it easy to query. Unfortunately, the information has to be converted back into a matrix before statistical software can access the data. This was a very time-consuming problem to fix a decade ago because neither the spreadsheet nor

3 Data quality flags include: detected, not detected, below the detection limit, rejected as unusable, found in QA/QC samples, and a few others. They are important for deciding if, and if so, how, a result should be included in an analysis.

4 There's also a lot of other metadata, like units and the dates the sample was processed, that don't usually enter into a data analysis and so don't have to be included in the dataset.

5 Excel 2007 was enhanced to accommodate 16,384 columns and 1,048,576 rows. Econometricians laugh at Microsoft's naiveté.

the database software was user-friendly enough to do such extensive data manipulations. Over time, this issue has become less of an obstacle as software capabilities improve.

Professional statisticians routinely use all these formats, though only to get to a matrix. Which format should you use? Well, if you're familiar with databases, by all means use them. If not, stick with spreadsheets. They are by far the easiest means to conceptualize, create, and scrub data matrices. If you have too much data for a spreadsheet and aren't familiar with databases, get help.

4

Hellbent on Measurement

Any variable that you record in a dataset will have some scale of measurement. Scales are properties of numbers, not the objects being measured. You could measure the same attribute of an object using more than one scale. For example, say you were doing a study involving cats and wanted to have a measure of each cat's age. If you knew their actual birth dates, you could calculate their real ages in years, months, or even days. If you didn't know their birth dates, you could have a veterinarian or other knowledgeable individual estimate their ages in years. If you didn't need even that level of precision, you could simply classify the cats as kittens, adult cats, or mature cats.

Measure twice; cat once.

Understanding scales of measurement is important for a couple of reasons. Use a scale that has too many divisions and you might be criticized for creating the illusion of precision. Use a scale that has too few divisions and you might be criticized for dumbing down the data. Most importantly, though, scales of measurement determine, in part, what statistical methods *can* be applied to a set of measurements. If you want to do a certain type of statistical analysis on a variable, you better plan on using an appropriate scale for the variable.

Simple Classification of Scales

The creator of the universe works in mysterious ways. But he uses a base ten counting system and likes round numbers.

From "Dilbert" comic strip on March 25, 1994, by Scott Adams

Qualitative versus Quantitative Scales

A very simple classification of measurement scales involves whether or not the scale defines groups that have no mathematical relationship to each other. Grouping scales are considered to be qualitative. All other scales are considered to be quantitative because they define some mathematical progression.

Consider this example. Soil texture is usually a qualitative variable that defines groups such as loam, sandy loam, clay loam, and silty clay. The information can be made quantitative by recording the percentages of sand, silt, and clay (which define the texture) instead of just the classification. The qualitative measure is much easier to collect in the field and is one variable to manage rather than three. On the other hand, the qualitative variables can be analyzed in more ways. Correlating the clay content of a soil to crop growth, soil moisture, or a pollutant concentration can be done only if soil texture is measured on a quantitative scale.

Discrete versus Continuous Scales

This scale classification is based on the number of levels in the scale. Discrete scales have a finite number of levels. For example, sex has two levels, male and female. Discrete scales with two levels are also called binary or dichotomous scales. Discrete scales with more than two levels are called categorical scales. Sampling round, for example, may have quite a few levels depending on how many sampling rounds there are, but there are no intermediate levels such as sampling round 1.25.

Continuous scales have an infinite number of possible levels. They can use any integer number as well as any number of decimal points after the integer. Another way to view the difference is this: Discrete scales are like steps. You can only be on a step not between steps. Continuous scales are like a ramp. You can be anywhere along the ramp. In wrestling and boxing, weight class is a discrete scale while weight, on which weight class is based, is a continuous scale. This classification of scales is especially useful for selecting appropriate statistical methods.

Stevens's Scales

Dante tells us that Divine justice weighs the sins of the cold-blooded and the sins of the warm-hearted on different scales.

Franklin D. Roosevelt, former U.S. president

Stevens (1946)[1] categorized scales of measurement into four major groups—nominal, ordinal, interval (and log-interval), and ratio. These categories are mentioned in almost every college-level introduction to statistics, so it's good to be aware of them.

Nominal Scales

Nominal scales are discrete qualitative scales that represent simple classes or names. None of the levels within a nominal scale have any sequential relationship to any of the other levels. One level isn't greater than or less than another level. Only statistics based

1 Stevens, S. S. 1946. On the theory of scales of measurement. *Science* v. 103, No. 2684, p. 677–680.

on counts should be calculated. Examples of properties that would be measured on a nominal scale include:

- Names—Kyle, Stan, Eric, Kenny
- Sex/Gender[2]—female, male, transgendered, ambiguous
- Identification—PINs, product serial numbers, Johnny5
- Locations—Wolf Creek, Area 51, undisclosed secure location
- Car styles—sedan, pickup, SUV, limo, station wagon

Ordinal Scales

Ordinal scales have levels that are ordered. The levels denote a ranking or some sequence. One measurement may be greater than or less than another. However, the intervals between the measurements might not be constant. Moh's scale of mineral hardness, for example, consists of ten levels.[3] However, the interval between levels 1 and 8 is about the same as the interval between levels 8 and 9. The interval between levels 9 and 10 is four times greater than the interval between levels 8 and 9. Counts and statistics based on medians and percentiles can be calculated for ordinal scales. This includes most types of nonparametric statistics.[4] Examples of properties that would be measured on an ordinal scale include:

- Events—sampling round 1, sampling round 2, sampling round 3…
- Time—business quarter, geologic period
- Rankings—first place, second place, third place…
- Distance—geologic strata, atmospheric layers
- Survey responses—very bad, somewhat bad, average, somewhat good, very good

2 Sex tends to refer to biological characteristics; gender tends to refer to social characteristics. The two terms are often used interchangeably to refer to differences between males and females, but both also have alternative meanings. Some products are considered to be sexy if they create a social statement even without appealing to anatomical differences, such as a sexy watch or car. Pipes and electrical connections have a gender when their ends mimic either a male or female sex organ. And people say statistics is confusing….

3 The hardness of a mineral is determined by what it will scratch and what scratches it. In 1822, the Austrian mineralogist Moh devised a scale based on ten common minerals of increasing hardness: 1=talc, 2=gypsum, 3=calcite, 4=fluorite, 5=apatite, 6=feldspar, 7=quartz, 8=topaz, 9=corundum, and 10=diamond.

4 Nonparametric statistics are based solely on the sample's frequency distribution. In contrast, parametric statistics are based on a theoretical model of a frequency distribution, such as a Normal distribution.

Interval Scales

Interval measurements are ordered like ordinal measurements *and* the intervals between the measurements are equal. However, there is no natural zero point and ratios have no physical meaning. The classical example of an interval scale is temperature in degrees Fahrenheit or Centigrade. The intervals between each Fahrenheit degree are equal, but the zero point (-32 degrees) is arbitrary. Elevation is sometimes considered to be an interval scale because the choice of sea level as the zero elevation is arbitrary. Time can also be thought of as an interval scale. Some statisticians consider log-interval scales of measurement, in which the intervals between levels are constant in terms of logarithms, to be a subset of interval scales. Earthquake intensity (Richter and Mercali scales) and pH are examples of log-interval scales. Statistics for ordinal scales and statistics based on means, variances, and correlations can be calculated for interval scales.

Forget sea level. Cat level is the new standard.

Survey responses can be considered to be one of several types of scales depending on how the levels are defined. For example, a scale in which 5=very good, 4=good, 3=no opinion, 2=poor, 1=very poor would be considered to be a nominal scale. If the 3 level were defined as fair or average, it could be considered to be an ordinal scale if it were believed that the intervals between levels were not constant. If the intervals between levels were believed to be fairly constant, it could be considered to be an interval scale. This is one of the points of contention with Stevens's categories of scales. A given measurement's scale might be perceived differently by different users.

Consider the example of color. To an auto manufacturer, color is measured on a nominal scale. You can buy one of their cars painted red *or* blue *or* silver *or* black. To a gemologist, the color of a diamond is graded on an ordinal scale from D (colorless) to Z (light yellow).[5] To an artist, color is measured on an interval scale because their color

5 Diamond color is based on a scale developed by the Gemological Institute of America. The scale consists of twenty-three letter grades: D-F (colorless), G-J (near colorless), K-M (faint yellow), N-R (very light yellow), and S-Z (light yellow). Color is determined by comparing the sample to a reference set of diamonds, called masterstones.

wheel contains the sequence: red, red-orange, orange, orange-yellow, yellow, yellow-green, green, green-blue, blue, blue-violet, violet, and violet-red. To a physicist, colors are measured by a continuous spectrum of light frequencies, which employ a ratio scale.

Ratio Scales

Ratio scales have all the properties of the preceding scales plus zero is not an arbitrary number and ratios are meaningful. Measurements made by most kinds of meters or other types of measuring device are probably ratio scales. Any type of statistic can be calculated for variables measured on a ratio scale. Examples of variables measured on ratio scales include:

- Concentrations, densities, masses, and weights
- Durations in seconds, minutes, hours, or days
- Lengths, areas, and volumes

Other Scales of Measurement

How do you measure—measure a year?
In daylights—In sunsets
In midnights—In cups of coffee
In inches—In miles
In laughter—In strife
In—Five hundred twenty-five thousand six hundred minutes
How do you measure a year in the life?
"Seasons of Love" from the musical *Rent* by Jonathan Larson

Understanding different types of measurement scales can help you select appropriate techniques for an analysis, especially if you're a statistical novice. Some statisticians take issue with Stevens's classification of scales, though, because they believe it:

- Restricts choices of statistical methods that can be applied
- Leads to unnecessarily resorting to nonparametric statistics[6]
- Is too strict to apply to real-world data

The first two reasons are pretty much the same idea. Stevens's scheme advises against using certain statistical procedures with some scales even though the procedures might actually be appropriate. The taxonomy still has merit, though, so long as it is viewed as

6 The statement that Stevens's classification system forces "unnecessarily resorting to nonparametric statistics" is an oblique criticism of nonparametric statistics. This is a controversy in itself. Some statisticians actually prefer using nonparametric statistics to parametric statistics. Go figure . . .

guidance rather than gospel. Statistics isn't always a clear-cut science; there is a component of art involved, too.[7]

The third reason is pretty much true. The classification system accounts for many but not all scales. The following sections describe a few scales that don't quite fit into Stevens's taxonomy.

Counts

Counts are like ratio scales in that they have a zero point, constant intervals and ratios are meaningful, but there are no fractional units. Any statistic that produces a fractional count is meaningless. The classic example of a meaningless count statistic is that the average family includes 2.3 children. Counts are usually treated as ratio scales, but the result of any calculation is rounded off to the nearest whole unit.

Restricted-Range Scales

A constrained or restricted-range scale is a type of scale that is continuous only within a finite range. Probabilities are examples of constrained scales because any number is valid between the fixed endpoints of 0 and 1. Numbers outside this range are not possible. Percentages can be considered constrained or unconstrained depending on how the ratio is defined. For example, percentages for opinion polls are restricted to the range 0 to 100 percent. Percentages that describe corporate profits can be negative (i.e., losses) or virtually infinite (as in windfall profits). Restricted-range scales must be handled with special statistical techniques, such as logistical regression, that account for fixed scale endpoints.

Cyclic Scales

Cyclic scales are scales in which sets of units repeat.

Repeating Units

Some cyclic scales consist of repeating levels for measuring open-ended quantities. Day of the week, month of the year, and season are examples. Time isn't the only dimension with repeating scales, either. Musical scales, for instance, repeat yet have very different properties compared to time scales.

Repeating scales can be analyzed either by (1) treating them as an ordinal scale or (2) ignoring the repeating nature of the measure and transforming them into nonrepeating linear units, such as day 1, day 2, and so on. The objective of the statistical analysis dictates which approach should be used. The first approach might be used to identify seasonality or determine if some measurement is different on one day or month rather than another. For example, this approach would be used to determine if work done on

7 Some would argue, a black art.

Fridays had higher numbers of defects than work done on other days. The second approach might be used to examine temporal trends.

Orientation Scales

Orientation scales are a special type of cyclic scale. Degrees on a compass, for example, are a cyclic scale in which 0 degrees and 360 degrees are the same. Special formulas are required to calculate measures of central tendency and dispersion on circles and spheres.

Concatenated Numbers and Text

Concatenated numbers and text are not scales in the true sense of variable measurement, but they are part of every data analysis in one way or another. Concatenated numbers contain multiple pieces of information, which must be treated as a nominal scale unless the information can be extracted into separate variables. Examples of concatenated numbers include social security numbers, telephone numbers, sample IDs, date ranges, latitude/longitude, and depth intervals. Likewise, labels can sometimes be parsed into useful data elements. Names and addresses are good examples.

It's time to pay attention to me.

Time Scales

Time scales have some very quirky properties. Time (as opposed to duration) is like a one-dimensional location coordinate.[8] You might think that time is measured on a ratio scale given its ever finer divisions (i.e., hours, minutes, seconds). Yet it doesn't make sense to refer to a ratio of two times any more than the ratio of two location coordinates. The starting point is also arbitrary.

Most measurement scales are based on factors of ten. With time, there are 60 seconds per minute, 60 minutes per hour, and 24 hours per day. Blame the Babylonians for starting this and every civilization for the next 4,000 years for being content with the status quo.[9] In contrast, calendars have

8 Someday, we may discover that time is multidimensional, like location. Imagine a road. The first spatial dimension is parallel to the road. The second and third dimensions are perpendicular to the road, one on the land surface and the other above/below the road. Like these other spatial dimensions, perhaps there are other time dimensions that locate a point in time. Maybe other time dimensions define alternate realities or timelines and the speed at which you experience time. Time will tell.

9 Everybody knows about seconds, minutes, hours, days, months, years, and even decades, cen-

evolved from the Hellenic calendar (~850 BC), the Roman calendar (~750 BC), the Julian calendar (46 BC), to the Gregorian calendar (1582).

Time measurements can be linear or cyclic. Year is linear, so it's at least an ordinal scale. For example, 1953 happened once and will never recur.[10] Some time scales, though, repeat. Day 8 is the same as day 1. Month 13 is the same as month 1. So time can also be treated as being measured on a nominal scale.

Time units are also used for durations, which are measured on a ratio scale. Durations can be used in ratios, they have a starting point of zero, and they don't repeat (eight days aren't the same as one day).

Time formats can be difficult to deal with. Most data analysis software offer a dozen or more different formats for what you see. Behind the spreadsheet format, though, the database has a number, which is the distance the time is from an arbitrary starting point. Convert a date-time format to a number format, and you'll see what I mean. The software formatting allows you to recognize values as times while the numbers allow the software to calculate statistics. This quirk of time formatting also presents a potential for disaster if you use more than one piece of software. Always check that the formatted dates are the same between applications.

Selecting an appropriate time scale is especially important because the scale can dictate the resolution and types of analyses that can be done. Resolution is an important matter. Select an interval that is too small, and your database may become unmanageably large. Select an interval that is too large, and you may not have enough resolution to investigate the time unit you are interested in. A good rule-of-thumb is to select an interval that is at least one time unit smaller than your unit of interest. For example, if you are interested in yearly trends, collect measurements every month. If you only collect measurements yearly, you won't be able to assess the variability that occurs within a year. If you collect measurements more often than daily, you may have to rollup the data to make it manageable.

---

turies, and millennia, but there are many other scales used for time. A *jiffy* is either one tick of a computer's system clock (about 0.01 second) or the time required for light to travel one centimeter (about 33.3564 picoseconds). A *warhol* is being famous for fifteen minutes; a *kilowarhol* is being famous for approximately ten days. A moment is a medieval unit of time equal to about a minute and a half. A *fortnight* is two weeks and a microfortnight is about 1.21 seconds. A *platonic year* is an astronomical unit measuring the time required for planets to align (about 26,000 calendar years). It has also been proposed that a day be divided into 10 hours, each hour into 100 minutes, and each minute into 100 *blinks*. A blink works out to be 0.864 second, which ironically is twice the time it takes for you to blink your eye (from www.neatorama.com/2009/01/30/fun-and-unusual-units-of-measurements/, see also en.wikipedia.org/wiki/List_of_humorous_units_of_measurement and en.wikipedia.org/wiki/List_of_unusual_units_of_measurement). These are all units of measurement, most of which fit into classification systems for scales. Then there's *geologic time*, which is subdivided into eon, eras, periods, epochs, and ages. But the subdivisions aren't the same lengths. Some periods are four times longer than others and the lengths can change as more is learned about the history of Earth. The units of the scale are also different in different parts of the world. I won't go into astronomical time. You get the idea. Measuring time can be complicated.

10 OK, without getting too metaphysical, some theories of the universe suggest that time is like a river. You start at some point and go with the flow. You can't get back to your starting point, but it still exists.

Location Scales

Just as there is time and duration, there is location and distance (or length), but there are a few twists. Time is one-dimensional, at least as we now know it. Distance can be one-, two-, or three-dimensional. Distance can be in a straight line ("as the crow flies") or along a path (such as driving distance). Distances are usually measured in English inches, feet, yards, and miles or metric centimeters, meters, and kilometers. Locations, though, are another matter.

I'm up here.

Defining the location of a unique point on a two-dimensional surface (i.e., a plane) requires at least two variables. The variables can represent coordinates (northing/easting, latitude/longitude) or distance and direction from a fixed starting point. Of the coordinate systems, only the northing/easting scheme is a simple, nonconcatenated scale that can be used for classical statistical analysis. However, this type of scale is usually not used for published maps. This can be a problem because virtually all environmental data are inherently location-dependent and multidimensional. Thus, coordinate systems usually have to be converted for one or the other use. Geostatistical applications[11] are based on distance and direction measurements, but these measurements are usually calculated from spatial coordinates.

At least three variables are needed to define a unique point location in a three-dimensional volume, so a variable for depth (or height) must be added to the location coordinates. Often, however, a property of an object occurs over a range of depths (or heights) rather than a finite point. Unfortunately, depth range is a concatenated number. It's always better to use two variables to represent starting depth and ending depth. Thus, it may take four variables to define an environmental space, such as the sampled interval of a well or soil boring.

Tipping the Scales in your Favor

The best measure of a man's honesty isn't his income tax return. It's the zero adjust on his bathroom scale.

Arthur C. Clarke, British physicist and writer

You probably haven't thought much about scales. Most people think that an attribute

11 Geostatistical software is used to analyze the spatial correlation between sample results to produce contour maps.

can be measured in only one way. This is untrue more often than it is not. If you can't measure an attribute on a ratio or interval scale, think about how an ordinal scale could be applied. You can almost always devise an ordinal scale to characterize an attribute; you just have to be creative. Think of opinion surveys. If you can measure opinions, you can measure anything. If you don't have a scale to measure weight, for instance, you could create a scale of:

- Too heavy to move
- Movable but too heavy to lift
- Liftable but can't be thrown
- Can be thrown at least twenty feet
- Too light to throw without blowing away

It's crude, but it's better than nothing. Think back to the example involving color. There are always alternatives to consider.

In parts IV and V of this book, you'll see how the choice of variable scales can influence a data analysis. With experience, you'll find that clever manipulation of scales can be the difference between a revealing data analysis and disappointment.

5 Catch an Error by the Tail

Imagine practicing hitting a target using darts, bow and arrow, pistol, cannon, missile launcher, or whatever. You aim for the center of the target. If your shots land where you aimed, you are considered to be accurate. If all your shots land near each other, you are considered to be precise. The two properties are not linked. You can be accurate but not precise, precise but not accurate, neither accurate nor precise, or both accurate and precise.

Accuracy and precision also apply to statistics calculated from data. If you're trying to determine some characteristic of a population (i.e., a population parameter), you want your statistical estimates of the characteristic to be both accurate and precise.

The same also applies to the data themselves. When you start measuring data for an analysis, you'll notice that even under similar conditions, you can get dissimilar results. That lack of precision is called variability. Variability is everywhere. It's a normal part of life. In fact, it is the spice in the soup. Without variability, all wines would taste the same. Every race would end in a tie. Even statistics might lose its charm. Your doctor wouldn't tell you that you have about a year to live, he'd say don't make any plans for January 11 at 6:13 PM EST. So a bit of variability isn't such a bad thing. The important question, though, is what kind of variability?

The Inevitability of Variability

> *Great numbers are not counted correctly to a unit, they are estimated; and we might perhaps point to this as a division between arithmetic and statistics, that whereas arithmetic attains exactness, statistics deals with estimates, sometimes very accurate, and very often sufficiently so for their purpose, but never mathematically exact.*
>
> Arthur L. Bowley, *Elements of Statistics*, 1901

Before going further, let me clarify something. Statisticians discuss variability using a variety of terms, including *errors, deviations, distortions, residuals, noise, dispersion, scatter, spread, perturbations, fuzziness,* and *differences*. To nonprofessionals, many of these terms hold pejorative connotations. But variability isn't bad ... it's just misunderstood.

Suppose you're sitting in your living room one cold winter night contemplating the high cost of heating oil. The thermostat reads 68°F, but you're still shivering. Maybe it's broken. Maybe you need more insulation. Maybe there's a warmer place to sit while you read *An Inconvenient Truth*. You grab a thermometer from the medicine cabinet and start measuring temperatures around the room. It's 95° at the radiator, 68° at your chair, 59° at the window, and 69° at the stairs. You keep measuring. It's 73° at the fish tank, 67° at the couch and bookcase, 76° at the TV, and 62° at the door. That's variation!

Think of those temperature readings as the summation of five components:

- **Characteristic of Population—**the portion of a data value that is the same between a sample and the population. This part of a data value forms the patterns in the population that you want to uncover. If you think of the living room space as the population you're measuring, the characteristic temperature would be the 68° at your chair where you want to read.

- **Natural Variability**—the inherent fuzziness, uncertainty, uniqueness or differences between a sample and the population. This part of a data value is the uncertainty or variability in population patterns. In a completely deterministic world, there would be no natural variability. You would read the same value at every sampling point you measured. But in the real world, if you recorded the same measurements again and again, you probably would get different values. If all other types of variation were controlled, these differences would be the natural or inherent variability.

- **Sampling Variability**—differences between a sample and the population attributable to how uncharacteristic (nonrepresentative) the sample is of the population. Minimizing sampling error requires that you understand the population you are trying to evaluate. The sampling variability in the living room would be attributable to where you took the temperature readings. For example, the radiator and TV are heat sources. The door and window are heat sinks. Furthermore, if all the readings were taken at eye level, the areas near the ceiling and floor would not have been adequately represented.

- **Measurement Variability**—differences between a sample and the population attributable to how data were measured or otherwise generated. Minimizing measurement error requires that you understand measurement scales and the actual process and instrument you plan to use to generate data. Using an oral thermometer for the living room measurements may have been expedient but not entirely appropriate. The temperatures you wanted to measure are at the low end of the thermometer's range and may be less accurate than around 98°. Also, the thermometer is slow to reach equilibrium and can't be read with more than one decimal place of precision. Use a digital infrared laser-point thermometer next time.

- **Environmental Variability—**differences between a sample and the population attributable to extraneous factors. Minimizing environmental variance is difficult because there are so many causes and because the causes are often impossible to anticipate. For example, the heating system may go on and off unexpectedly. Your own body heat adds to the room temperature and walking around the living room taking measurements mixes the air and adds variability to the temperatures.

When you analyze data, you usually want to evaluate characteristics of some population and the natural variability associated with the population. Ideally, you don't want to be mislead by any extraneous variability that might be introduced by the way you select your samples (or patients, items, or other entities), measure (generate or collect) the data, or experience uncontrolled transient events or conditions. That's why it's so important to understand the ways of variability.

Natural Variability

Say you wanted to work on your golf game. You grab your pitching wedge and a bushel of balls and head for the practice range. Hours later you have a small Mount Titleist near the pin you were aiming for. Most of the balls are close to the pin, fewer are farther away, and a couple are at a that's-really-not-my-ball distance. Why didn't all the balls lie near the pin? Maybe your form was inconsistent, or the balls had different designs or defects, or there were periodic wind gusts, or all of those things. Even if you hit the same ball with the same club off the same tee with the same swing, you would get some scatter.

Natural variability comes from the interaction of all the things that affect the population you are trying to measure. It's what's left after you control all the variance that can be controlled. It's the variance you want to get at to characterize a population.

Try this experiment. Weigh yourself on a digital scale that provides a reading with at least one decimal place. Weigh yourself every hour for the next few days. Don't move the scale or wear different weights of clothing when you weigh yourself, in other words, control as much of the variation as you can. What you'll see is that over the course of a day, your weight fluctuates by a small amount. That's the natural variability associated with your body's weight.

Does this fur make me look fat?

Sometimes natural variability *appears* to be large, sometimes small. I say appears because you're never sure if the variability you measure is just natural variability or if it's contaminated by other types of variability that haven't been controlled. The more extraneous variability you have obscuring the natural variability, the harder it will be to analyze patterns in the population.

Sampling Variability

Sampling variability is attributable to the way samples are selected, collected, or processed before measurement. For example, randomly selected sampling locations are sometimes moved because they seem too close together or locations are added because the selected locations appear too far apart or are not in all the areas that the investigator thinks they should be. Sometimes locations are selected in the field on the basis of their physical appearance, such as being near dead animals or vegetation, discolored soils or water, or spills of waste materials. Such samples might be appropriate for identifying contaminants but not for conducting most statistical analyses.

Variability can also be added by the way samples are processed. Take, for example, samples that are collected for analysis by a laboratory, such as environmental samples, medical samples, or product samples (e.g., foods, pharmaceuticals). Causes of variability in the collection of such samples might include:

- **Collection**—lack of a consistent procedure for sample collection, inadequate training of sampler, leaching of chemicals from sample-collection device or container, inadequate decontamination of sampling devices.[1]
- **Preparation**—excessive agitation of liquid samples or exposure to air, incorrect or inadequate sample preservation, impurities in sample preservatives.
- **Transport**—long transport times, elevated temperatures, undocumented chain-of-custody.
- **Analysis**—incorrect analytical protocols, contaminants in glassware or lab devices, inaccuracies in interpreting analytical results.

Surveys are especially susceptible to sampling problems although sometimes the problems are unintentional. Open-invitation web-based surveys, for instance, are susceptible to sampling irregularities because anyone who finds the survey site can answer it. Web surveys can be *freeped*, when respondents convince like-minded friends to also answer the survey. This happens frequently with political surveys. Sampling errors in surveys also occur when respondents are improperly selected from a restricted group and then the results are extrapolated to the larger, unrestricted group. For example, a business considering starting a work-at-home program decides to conduct a survey to see if employees would value the program. However, they survey only the main office, which has a disproportionably large number of managers and supervisors who do not want the program. The company reports that there is not enough interest to start the program, thus disappointing rank-and-file employees who do want to work at home.

Variability can also be introduced into surveys by the way questions are worded or

---

1 The medical profession really has their act together on this one. Vials for blood samples are color coded with the preservative already inside. Since the sample is introduced through a needle, there's no need to take the cap off, thus minimizing the possibility of contamination. Medical technicians are trained on the proper procedures to use, including the order to collect samples for various analyses. Aren't you glad?

asked in an interview. Respondents sometimes also give answers they think the experimenter wants based on subtle cues like body language.

Measurement Variability

Measure with a micrometer
Mark with a grease pencil
Cut with a chainsaw
Pound to fit
Paint to match.

old saying in the military

Measurement variability is introduced into a data value by the very process of generating the data value. It's like tuning an analog radio. Turn the tuning dial a bit off the station, and you hear more static. That's more variance in the station's signal.

Every measurement can be thought of consisting of three elements:

- **Benchmark**—the accepted standard against which a data value is made. Scientific instruments, meters, rulers, scales, comparison charts, and survey question response options are all examples of measurement benchmarks.
- **Processes**—repetitive activities that are conducted as part of generating a data value. Equipment calibration, measurement procedures, and survey interview scripts are all examples of measurement processes.
- **Judgments**—Decisions made by the individual to create the data value. Examples of measurement judgments include reading instrument scales, making comparisons to visual scales, and recording survey responses.

Understanding these three facets of measurements is important because it will help you decide how to control extraneous variability in data collection.

Consider the examples of data types shown in Table 4. For any particular data type, all three of these elements change over time. Benchmarks change when new measurement technologies are developed or existing meters, gauges, and other devices become more accurate and precise. Standardized tests change to safeguard the secrecy of questions. Likewise, processes change over time to improve consistency and to accommodate new benchmarks. Judgments improve when data collectors are trained and gain work experience. Such changes can create problems when historical and current data are combined because variance differences attributable to evolving measurement systems can produce misleading statistics.

Table 4. Examples of Components of Data Measurements

Data Type		Examples	Measurement Devices	Components of Data Measurement		
				Benchmark	Process	Judgment
Social Sciences	Sociology	Population, socioeconomic class	Census and survey data	Description of data and data collection procedures	Obtaining data from websites and publications and converting to electronic form	Interpreting data based on description of collection process
Social Sciences	Psychology	Opinion and preference surveys	Surveys	Response scales defined by experimenter	How the survey questions are asked	Construction of survey (what questions, what order, etc.)
Business	Money	Income, expenses	Personal observation	Business definitions and conventions	Value of money varies internationally and over time	Recognition of denominations, interpretation of business definitions
Business	Manufacturing	Defect rate	Personal observation	Engineering specifications	Selection and measurement of products	Comparison of products to specifications
Chemistry	Bench Analysis	Analyses involving glassware and manual procedures	Separations, calorimetry, titrations, etc.	Standard procedure for analysis	Process of preparing sample and following procedure	Interpretation of output and possible chemical interactions
Chemistry	Quantitative Analysis	Analyses involving instrumentation	Spectroscopy, crystallography, microscopy, etc.	Instrumentation standards (e.g., chromatograph retention times)	Process of preparing sample, instrument calibration, and usage	Interpretation of output and possible chemical interactions
Earth Sciences	Hydrology	Pump discharge, stream flow	Flow meter, weir, bucket and stopwatch	Accuracy of measurement device	Usage and calibration of device	Reading and recording increments on measurement devices
Earth Sciences	Geology	Rock unit description, soil, and rock classification	Measuring tape geologic compass, references for classification	Descriptions of rock formations and lithologic and soil classification procedures	Making observations for classification procedure, calibration, and use of devices	Reading and recording increments on measurement devices, comparison of observations to classification parameters
Earth Sciences	Meteorology	Wind speed, temperature, barometric pressure	Anemometer, thermometer, barometer	Instrumentation standards	Correct placement, usage, and calibration of instruments	Reading and recording increments on measurement devices

Data Type		Examples	Measurement Devices	Components of Data Measurement		
				Benchmark	Process	Judgment
Life Sciences	Biology	Species counts, habitat maps	Observation of species/habitats	Technical descriptions of species/habitats	Process of comparing observation to description	Identification of specific species and habitats
Life Sciences	Pharmacology	Drug effectiveness	Personal or lab observation of drug effects	Comparison to a reference drug or placebo control group	Processes of instructing patients and administering drugs	Patient/doctor assessment of drug effects or analysis of samples from patient
Life Sciences	Epidemiology	Disease clusters	Personal observation of community health	National statistics on disease rates	Compiling individual patient histories or selecting local statistics on disease rates	Interpretation of statistics from differing sources

There's a special type of analysis aimed at evaluating measurement variance called *Gage R&R*. The R&R part refers to:

- **Repeatability**—the ability of the measurement system to produce consistent results. The focus of repeatability is on the benchmark and process portions of the measurement system. Testing for repeatability involves using the same subject or sample, the same characteristic or other variable, the same measurement device or instrument, the same environmental setting or conditions, and the same researcher to make the measurements.
- **Reproducibility**—the ability of the measurement system *and* the people making the measurements to produce consistent results. The focus of reproducibility is on the entire measurement system. By comparing reproducibility to repeatability, the effects of the judgments made by the people making the measurements can be assessed. Testing for reproducibility involves using the same sample, characteristic, measurement instrument, and environmental conditions but using different researchers to make the measurements.

Gage R&R is a fundamental type of analysis in industrial statistics, where meeting product specifications requires consistent measurements, but it can be used for any measurement system from medical testing to opinion surveys.

Environmental Variability

Environmental variability refers to all the other variability in a value that isn't natural or attributable to sampling or measurement. Some of it you might anticipate and plan for, but there are also some things you might never think about.

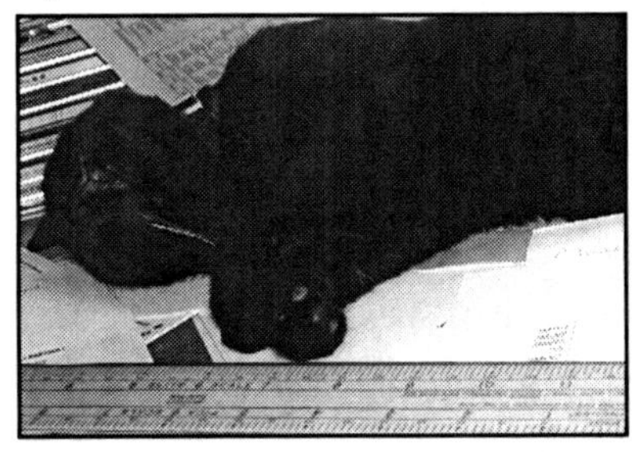
All cats are drawn to scale.

Remember when you were in school taking a

final exam? All kinds of things conspired to distract you. The room was too hot and stuffy. The class brain on your left is looking out the window, so you can see all her answers. The guy in front of you is farting the burrito he had for breakfast. Someone behind you is coughing and sneezing with something highly contagious you know you'll be coming down with when your vacation starts tomorrow. Your cell phone in your pocket is vibrating your leg into numbness, probably old Mrs. Wrinklebaum wanting to know when you were going to mow her lawn like you promised. And all you can think about is the party you want to go to later that night. Did you anticipate any of that might happen? Maybe the cell phone; you should have just turned it off. But at least you remembered to eat breakfast and bring your calculator and pencils, so they weren't a problem. Do you think, though, that any of those other things might influence your score on the test? That's how environmental variability works.

Environmental variability can swamp other types of variability, not to mention the population characteristic itself that you're trying to measure. Drive east on the Pennsylvania Turnpike from Pittsburgh, set your cruise control to the 60 miles-per-hour (mph) speed limit, the equivalent of a population parameter, and see what happens. The natural variations in the car's speed while the cruise control is active plus the speedometer's sampling and measurement variability might amount to 10 miles per hour. That's the margin the Pennsylvania State Police often allow. But the valleys and ridges of the Appalachians add a lot of environmental variability. You can lose 30 mph going up the long hills and or gain it when you're going down. So your speed could be 60 mph ± 30 mph if you leave the cruise control alone.

Here's one last example to get you thinking about environmental variability.

Say you want to estimate how long it will take to fill your new pool so that you can run some errands while it's filling. You plan to estimate the discharge from your garden hose by measuring the time it takes to fill a one-gallon bucket and then calculating the flow rate in gallons per minute. In calculating the hose discharge, for example, there might be some natural variability attributable to fluctuations in water pressure at the faucet. There's not too much you can do about that. There might be sampling error attributable to how the hose is coiled and kinked, and if the hose expands as it sits in the hot sun, but about all you can do for that is stretch out the hose and try to keep it in shady areas. There might also be measurement error attributable to filling the bucket to the same level and simultaneously trying to look at the second hand on your watch for the elapsed time. For that, you decide to use a stopwatch and a five-gallon carboy. Finally, knowing your pool can hold about 8,000 gallons, you can calculate how long it will take to fill the pool. So you make the measurements, do the calculations, and hop in the car planning to be back when the pool is about full.

Where is the level of the pool when you get home? If it rained, more water than expected would have entered the pool making the level higher. If your household water use caused the pressure to drop, less water than expected would have entered the pool making the level lower. If the environmental variability is small or random (i.e., the water additions are about equal to the water subtractions), the level would be where you expected it to be.

Variability versus Bias

A man should look for what is and not for what he thinks should be.
Albert Einstein, physicist

Remember target practice? If there is little variation in your aim, the deviations from the center of the target would be random in distance and direction. Your aim would be accurate and precise. But what if the sight on your weapon were misaligned? Your shots would not be centered on the center of the target. Instead there would be a systematic deviation caused by the misaligned sight. Your shots would all be inaccurate, roughly the same distance and direction from the center. That's called bias. You may not even have known there was a problem with the sight before shooting, although you would probably suspect something after all the misses.

Bias usually carries the connotation of being a bad thing. If the systematic deviation is a good thing because it fixes another bias, it's called a correction. For example, you could add a correction, an intentional bias in the direction opposite the bias introduced by the sight, to compensate for the inaccuracy. So bias can be good (in a way) or bad, intentional or not, but it's always systematic. On the other hand, a bias applied to only selected data is a form of exploitation, and is nearly always intentional.

So the relationships to remember are:

Variance ↔ Imprecision
Bias ↔ Inaccuracy

Analysis Bias and Variability

Most statistical techniques are unbiased themselves, as long as you meet their assumptions. If something goes wrong, you can't blame the statistics. You may have to look in the mirror, though.

During the course of any statistical analysis, there are many decisions that have to be made. Most of these decisions involve data. Each decision has the potential to slant the results. Whatever the decisions are, there will be some impact on precision and perhaps even accuracy. In an ideal world, the sum of the decisions won't add appreciably to the variability. Often, though, data analysts want to be conservative, so they make decisions they believe are counter to their expectations. Sometimes they are too conservative and when they don't get the results they expected, they go back and try to tweak the analysis. At that point they have lost all chance of doing an objective analysis and are little better than analysts with vested interests who apply their biases from the start. Avoiding analysis bias requires no more than to make decisions based solely on statistical principles. This sounds simple, but isn't always so.

Analysis variability is a bit more subtle than analysis bias. It stems from there being many options in conducting the same analysis. Say you wanted to calculate descriptive statistics (just to keep the example simple) for a dataset. Here are the data you're given:

37, 73, , 96, <10, 88, 44, 79, 888, 17, , 23, 47, 24, 32, <10, 56, 52, 60, 76

The first thing you notice is that you have two missing values and two censored values (the values with the less than sign, <).[2] You also wonder about whether the 888 value is an error since it is so much higher than the other values. So you call the client and find out that the missing values and the censored values are correct. As you suspected, the 888 should be 88. And it's not a random typo. It seems that the keyboard the data entry person uses has a sticky 8 key. So you ask: what about the 88 value? Is that an error, too? Oops. Yes it is. It should be an 8. That's an example of why it's good to find out why an outlier occurred. So you make those changes.

Yyyyyyyeah, I'm sitting on yyyyyyour keyyyyyboard. Whyyyyyyyy?

Then you look at the censored data. If you ignore the censored data, the mean is 50.8. That is the highest the mean could be because the censored data are at the low end of the data. Ignoring censored data, though, is not an acceptable method because it introduces bias into the dataset. There are several statistical methods that you can use, however. One method is to replace the missing values by a constant. If you substitute zero for the two censored data values, the mean is 45.1. If you substitute the censoring limit for the two censored data values, the mean is 46.2. So if the censored data below 10 were available, the mean would be somewhere between 45.1 and 46.2. There would still be some bias in the mean, but it wouldn't be as large as it would be by ignoring the censored data. So what should the replacement value be? Some statisticians use the censoring limit, some use half of the limit, and others use something else. You decide to use half of the censoring limit, so you replace the <10 values by 5. That results in a mean of 45.7.

---

2 Censored data are measurements that aren't exact; they are less than or greater than some limit of measurement. Censoring occurs when the measurement instrument can't measure part of the range of possible data values. For example, I once owned a car which had a gas gauge that did not measure below half a tank. When the gauge reached half a tank, it stopped moving. At that point, I would know only that I had less than half a tank of gas. That's a censored measurement. Chapter 18 discusses ways to deal with censored data.

Table 5. Summary of Example Datasets

	Dataset Given to Statistician	**Errors Corrected**	**Censored Data Replaced by Half of Limit**	**Missing Data Replaced by Mean**
	37	37	37	37
	73	73	73	73
	(missing)	(missing)	(missing)	45.7
	96	96	96	96
	<10	<10	5	5
	88	8	8	8
	44	44	44	44
	79	79	79	79
	888	88	88	88
	17	17	17	17
	(missing)	(missing)	(missing)	45.7
	23	23	23	23
	47	47	47	47
	24	24	24	24
	32	32	32	32
	<10	<10	5	5
	56	56	56	56
	52	52	52	52
	60	60	60	60
	76	76	76	76
Count	16	16	18	20
Average	105.8	50.8	45.7	45.7
Standard Deviation	203.3	25.7	28.1	26.7

Finally, you look at the missing values. Should you ignore them and not use them in any calculations? No, that wouldn't be representative of the population for the same reasons as for censored data. Replacing the missing values is a legitimate procedure, but you can't assume the sample limits are the same as the population's limits. Thus, your best bet is to replace the missing data by the estimated mean. That approach will introduce the least amount of bias of any replacement value.

What about the standard deviation? That's a bigger problem because the standard deviation is much more sensitive to changes in data values than the mean. You know it'll be somewhere between the dataset that replaces the censored data by 0 and the dataset that ignores the censored data, which would be between 25.7 and 29.0. Replace the missing values by the mean of the available values, and the standard deviation is only 26.7.

Table 5 summarizes the changes made to the dataset. Can you see why there is variability associated with how you do an analysis? Another statistician may have used different methods to address the censored and missing data. Now imagine a number of statisticians[3] doing a complex multivariate statistical procedure with scores of options beyond this simple data scrubbing example. At best, they'll all get the same interpretation, but the chances of getting all the same values for specific statistics would be low. What's important to remember, then, is that even if you are totally unbiased, there is going to be a certain amount of fuzziness in the results because of the decisions you make in scrubbing the data and implementing the statistical procedures. So if a reviewer of your analysis wants to replicate the analysis to see if you made any errors, beware. The two analyses will probably never match.

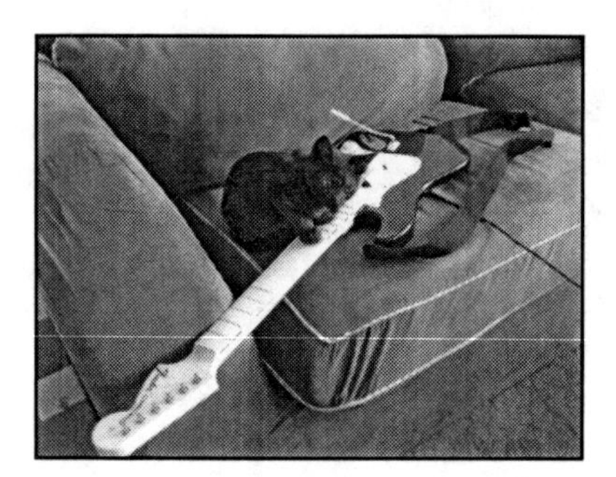

"Muscled, black with steel-green eyes / swishing through the rye grass / with thoughts of mouse-and-apple pies / Tail balancing at half-mast / And the mouse police never sleeps / lying in the cherry tree / Savage bed foot-warmer." (From "And the Mouse Police Never Sleeps" by Ian Anderson.)

3 Does anybody know what the collective noun is for statisticians? Is it a sample of statisticians? A population? A census? I think it should be a variance of statisticians.

Reporting Bias and Variability

A statistician visiting a foreign country was arrested by the secret police as a spy because they thought the random number tables he was carrying were secret codes. They threw him into a stifling, windowless prison cell with a disheveled and dirty old derelict. After a few days, the statistician found out that his cellmate was a psychologist who was imprisoned by the secret police as a spy because he answered their questions with his own questions.

After a few months in the prison, the statistician decided he had to escape, so when the guard brought his bread and water, the statistician quickly began lecturing on the central limit theorem. Before the guard could cover his ears, he soundlessly slipped into a sweet statistical slumber. The psychologist didn't want to go along, so the statistician took the sleeping guard's keys and made his way out of the prison.

Once he got out of the prison, the statistician was surprised to find himself in the middle of a desert, with nothing but sand for as far as he could see. He set off toward the sunset, but after staggering around lost for days, he crawled back to the prison and was again thrown into the jail cell.

When the psychologist saw his cellmate return, he told him, "I knew you'd be back. I tried the same thing last year and ended up crawling back after a week, too." The sun-burnt and emaciated statistician stared in amazement at the psychologist and yelled: "Why didn't you tell me that before I left?" The psychologist shrugged, stroked his pet rat, and said:

"So who reports negative results?"

You can see reporting bias and variability every day in any media outlet with a political beat. For example, compare stories on the Drudge Report (a conservative site at drudgereport.com) and the Huffington Post (a liberal site at huffingtonpost.com). Same news; different presentations. The same can and does happen in statistics. Some variation stems from ignorance, such as when a reporter does not understand the subtleties of a complex statistical analysis. Other examples might not be so innocent.

Reporting bias and variability is where the saying about lies, damn lies, and statistics is most applicable. In professional circles the most common form of reporting bias is probably not reporting nonsignificant results. Some investigators will repeat a study again and again, continually fine-tuning the study until they reach their statistical nirvana of significance. Seriously, is there any real difference between probabilities of significance of 0.051 versus 0.049? But you can't fault the investigators alone. Some professional journals won't publish negative results, and professionals who don't publish perish. Can you imagine the pressure on an investigator looking for a significant result for some new business venture, like a pharmaceutical? He might take subtle actions to help his cause then not report *everything* he did.

Perhaps the most common form of reporting bias in nonprofessional circles is cherry picking, the practice of reporting just those findings that are favorable to the reporter's position (en.wikipedia.org/wiki/Cherry_picking). Cherry picking is very common in studies of controversial topics such as climate change, marijuana, and alternative medicine.

Given that someone else's reporting bias is after-the-analysis, why is it important to your analysis? The answer is that it's how you can be misled in planning your statistical study. Never trust a secondary source if you can avoid it.[4] Never trust a source of statistics or a statistical analysis that doesn't report variance and sample size along with the results. Remember: *statistics don't lie; people do.*

---

4 Sometimes primary references may be unavailable or in a foreign language.

6 The Zen of Modeling

What's the first thing you think of when you hear the word *model*? The plastic model airplanes you used to build?[1] A fashion model?[2] The model of the car you drive? The person who is your role model? But what do any of those things have to do with data analysis? Read on; you're about to find that statistical analyses begin and end with models.

By Any Other Name

...statistical techniques are tools of thought, and not substitutes for thought.
Abraham Kaplan, *The Conduct of Inquiry: Methodology for Behavioral Science*

What do a Ford Focus, a plastic airplane, and Tyra Banks have in common? They are all called models. They are all representations of something, usually an ideal or a standard.

Models can be true representations, approximate (or at least as good as practicable), or simplified, even cartoonish compared to what they represent. They can be about the same size, bigger, or most typically, smaller, whatever makes them easiest to handle. They usually represent physical objects but can also represent a variety of phenomena, including conditions such as weather patterns, behaviors such as customer satisfaction, and processes such as widget manufacturing. The models themselves do not have to be physical objects either. They can be written, drawn, or consist of mathematical equations or computer programming. In fact, using equations and computer code can be much more flexible and less expensive than building a physical model.

Customarily, models are used either to:

- **Display** what they represent (e.g., model airplanes) or are associated with (e.g., fashions)
- **Substitute** for incomplete real world data, such as using the Normal distribution as a surrogate for a sample distribution.

1 That is, until your mom found out you were enjoying the glue more than the model.
2 That is, until your spouse found out you were enjoying the model more than the fashion.

- **Manipulate** their components to learn more about the things they represent (e.g., scientific models for planetary motion).

Whether you know it or not, you deal with models every day. Your weather forecast comes from a meteorological model, maybe several. Mannequins are used to display how fashions may look on you. Blueprints are drawn models of objects or structures to be built. Examples are plentiful. Table 6 provides some additional examples of physical models.

Table 6. Examples of Physical Models

Accuracy of Representation		Relative Size of Model		
		Larger than Actual	**Same as Actual**	**Smaller than Actual**
	True	Oversized exhibition models for industrial trade shows	Cadavers used in medical schools	Ant Farms
	Approxi-mate	Anatomical models used in colleges and medical schools	NASA flight simulators	Army Corps of Engineers models of waterways
	Simplified	Molecular models used in education	ResusiAnnie (dummies) used for CPR instruction	Architectural models

Humans, in particular, are modeled all the time because of our complexity. Children play with dolls as models of playmates. Mannequins are simplified models of fashion models, who in turn, are models of people who might wear a fashion designer's wares. Posing models provide reference points for artists. Crash test dummies reveal how the human body might react in an automobile accident. Medical researchers use laboratory animals in place of humans for basic research. Medical schools use donated cadavers as models, very good ones as it turns out, of the human anatomy. So, there should be nothing unfamiliar or intimidating about models.

Whether it is a physical scale-model of a hydroelectric dam or a mathematical model of weather patterns, a model is nothing more than a tool used to stimulate the imagination by simulating a phenomenon. The model airplane takes its young pilot looping through the blue skies of a summer day. Globes teach geography and orreries teach planetary motion. The mannequin shows the bride-to-be how beautiful she'll look in the gown at her wedding. The concept car unveiled today gives consumers an idea of what they may be driving in a few years. The National Hurricane Center uses over a dozen mathematical models to forecast the intensities and paths of tropical storms and help understand the complex dynamics of hurricanes.

It should come as no surprise, then, that scientists, engineers, and mathematicians use models, especially virtual models, all the time. It may be surprising, though, that virtual models are also used extensively in business, economics, politics, and many other fields. Nevertheless, there is a mystique associated with modeling, especially the mathematical variety. Some believe that models are infallible and unchanging. Some believe that models are impossibly complex and necessarily unfathomable. Some believe that models are sophisticated delusions for obfuscating real data. In reality, none of these opinions is correct, at least entirely.

A Medley of Numbers

In theory, there is no difference between theory and practice. But, in practice, there is.

Jan L. A. van de Snepscheut,
former executive officer of the
computer science department at the
California Institute of Technology.

Mathematical models can be either theoretical (i.e., derived mathematically from scientific principles) or empirical (i.e., based on experimental observations). For example, celestial movements and radioactive decay are phenomena that can be evaluated using theory-based models. To calibrate a theoretical model, the model (i.e., the equation) is fixed and the inputs are adjusted so that the calculated results adequately represent actual observations.

Empirical models differ from theoretical models in that the model (i.e., the equation) is not necessarily fixed for all instances of its use. Rather, empirical models are developed for specific situations from measured data. Model formulation and calibration are simultaneous. However, the selection of the form of the equation and the inputs used in an empirical model are usually based on related theories. Models developed using statistical techniques are examples of empirical models.

Empirical models can also be deterministic, stochastic, or sometimes a hybrid of the two. Deterministic empirical models presume that a specific mathematical relationship exists between two or more measurable phenomena (as do theoretical models) that will allow the phenomena to be modeled without uncertainty under a given set of conditions (i.e., the model's inputs and assumptions). Biological growth models are examples of deterministic empirical models.

Both theoretical models and deterministic empirical models provide solutions that presume that there is no uncertainty. These solutions are termed "exact" (which does not necessarily imply "correct"). Conversely, stochastic empirical models presume that changes in a phenomenon have a random component. The random component allows stochastic empirical models to provide solutions that incorporate uncertainty into the analysis.

Statistical models are examples of stochastic empirical models in which the model equation is generated by quantifying and minimizing errors (i.e., uncertainty). Statistical models place great emphasis on examining and quantifying uncertainty, whereas theoretical models generally do not.

This is supposed to be a model of ME?

OK, that's way more than you need to know. Let me simplify. Models and statistics are closely intertwined. Statistical analyses involving descriptive statistics and testing rely on mathematical models like the Normal distribution to represent data frequencies and error rates. Statistical analyses involving detecting differences, prediction, or exploration usually involve using statistics to estimate the mathematical coefficients of a conceptual model. So models serve as both inputs and outputs of statistical analyses. You can't do without them, so you might as well understand what they are.

7 Assuming the Worst

Before you go poking around your data looking for patterns and anomalies, you should know that there are a few fundamental statistical requirements that you can't ignore. Unfortunately, you can't always verify that you meet all these requirements so you have to make assumptions. All models make assumptions, an evil necessity for simplifying complex analyses. If your statistical analysis deals in probabilities, and what statistical model doesn't, you'll be making at least five assumptions:[1]

- **Representativeness**—The samples or measurements used to develop the model are representative of a population of possible samples or measurements that is the focus of the analysis.
- **Linearity**—The model can be expressed in an intrinsically linear, additive form. This sounds intimidating, but it's probably the easiest of the assumptions to deal with.
- **Independence**—Errors in the model are independent of each other (i.e., are not correlated).
- **Normality**—Errors in the model are Normally distributed.
- **Homogeneity of Variances**—Errors in the model have equal variances for all values of the dependent variables (i.e., the measure you are investigating).

The representativeness assumption means that you have the right data from the population you want to create a statistical model of. The independence assumption means you didn't screw up anything in the way you selected and generated the data. The linearity assumption means you have the right statistical model for the data. The Normality and homogeneity-of-variance assumptions mean the Normal distribution is a good model for the frequency distribution of the dependent variable you are trying to model. In a nutshell, the assumptions mean you have the right data and the right models.

In practice, the assumptions are almost always violated to some degree, which does

1 Textbooks on statistics don't include the assumption of representativeness. Those authors probably know better than I do, but this is my book. I include it because if the assumption isn't valid, your model will lie to you and you won't know it. Some authors describe many more assumptions, particularly those associated with specific statistical procedures. Drive off that bridge when you get there.

not necessarily mean that the model is invalid. It does mean, however, that additional judgment must be used in interpreting the results. Sometimes, the model just needs another component or two. Other times, there may be something more fundamentally wrong that will cause you to rethink your whole approach. In either case, your calculated probabilities will be wrong, maybe by a lot or maybe by a little, but you'll usually be able to recognize the problem. If the data aren't representative, on the other hand, your model is crap, and you won't even know it.

Representativeness

[The Horta] found humanoid appearance revolting, but she felt she could get used to it. . . . I did get the distinct impression she found [Spock's pointed ears] the most attractive human characteristic of all. I didn't have the heart to tell her that only I have them.

Spock on *Star Trek* episode "Devil in the Dark"

The most important assumption common to all statistical models is that the samples used to develop the model are representative of a population of possible samples that are being investigated. Some statistics books don't discuss this as a basic assumption because it is viewed as more of a requirement than an assumption. But obtaining representative samples of populations can be a challenge. Unlike the other assumptions, failure to obtain a representative sample from a population under study would necessarily be a fatal flaw for any statistical analysis. You might not know it, though, because there isn't always a good way to determine if a sample is representative of the underlying population. To do that, you would need to know a lot about the population.

Hmmm. Doesn't taste like sushi. Must not be a representative sample.

For example, say you were going to conduct a survey of doctors about their preferences in prescribing drugs for reducing cholesterol. You buy a mailing list from a market research firm and select a sample in which you control for the doctors' age and sex. Would the sample be representative? Maybe not.

The list included ophthalmologists, dentists, proctologists, chiropractors, veterinarians, doctors of philosophy, acupuncturists, homeopaths, faith healers, and psychiatrists. You really just wanted cardiologists, so you go back and try again. This time you don't buy a list until you answer the question, who knows the answers to the questions I want to ask? If somebody can't answer your questions, there's no sense asking them. Even cardiologists might not be picky enough. You need cardiologists who do not work in a hospital or in emergency care because you don't want the guys who are zapping heart attack victims back to life. They see patients for a day and then send them on their way, maybe with a thirty-day supply of something. You want the guys who are following heart attack patients long-term and deciding if they should be on drug C, drug L, drug Z, or one of the other dozen or so long-term cholesterol reduction drugs. You may think you want cardiologists, but actually you might be better off talking to primary care physicians (i.e., family doctors). So you buy a list of PCPs. Is that a representative sample? Maybe not.

You realize that doctors who practice in low-income neighborhoods may not prescribe drugs in the same way as doctors who service patients of greater means because of differences in drug costs and insurance reimbursements. More educated patients may also request a specific drug they saw advertised in *National Geographic*. Your results may be skewed if you don't address demographic and socioeconomic factors. So you do that. Is that a representative sample? Maybe ... not.

Sometimes there are even more insidious divisions in a population, ones you might never have thought about. Prescribing could be greatly influenced by promotional efforts of the companies selling cholesterol reducing drugs. One pharmaceutical company may be promoting their cholesterol drug heavily in Fort Lauderdale, Florida because there is a large population of elderly people. But where the population is younger, like in Seattle, Washington, the PCPs may not have heard anything about the drug the PCPs in Fort Lauderdale are handing out like Halloween candy. You'd have no way of knowing where all the companies have their sales forces or what promotions/incentives are going on. Representative sampling isn't easy.

So here's the problem. If you knew about all the quirks of the population that might affect the results of your study, you probably wouldn't need to bother conducting the study. So representativeness has to be addressed indirectly by building randomization and variance control into the sampling program before it is undertaken. If randomization cannot be incorporated into the sampling procedure in some way, the only alternative is to try to evaluate how the sample of the population might not be representative. This is seldom a satisfying exercise. Making statements like "the results are conservative because only the worst cases were sampled" are usually conjectural, qualitative, and unconvincing to anyone who understands statistics.

Linearity

I shall not go nonlinear. It is a model killer. Those curvy patterns bring no small complexities. When they confront me, I shall transform the data through roots or logs or degrees. Then I shall turn my inferring eye to look upon the data's path and see that it is straightly drawn. Only answers will then remain.

Bene Gesserit, litany against nonlinear models

The linearity assumption requires that the statistical model of the dependent variable being analyzed can be expressed by a linear mathematical equation consisting of sums of arithmetic coefficients times the independent variables. The effects of nonlinear relationships are usually substantial. Applying a linear model to a nonlinear pattern of data will result in misleading statistics and a poor fit of the model to the data. Evaluating the linearity assumption is usually straightforward. Simply plot the dependent variable versus the independent variables. If the plotted data do not approximate a straight line, you may have a problem.

This assumption is seldom a major problem for three reasons:

- First, in practice, most models of dependent variables *can* be expressed as linear mathematical equations consisting of arithmetic sums of coefficients times the independent variables. Mother Nature seems to like this simplicity. Just plot the dependent variable versus the independent variable and if there is any pattern at all it will probably be a straight line.
- Second, the assumption will still be met when one (or more) of the independent variables has a nonlinear relationship with the dependent variable *if* a mathematical transformation can be found to make the relationship linear. Put another way, you can legitimately do just about anything to the data for your independent variables (raise them to powers, take square roots, logs, or reciprocals) to make the relationship linear. The only catch is that the coefficients (termed the *parameters* of the model) must still be linear. Models that have parameters that are linear (e.g., sums of constants) are termed *intrinsically linear.* In contrast, *intrinsically nonlinear* models have coefficients that are nonlinear (e.g., powers or roots). So if the plot of the dependent variable versus the independent variables does not form a straight trend, find a transformation of the independent variable that will make the pattern straight. Chapter 20 will give you some tricks for doing this.
- Third, if a transformation cannot be found to correct a nonlinear relationship, you can still resort to using statistical methods for intrinsically nonlinear models. Nonlinear modeling uses different terminology and optimization processes than linear regression and requires specialized software. You won't get nonlinear parameters by accident.

Independence

People who wish to salute the free and independent side of their evolutionary character acquire cats. People who wish to pay homage to their servile and salivating roots own dogs.

Anna Quindlen, U.S. writer

The third assumption common to statistical models is that the errors in the model are independent of each other. Some introductory statistics textbooks describe this assumption in terms of the measurements of the dependent variable. There are two reasons for this. First, it's a lot easier for beginner students to understand, especially if they aren't familiar with the mathematical form of statistical models and the concept of model errors, called residuals. Second, and more importantly, the two approaches to describing the independence assumption are equivalent. This is because a data value can be expressed as the sum of an inherent "true" value and some random error. If you have controlled all those sources of extraneous variation, the data and the model errors should be identically distributed. That's another reason why variance control is important.

Am I independent? I'm a CAT!

Say you were conducting a study that involved measuring the body temperature of human subjects. Without your knowledge, a well-meaning assistant provides beverages in the waiting room—piping hot coffee and iced tea. When you plot a histogram of the body temperature data, you see three peaks (called modes), one centered at 98.6°F, another peak a degree or so higher, and a third a degree or so lower. Your data has violated the independence assumption. The subjects who drank the coffee all had their temperatures linked to the higher temperature of the coffee. The subjects who drank the iced tea all had their temperatures linked to the lower temperature of the iced tea. What are the chances you might notice this dependency? If you had lust a few subjects, the chances wouldn't be good unless you caught the assistant in the act. With 100 subjects, you might notice something. With 1,000 subjects, you would certainly notice the effect.[2] That's one of the reasons why statisticians prefer more data than most clients want to pay to collect. Better training of your assistants and more supervision of the patients, and this wouldn't have been a problem.

Sometimes data dependencies cannot be avoided. Environmental data, for instance, usually always have some inherent spatial or temporal dependence. For example, two soil samples located very close together are more likely to have similar attributes than two samples located very far apart. Likewise, two well water samples collected a day apart are more likely to have similar attributes than two samples collected two years apart.

Assessing independence involves looking for serial correlations, autocorrelations, and spatial correlations. The terms *serial correlation* and *autocorrelation* are often used inter-

2 Although if you're providing beverages to 1,000 subjects, you might consider getting out of research and opening a coffee shop.

changeably. A serial correlation is the correlation between data points with the previously listed data points (termed a *lag*). A second-order lag would involve a data point and the data point before the previously listed data point. If the data points are collected at a constant time interval, the term *autocorrelation* is more typically used. Serial correlations can be calculated for any number of lags, although usually only the first few are important.

To check a variable or residuals from a model for sample independence, arrange the data in the order in which the samples were collected or generated, and then conduct a Durban-Watson test. The Durban-Watson test statistic ranges from 0 to 4. If the statistic is close to 2.0, then serial correlation is not a problem. Most statistical software will allow you to conduct this test as part of a regression analysis.

For temporally related data, examine correlograms for autocorrelations and partial autocorrelations. Correlograms are plots of the temporal correlation (autocorrelation or partial autocorrelation) versus the order of the data lag. An autocorrelation is the correlation between the data value and the order of the lag. A partial autocorrelation is the correlation between the data value and the order of the lag with the effects of previous lags held constant.

For spatially related data, create variograms, which are plots of the spatial variance versus the distances between samples. Correlograms and variograms require specialized software to produce and some experience to interpret. If these statistical procedures suggest that violations of the independence assumption may be significant, more sophisticated methods of evaluating the dependence can be used, but you'll probably need professional assistance.

When the independence assumption is violated, the calculated probability that a population and a fixed value (or two populations) are different will be underestimated if the correlation is negative or overestimated if the correlation is positive. The magnitude of the effect is related to the degree of the correlation.

Some people confuse the independence assumption, which refers to model errors or measurements of the dependent variable, with the assumption that the independent variables (a.k.a., predictor variables) are not correlated. Correlations between predictor variables, termed *multicollinearity*, are also problematical for many types of statistical models because statistics associated with such models can be misleading. But that's a story for another time.

Normality

Nobody realizes that some people expend tremendous energy merely to be normal.
Albert Camus, French writer

The Normality assumption requires that model errors (or the dependent variable) mimic the form of a Normal distribution (viz., a bell-shaped curve, another example of a mathematical model). This assumption is important because the Normal model is used as the basis for calculating probabilities related to the statistical model. If the model errors don't at least approximate a Normal distribution, the calculated probabilities will be wrong. It would be like trying to calculate the circumference of a round hole from a square peg.

There are many methods for evaluating the Normality of a distribution, which fall into one of three categories (these are described in more detail in Chapter 19):

Normal is really such a matter of opinion.

- **Descriptive Statistics**—These include the coefficient of variation (the standard deviation divided by the mean), the skewness (a measure of distribution symmetry), and the kurtosis (a measure of relative frequencies in the center versus the tails of the distribution). If the coefficient of variation is less than 1 (or 1.2 in some references), and the skewness and the kurtosis are close to zero, a Normal distribution of the errors is probably a reasonable assumption.

- **Statistical Graphics**—Statistical graphics are more revealing than descriptive statistics because they indicate visually what data deviate from the Normal model. Interpreting these graphics can be somewhat subjective, however. The most commonly used statistical graphics are histograms, box plots, and probability plots (examples of these plots are provided in Chapter 19). Other statistical graphics sometimes used to evaluate Normality include stem-and-leaf diagrams, dot plots, and Q-Q plots.

- **Statistical Tests**—Statistical tests are more rigorous than either descriptive statistics or statistical graphics. Commonly used tests of Normality include the Shapiro-Wilk test, the Chi-squared test, and the Kolmogorov-Smirnov test. One of the problems with statistical tests of Normality is that they become more sensitive to departures from Normality as the sample size gets large. So a statistical test on a large number of subjects may indicate a departure from Normality that is unimportant to a statistical model. Thus, tests of Normality may be definitive but irrelevant.

So how should you evaluate Normality? Focus on one method or decide on the basis of a preponderance of the evidence? First, you have to understand that statistical tests, statistical graphics, and descriptive statistics are like advisors. They all have an opinion, none is always correct, and they often provide conflicting advice. Unfortunately, you are the one responsible for making the correct determination.

One approach to consider is to first look at a histogram to get a general impression of whether the data distribution is even close to a Normal distribution. If it is, look at a test of Normality, preferably a Shapiro-Wilk test. The test assumes Normality, so if there's no significant difference, you can conclude that the data came from a Normally distributed population. If there is a significant difference, then your decision becomes problematical. Look at a probability plot to determine where the departures from Normality are. If you can't generate a probability plot, look at the skewness and kurtosis. If there is an appreciable deviation from Normality, you might consider pursuing a data transformation or a different type of analysis, such as a nonparametric procedure.

Practically, your Normality evaluation will come down to what your software can do and what the reviewer of your analysis will accept. If you have software that will prepare a probability plot or conduct a statistical test, you're in great shape. If all you can calculate is a coefficient of variation, then start there. It's better than nothing. But beware; if all the reviewer wants to see is a test, it'll be unlikely that he or she would be willing to consider a statistical graphic or other method of evaluating Normality.

The Normality assumption receives a lot of attention, perhaps in part because there are so many methods for assessing its validity. In practice, though, statistical modeling tends to be relatively insensitive to violations of the Normality assumption.

Equal Variances

You know, they want everything to be equal, everything! But when the check comes, where are they?

George Costanza, on *Seinfeld* episode "The Handicap Spot"

The last assumption is called equal variances, equality of variances, homogeneity of variances, or homoscedasticity. They all mean the same thing, that the errors in a statistical model have the same variance for all values of the dependent variable. For models involving grouping variables, the assumption means that all groups have about the same variance. For models involving continuous-scale variables, homoscedasticity means that the variances of the errors don't change across the entire scale of measurement. In the temperature measurement example mentioned in Chapter 5, for example, the variances of the lower temperatures would have to be about the same as the variances of the higher temperatures. In the case of analyte concentrations from laboratory analysis, homoscedasticity requires that the error variance be about the same for measurements at both the low and high ends of the analytical instrument's range, which could be a difficult requirement to meet. Another example would be measurements made over many years. Improvements in measurement technologies could cause more recent measurements to be less variable than historical measurements.

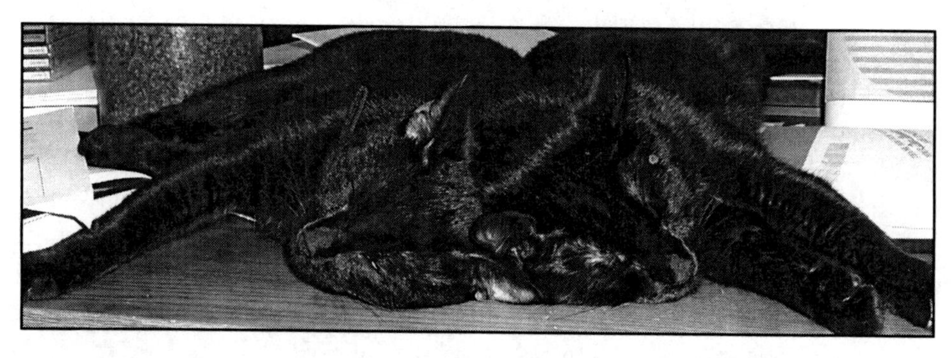

We have equal variances.

Assessing homoscedasticity is more straightforward for discrete-scale variables than for continuous-scale variables because there are usually more than a few data points at each scale level. A simple qualitative approach is to calculate the variances for each group and look at the ratios of the sample sizes and the variances. There are more sophisticated ways to evaluate homoscedasticity, such as Levene's test, but these tests are subject to false negative results when the variances are in fact unequal.

Violations of the homoscedasticity assumption tend to affect statistical models more than do violations of the Normality assumption. Generally, the effects of violating the homogeneity-of-variances assumption will be small if the ratio of the variances is near 1 and the sample sizes are about the same for all values of the independent variables. However, as differences in both the variances and the numbers of samples become large, the effects can also be great. Violations of homoscedasticity can often be corrected using transformations. In fact, transformations that correct violations of Normality will often also correct heteroscedasticity. Nonparametric statistics also have been used to address violations of this assumption.

8 Perspectives on Objectives

[Knowing what to do] ... is a matter of having the right purpose, the purpose appropriate to the situation in hand. ... The one who "knows what to do" is the one on whom you can rely to make the best shot at success, whenever success is possible.
Roger Vernon Scruton, English philosopher

Once you understand data, samples, variables, and scales, you can start thinking about what you could do in a statistical analysis. Statistical analyses usually fall into one of five categories of objectives:

- **Describe**—characterizing populations and samples using descriptive statistics; statistical intervals, correlation coefficients, graphics, and maps.h
- **Identify or Classify**—classifying and identifying a known or hypothesized entity or group of entities using descriptive statistics; statistical intervals and tests, graphics, and multivariate techniques such as cluster analysis.
- **Compare**—detecting differences between statistical populations or reference values using simple hypothesis tests and analysis of variance and covariance.
- **Predict**—predicting measurements using regression and neural networks, forecasting using time-series modeling techniques, and interpolating spatial data.
- **Explain**—explaining phenomena using regression, cluster analysis, discriminant analysis, factor analysis, and other data mining techniques.

Don't feel constrained by these five categories. There are probably other objectives and other classification schemes, but this is a reasonable place to start. Table 7 provides some examples of data analysis tools that can be used for addressing the objectives.

Table 7. Examples of Tools and Uses of Statistical Objectives[1]

Objectives	Commonly Used Tools	Examples of Applications
Describe	Text and images Graphs Descriptive statistics	Opinion surveys Demographic surveys
Compare	Text and images Graphs Descriptive statistics Statistical tests	Pharmaceutical effectiveness Educational methods
Identify or Classify	Visual scans Filters, queries, and sorts Graphs Discriminant analysis Association rules Classification trees Data mining	Biological species Tax return audits Possible criminals or terrorists
Predict	Graphs Regression Neural networks Data mining Unsupportable methods*	Credit worthiness Student success in college
Explain	Regression Analysis of variance (ANOVA) Other multivariate statistics	Academic research

* Unsupportable methods include: intuition, rules-of-thumb, scientific guess, best engineering judgment, punditry, biorhythms, *Old Farmer's Almanac*, folk tales, Nostradamus quatrains, crystal balls, magic-8 balls, oracles, divining rods, pendulums, runes, tea leaves, entrails, tarot cards, **astrology, palmistry, phrenology, numerology,** psychometry, **and the CIA's** "enhanced interrogation techniques."

This classification scheme has three features. First, it's easy to figure out so that nonstatisticians can decide in which category their project fits. Second, the major statistical techniques tend to be used primarily in just one of the classifications. And third, the scheme can be thought of as an index of the professional peril a statistician could face in doing the analysis. Here's why.

Description is relatively easy. You can do the calculations on spreadsheet software. All you have to be aware of are measurement scales, distributions, sampling schemes, and methods for dealing with outliers and missing data. Most description creates information from data, and sometimes knowledge, but never wisdom.

Identification and classification range from simple visual recognition to the exploration of arcane mathematical dimensions where only bold number crunchers venture. You need both capabilities. Say you were trying to find Waldo. At a convention of funeral di-

1 If you want to know what these statistical methods involve, visit the glossary at the end of the book, or even better, search the Internet for several definitions.

rectors, one look would be all you needed. If he were hiding in a candy cane forest holding American flags, though, you would need some nonvisual clues. Here's another example. You can determine a person's sex (usually) by looking at him or her but not from a table of eye and hair color. On the other hand, you couldn't tell who the best players were on a sports team from their pictures, but you could from their performance statistics. Identification is the gateway to classification. If you can do one, you can probably do the other. Most identification and classification creates information or knowledge but not wisdom.

Comparison is tougher even though there is ample software available for most analyses. You need to know what test to run or ANOVA design to use as well as understand probability, effect size, and violations of assumptions. There's a much greater chance of something going wrong that a reviewer will catch. Most analyses aimed at comparison create knowledge but not wisdom.

Prediction is next. In addition to all the description and comparison techniques, you'll need to know how to use a variety of model building and assessments methods and understand the morass of prediction error. In prediction, even if you did nothing wrong statistically, a poor prediction will eventually be noticed. It's easy to make a prediction. It's hard to make an accurate prediction. It's damn near impossible to make an accurate prediction that is also precise. One really good prediction, and a psychic is famous; one really bad prediction, and a statistician is relegated to selling insurance. Prediction may create knowledge, but seldom does it become wisdom.

Finally, explanation is the toughest of all objectives. Not only do you need to understand some of the more esoteric statistical methods, like factor analysis and canonical correlation, but you also have to understand the conceptual framework of the systems the data come from. Then you have to have the talent to apply the knowledge creatively. You can't explain your statistical model of stream contamination without knowing something about stream hydraulics, hydrogeology, meteorology, and environmental geochemistry. You can't explain customer satisfaction without knowing something about demographics, marketing, business, and even psychology. You'll also probably have to integrate the information and think of it in ways that have never been thought of before. Explanation creates wisdom, although most of the time, your results will be humdrum. If you do come up with something truly consequential, though, some people will believe your results are erroneous, coincidental, or faked. Some people will claim that your finding is old news, having discovered it themselves years before. Some people will down vote your findings on Reddit (www.reddit.com) then post them on Digg (www.digg.com). Most people, though, will just ignore you.

Explain to us again why we can't go out there.

So if you are contemplating doing a statistical analysis, know where you're going but be prepared for where you might eventually end up.

PART 2

Frisky Business

Have you ever studied for an exam only to find that a good chunk of the test concerned topics you didn't think would be on the exam, and in fact, know nothing about? It's a bad situation. Well, guess what? After spending all that time studying about statistics, you may find that the toughest part of a data analysis project is planning and managing it. That's what this part is about.

Part II consists of four chapters:

Chapter 9—The Statistical Do-It-Yourselfer has a single purpose—to help you decide if you should analyze your data yourself or hire a professional to do it.

Chapter 10—Manage to Get It Right describes eight aspects of a project that you'll need to know about if you want to analyze data for a living.

Chapter 11—Weapons of Math Production summarizes the resources that you'll want to have available to you for your analysis, including software and information sources.

Chapter 12—Tales of the Unprojected provides examples of some of the things that can go wrong on a data analysis project and suggestions for how the problems can be addressed.

If you have management experience, feel free to skip ahead to Part III. But don't be overconfident. You might find some of this information as valuable as all the poppies in Afghanistan.

9 The Statistical Do-It-Yourselfer

The paradigm shift from mainframe-based statistics to PC-based statistics has led to an even greater paradox, namely that there are people—bosses, reviewers, advisers, teachers, and government regulators—who expect *you* to be able to do all kinds of data analyses, regardless of your training and experience, even when *they* have little notion of what the analysis might involve. Many of these same people also expect that people have a professional license before they exterminate termites or cut hair. Go figure.

You build. I'll supervise.

There are times in all our lives when we make decisions that require us to learn new things. Buy a car, and you have to learn about licensing, insurance, maintenance, maps, and road rage. Buying a house is the ultimate reality show for wannabe interior designers, carpenters, plumbers, electricians, and landscapers. So when that touchy-feely job you took after college suddenly compels you to interact with data instead of people, it's time to pick up some new skills.

This chapter is about one thing—deciding how to get a data analysis done. More and more people have been doing their own analyses over the past decade or two because of improvements in computer hardware and statistical software. You may or may not be confident enough to do an analysis yourself. Only you can answer that question. If you decide you will do an analysis, you need to be sure you have the time and resources to devote to the project. If you decide you need some help, you need to find a person who will actually *help* and not just criticize. If you decide you need someone to do the work for you, you may have to go through an extensive screening process, but if you care about the analysis, you have to do it. Quality begins with the person doing the work.

Begin with a Blueprint

Intuition becomes increasingly valuable in the new information society precisely because there is so much data.

John Naisbitt, U.S. writer

Statistics are like power tools. If you know how to use them, they are incredibly valuable and fun to use. They help you do your job better, more thoroughly, and more quickly. But if you are careless, they can cause great damage. Think of an expert carpenter like Norm Abram on *This Old House*. Norm has a different tool for every possible job he might need to do in his workshop. Statistical methods are like that. There are many different types of statistical analysis. Some perform a single function, and some perform many. In the same way that there are several different types of saws, there are different statistical methods for doing exactly the same thing. And just as Norm knows when to use his table saw and when to use his band saw, a statistician knows when to use different types of statistical analysis.

The proliferation of personal computers has changed how statistical analyses are done and who does them. Many of the specialized software tools that statisticians use now have easy-to-learn GUIs and are inexpensive enough that do-it-yourself statistics is commonplace. But it's one thing to own the same band saw that Norm Abram does and another to demonstrate the same woodworking prowess. The tools will help you build a condo for your cat, but you'll need more skills to build a house for yourself. Likewise, there are many statistical analyses you can do on your own, but also, others that call for greater skills. The more you use the tools, the better you'll get at using them.

If you're planning to do a statistical analysis yourself, you'll need a blueprint for what you plan to do. First you'll need to identify what has to be done and what questions have to be answered. You'll need to be clear on why you're using statistics. You'll need a schedule so that you know if the data processing is proceeding as planned. From that, you'll need to decide how extensive the statistical analysis should be. A statistical analysis can be simple and quick, or it can involve complex data modeling or data mining. You can do a "mahogany" interpretation or a "balsa wood" interpretation. You'll need to ascertain whether the necessary tools are available to do the analysis. Tools can include a fast computer, good statistical software, and knowledge resources such as technical books and journal articles. But the first big decision you'll have to face is: *who should do the work?*

Who Builds the House?

If you didn't want them to think, you shouldn't have given them library cards.
Harry Bailey in *Getting Straight*

Doing a statistical analysis is like building a house. You identify study objectives and select an experimental design for your statistical analysis like you would prepare designs and floor plans for your house. You specify a sampling strategy and collect data like you would specify and purchase building materials. You prepare the dataset, the foundation of the analysis, construct statistical hypothesis and tests to frame your analysis, and build stories step by step that manifest your results. Finally, you live with the decisions you make based on the analysis as you would live in the house you built.

Faced with the responsibility of getting a data analysis done, you have two options. You can do it yourself or you can employ a professional. It's not a good idea to jump into a statistical analysis without first thinking about how you intend to get the analysis done and use the results. Assuming you have an adequately defined problem and work scope, there are a number of considerations in deciding whether you can do the statistical analysis yourself or whether you need to hire a professional.

First, there are your own constraints. Will you feel comfortable doing the calculations (i.e., using the software) yourself or having an associate do them? Do you have enough time available to invest in doing the analysis, knowing that you will probably have to learn some new things? In this regard, it is worthwhile to think ahead to the ultimate use of the analysis. How visible is the project? Will it be reviewed by outside experts? Might the results be the focus of a court proceeding? Will your results be picked up by the media? Is there any risk to human health or the environment? Are there financial or business risks? In other words, what would the ramifications be if you made a few mistakes? How about if you really screw up?

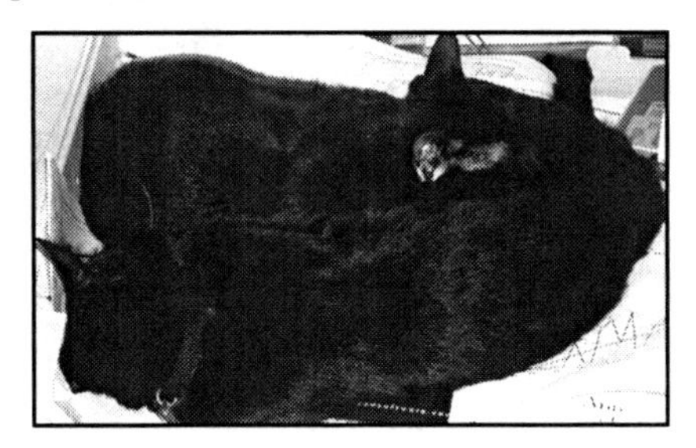

Make sure you know whether you're coming or going.

Second, there are your resources. Do you know what kind of statistical analysis needs to be done? Do you have reference material that can help you in case you run into problems? Do you have appropriate hardware and software for conducting the analysis and preparing the results? Finally, do you have a contingency plan? Is there someone you can rely on for advice, or call in to replace you if need be? Remember, there are no Home Depots for statistics.

Measure Twice; Cut Once

Statistics is, or should be, about scientific investigation and how to do it better, but many statisticians believe it is a branch of mathematics.

George E. P. Box, U.S. statistician

If you can answer all of those questions and still feel comfortable that you can do the analysis (or oversee it to someone who works for you), then go for it. Be sure you will be able to obtain five essential elements:

1. **An *appropriate* dataset**. This turns out to be a huge issue that is discussed in Parts III and IV of the book.
2. **Basic knowledge**. You might have taken a statistics course or two in college and feel like you have a good basis for conducting a statistical analysis. Maybe you do. But this can be like saying you're ready to build a house after assembling an IKEA bookcase. If the statistical analysis is simple enough and you have suitable technical resources to fall back on, you may be safe. If not…
3. **Technical resources**. Access to a library of statistical textbooks, articles, and Internet sites, and most importantly, *having the time and the patience to use these re-*

sources are necessities. There is no type of statistical analysis that doesn't require a more sophisticated understanding of how statistics work than is provided in an introductory college course. Take for example, calculating central tendency statistics, one of the most elementary topics covered in Statistics 101. You may also need to know about censored data, Normality, and Lognormality, and even spatial and serial correlation. So consider whether you're prepared to teach yourself some new tricks.

4. **Computer and software**. Statistical programs are pretty efficient, so you won't need a state-of-the-art machine unless you want to play World of Warcraft when your boss isn't looking. Just don't plan on using the hand calculator you got when you opened your checking account. The chance of making mistakes is far too great. You might be able to calculate the statistics you want in a spreadsheet program such as Microsoft Excel if you know how to use its formulas and tools. Be aware though, some of the statistical functions in spreadsheet programs are notably limited. More advanced statistics require the use of either a general statistical package or specialized software for a particular type of analysis. This topic is discussed in Chapter 11.
5. **A backup plan**. If it turns out that you or your associate can't complete the analysis, you'll need to have a plan for what to do. Remember, if there's barely enough time to do the work, there sure won't be any time to do it over.

If there are aspects of the work that you're not comfortable with, it's probably better to call in a professional statistician. In that case, you'll need to know how to find the right professional, hopefully without having to resort to Craig's List (www.craigslist.com).

Hired Hands

To consult the statistician after an experiment is finished is often merely to ask him to conduct a post mortem examination. He can perhaps say what the experiment died of.

Sir Ronald A. Fisher, *The Design of Experiments,* 1935

A statistician is more than just a machine for converting raw data into usable knowledge. If you decide to use a professional statistician, you first need to consider what role you might want him or her to play in your analysis. Sometimes, you may only need a statistician to help you design the study. You can then collect and analyze the data and report the results. This happens a lot with simple surveys. Alternatively, you may want the statistician to both design the study and conduct the analysis. Either of these approaches is perfectly acceptable.

A less acceptable approach would be where you design the study and collect the data, and then expect the statistician to conduct the analysis. This can present problems because the type and number of data points can have a big impact on the statistical analysis. For example, you might want to develop a statistical model of how some measurement

Cats like us, baby we were born to rub.

changes over time. If you don't collect your data at constant time intervals, you won't be able to use some types of statistical analysis (i.e., autoregressive modeling). If the time interval between your sampling rounds is too large, you may miss important sources of temporal variation.

The least acceptable approach to using a professional statistician would be for you to design and conduct the analysis, and then ask a statistician to review your product. This approach invariably results in disaster. Trust me on this one; you won't be able to find a statistician who'll take a bullet for you if you screw up the analysis.

Handyman or Specialist

For every expert there is an equal and opposite expert.

Arthur C. Clarke, British writer

If you decide that it would be better to employ a professional, there are several considerations for deciding what type of statistician would be best suited for your work, including:

- Diversity of expertise
- Educational background
- Data versus method focus
- Conservatism
- Experience
- Connection
- Cost

Diversity of Expertise

Professional statisticians are like house builders. There are general contractors, and there are specialty contractors. Some statisticians specialize in a certain type of data, for example, pharmaceutical or economic statistics. Some statisticians specialize in a certain type of analysis, for example, factor analysis, geostatistics, or Bayesian statistics. These statisticians would be like stonemasons or plasterers in that they are extremely knowledgeable of their particular craft and seldom venture outside it. Some statisticians, on the other hand, or more like general contractors. They can use a wide variety of statistical methods to analyze many types of data, although perhaps not as thoroughly as specialists.

As in the construction industry, there are trade-offs between hiring specialists and generalists. A carpenter can build or fix many of the parts of a house. But if you have a unique problem like safely removing asbestos insulation, even a master carpenter like Norm Abram would not be able to provide the best solution. Statistical professionals are similar. A statistician who is a generalist will be able to conduct most of the kinds of analysis that may be needed for your work. However, if your data or the problems you are trying to solve have some special quirks, more specialized knowledge and experience might be needed. Beware, though, if you employ a statistician who, for example, specializes in nonparametric statistics, your answers will all likely involve nonparametric statistics. When all you have is a hammer, everything looks like a nail.

Educational Background

You might think that all statisticians have degrees in mathematics or statistics. This is not true. Many statistical professionals have degrees in engineering, a natural or social science, or business or economics. The U.S. Office of Personnel Management requires that statisticians (job series 1530) in the federal government have *any* degree that includes at a minimum:

A. 15 semester hours in statistics, or 6 semester hours in statistics and 9 semester hours in mathematics, and

B. 9 semester hours in the natural or social sciences; engineering; or education (and just about everything else except fine arts, languages, and religion).

That's about five courses in statistics if you take no other math. They even give credit for courses in which 50 percent of the syllabus *appears* to involve statistical methods. Not very rigorous, huh? For comparison, meteorologists have to take 24 semester hours just in meteorology. Chemists have to take 24 semester hours in chemistry. And engineers, well, don't ask. It's no wonder that everybody thinks they're qualified to do their own statistical analyses, the government's Office of Personnel Management[1] thinks you can too.

Data versus Method Focus

Educational background often does influence a statistician's technical focus. Some statisticians focus on the statistical methods that they use. For example, mathematical

1 I'm sure it's just a coincidence that the office's acronym is pronounced "opium."

statisticians specialize in the theoretical calculations that go into a statistical method. They tend to be more strict about violations of statistical assumptions, testing for outliers, and other methodological issues. Applied statisticians, on the other hand, are more focused on the data that they are analyzing. They tend to be more strict about data quality. This distinction can be important in a statistical analysis where the reasons for an influential observation must be evaluated or deviations of the statistical assumptions must be addressed.

Conservatism

Another important personality trait to consider is a statistician's methodological flexibility or orthodoxy, their willingness or aversion to taking risks in their analysis, their flamboyance or conventionality in the use of statistical techniques. For lack of a better word, I'll say *conservatism.* What I mean is someone described as orthodox, conventional, cautious, prudent, careful, or guarded versus someone described as flexible, original, creative, innovative, imaginative, or enterprising. Long before the ascendancy of neoconservatives in politics, the term *conservative* referred to someone who tries to "preserve from ruin, injury, innovation, or radical change," that is, someone who is "a preserver; a conserver." In this context, a conservative statistician is one who would decline to conduct analyses that he or she feels will not rigorously meet all mathematical requirements. Small sample sizes, violations of statistical assumptions, and missing data will often bring conservative statisticians to a dead stop. In contrast, less conservative statisticians might resort to transformations, resampling, missing-value replacement, and other adventurous statistical techniques to try to complete the desired analysis. In selecting a professional statistician for your project, you might find that a conservative statistician might be better for projects likely to undergo close scrutiny, such as legal or regulatory reviews. Pharmaceutical testing might be one example. Less conservative statisticians might be preferable if the goal is knowledge discovery.

Experience

And then there's experience. How important is experience? Here are two opinions I agree with.

> Most [statisticians] really want to be helpful. If they aren't, it's because they are inexperienced in dealing with clients or don't have a proper respect for the field work. I once heard that it takes 200,000 casualties to properly train a major general. I suppose most [statisticians] have to personally kill or cripple at least fifteen projects before they can learn their craft.
>
> John Bell[2]

---

2 Bell, J. 1988. *Statistics for Practical People. Part II – How to Get Statistical Help.* Located at: www.pro-axis.com/~johnbell/sfpp/sfpp2.htm. Bell's original article referred to biometricians and statisticians.

> ... in statistical consulting, experience counts for a lot. You can have a bright young mathematician but there's no such thing as a bright young statistician unless he's doing mathematical statistics. Successful data analysts are old and grizzled and have been burned a few times.
>
> Peter Lewycky[3]

Some educators claim that any skill can be learned with enough practice. Remember the punch line to the old joke about the tourist asking the Manhattan cabbie how to get to Carnegie Hall—*practice, practice, practice.*[4] Remember homework? Whether you're learning math ("do all the even-numbered problems") or basketball ("shoot baskets for at least an hour after school") or piloting ("you'll need 1,000 hours of flight experience"), practice is the difference between those who do and those who watch. So the more experience your prospective statistician has the better. But, it has to be *effective* experience.

Geologists have an interesting concept called *effective porosity*. You know what porosity is. It's the holes in sponges and paper towels that allow them to absorb water. For geologists, the holes are in rocks. Geologists refer to the holes that are connected so that water can pass through as *effective* porosity. A rock can have a high porosity, but if the holes aren't effectively connected, they're pretty hard to get groundwater or oil from. In the same way, experience that doesn't move you closer to perfecting a skill is pretty useless. Practicing "Chopsticks" on the piano is a good place to start, but ten years of playing "Chopsticks" won't prepare you to play Rachmaninoff. A decade of painting buildings won't prepare you for painting murals on buildings. A thousand pie charts don't equal one analysis-of-variance. Every statistician has to start somewhere, whether it's calculating descriptive statistics, administering surveys, or creating charts and graphs. To be effective, though, experience must evolve over a professional's career. It's what job ads call progressive experience.

Remember taking your first class in statistics? Hopefully, it was an invaluable experience, or at least, didn't require more than a couple of visits to a therapist. Now, what if you had repeated that same class?[5] You might acquire some additional insights the second time around, but after taking the class again and again, the rewards of the experience would diminish. Even *Star Trek* reruns lose their appeal after the seventeenth viewing. What if your prospective statistician says he or she has a decade of experience conducting a certain type of statistical analysis? Does that make him or her an expert in what you need? Is that really ten years of experience or is it one year of experience, ten times?

And it's not just gaining progressively more diverse and enriching exposure to statistical techniques that makes experience effective. It's the application of the experience. Analyzing data is an essential component of all statisticians' training. Many statisticians also progress to planning and carrying out statistical studies. But to take the next step to

---

3 Lewycky, Peter. Excerpt from message posted in sci.stat.consult on Jan 10, 2003, available at http://groups.google.com/group/sci.stat.consult/browse_frm/month/2003-01.

4 Obviously, this joke dates back to a time when the cab drivers spoke English.

5 Maybe you did, but we won't go there.

become a statistical consultant, your prospective statistician will have to be able to solve problems using statistics, in particular, your problem.

So there's experience, and then, there's *experience*. It isn't just a matter of how much experience, it's how diverse and how well done it was, and above all, how well it was put to use in solving problems. So if you know nothing about statistics, what do you ask a prospective statistician so that you can decide if they're the right person for your project? It's actually pretty simple. Just tell them about your project, and then—listen.

Talk to me. I'm all ears.

Does the statistician ask you exactly what kind of analysis you want them to do? Beware. They might just do exactly what you asked for. Are you sure you can tell them the correct thing to do? Does the statistician expound on ANOVA designs or Bayesian statistics or other statistical techniques they might use? You'll probably get a type of analysis he or she knows very well, maybe even a very sophisticated type of analysis, but it may not solve your problem. But if he or she asks you about your objectives and how you will use the results, and seems to know a bit about the kind of data you want to analyze, you have your statistician.

Connection

After experience, the remaining considerations depend more on your personal preferences for connecting with your statistician. Some people prefer to work with people located close by so that they can interact face to face more often. With the availability of the Internet, cellular telephones, webcams, and other forms of communication, this doesn't have to be very important. Most statistical analyses can be conducted remotely without any degradation in service. I've personally conducted scores of analyses for clients who I have never met in person.[6] What's more important is personal chemistry. This is of course true of any professional interaction. If you cannot understand the explanations given by the statistician or are sensitive to a personal quirk, it's often better to find someone else.

Cost

As with all work efforts, there will be a trade-off between using a less experienced, less expensive individual who might take longer and do a less sophisticated job versus using a more experienced, more expensive professional who is more likely to complete the job satisfactorily in less time. Most professional statisticians charge fees in the range of $25/hour to $150/hour and up depending on credentials, the competitive market, the geographic location, the type of work, and other factors. For example, an independent statistician in Prairieland, Kansas, might charge you $30/hour to make some pie charts for your business's annual report while a San Francisco consulting firm might charge

6 In fact, I've never met many of my most loyal repeat customers. Hmmm, I wonder if this means anything?!?

$200/hour for one of their PhDs to support you as an expert witness in court. Usually, you get what you pay for, but there's no reason to pay top dollar for simple work.

The Right Tool for the Job

It is tempting, if the only tool you have is a hammer, to treat everything as if it were a nail.

Abraham H. Maslow, *The Psychology of Science*, 1966. p. 15.

Once you have an idea of what kind of statistician you want to look for, here's how you might go about the search:

Step 1. Write a description of the scope of your project including statements of the particular questions you want to answer. Don't worry about mathematical statements, like null hypotheses. Just write out what you need to know. Establish an acceptable schedule for the work being sure to include adequate time for report preparation, review, and revisions. Estimate what an appropriate budget would be for the efforts. If you don't know what acceptable schedules and budgets might be, you can let the statistician figure it out. If you have a limited budget or schedule, it's better to let the statistician know beforehand.

Step 2. Assemble the datasets that would be used in the analysis. Decide whether you will do the data formatting or expect the statistician to do it. Be aware that, no matter how well you scrub the datasets, the statistician will inevitably have to make changes to accommodate his software or methods of analysis, as well as quirks in the data.

Step 3. Conduct your search. Don't ignore the obvious; start by looking within your own organization. If you find no good candidates, search the Internet for statisticians using keywords from your scope statement. There are many independent statisticians and statisticians who work for consulting firms who can be found this way. You might also try visiting the web site of the American Statistical Association (www.amstat.org/). If you prefer, you can look for statisticians at local universities. Don't confine your search to the Math Department, unless of course you don't want any peers in your field snooping around your dataset.

Step 4. Obtain credentials from the candidate statisticians. Of particular importance would be the statistician's résumé showing what other projects he or she may have worked on involving the type of data you also want analyzed.

Step 5 – Evaluate the resume in terms of your project requirements. Here are a few considerations.

- **Level of Education**—If you are doing academic or high-profile research, find a PhD. With that exception, a candidate statistician's degree is usually only important for public perception. Courts don't typically differentiate levels of experts. You're an expert or you're not. Regulators usually don't care as much about the degree as the technical arguments.

- **Academic Discipline**—The training a statistician has will often influence how they approach a problem. For example, statisticians trained by mathematicians tend to focus on methods (i.e., the statistical tests and procedures) rather than data. Statisticians trained by economists tend to favor regression and time series analysis. Sociologists like opinion surveys. Geostatisticians, usually trained in mining geology or engineering, favor mapping solutions. These tendencies will become apparent on reviewing descriptions of projects they have completed. You might also review writing samples. Statisticians trained in mathematics tend to use more mathematical equations and notation than statisticians trained in business or the sciences.
- **Experience**—As with any discipline, the more years of experience a candidate has, the more likely it is that he or she has seen a similar problem and will know how to solve your unique problem. In the summaries of project experience, look for two things: the types of data that were analyzed and the statistical methods that were used. A statistician who analyzes many types of data would tend to be more comfortable analyzing your data even if he or she had never analyzed that type of data before. A professional who has analyzed only one type of data may be tentative in approaching a new type of data. For example, a statistician with a background in business who has only analyzed credit worthiness data may not be well-suited for evaluating biological diversity. Also look to see if the candidate has experience with a variety of statistical methods. If all of the candidate's experience involves preparing geostatistical maps, don't expect them to necessarily be proficient with factor analysis, and vice versa.

Step 6. Interview each of your candidates to get a feeling for how it would be to work with them. Be sure to broach any conflict-of-interest issues and special requirements you may have. You should also be prepared to discuss in general terms the data and the analysis goals. Then make your decision.

Plans and Specifications

Statistics is the grammar of science.

Karl Pearson, *The Grammar of Science,* 1892.

Once you have identified the person who will be your statistician, you will need to specify your project goals and scope to him or her in a clear manner. The budget and the schedule are the first and easiest things to convey. More difficult may be the exact study questions or goals of your project. Be as specific as possible. The statistician will need to translate your goals into mathematical hypotheses that can be tested or evaluated. It's not necessary for you to specify what statistical methods to use. That's the statistician's job.

However, you should specify whether you want the statistician to use only conventional methods or whether he or she is free to innovate.

After the scope, the most difficult item to describe is the dataset. You should be able to describe what type of dataset the information resides in, that is, a hierarchical dataset, or a flat file that is formatted or delimited, or a matrix like a spreadsheet. This has important implications for the statistician (discussed in Chapter 16). Statistical software requires that the data be arranged in a matrix. Consequently, hierarchical datasets have to be restructured before they can be processed. Spreadsheets are acceptable once they are formatted correctly. The statistician will need to know what data elements or variables are in the datasets and how many observations there are. Also, if there are replicate samples or quality assurance samples in the data, these will have to be addressed. Formatting the dataset before statistical analysis can easily be as little as half of the work to almost all of the work of the statistical analysis. Hence, it is important for the statistician to understand how much work will have to go into preparing the data.

"A computer and a cat are somewhat alike—they both purr, and like to be stroked, and spend a lot of the day motionless. They also have secrets they don't necessarily share." John Updike, U.S. writer

It is important not to forget to specify what type of output you will be expecting from the statistical analysis. For example, is a letter report sufficient or is a complete report with data records and methodological references and descriptions required? Do you have a specific format for the report? Will the statistician's report need to be incorporated in another document? Do you want to receive the formatted data and even the report itself as electronic deliverables? If you require an expanded report, who will be the audience for the report? Should the statistician write for a general audience or is the use of statistical terminology acceptable. Also it is important to specify how you want graphics to be produced. Should they be included in the report file or enclosed as separate files? Should they all be standard 8 1/2 by 11–inch or can oversized graphics be used? Is color acceptable or should everything be in black and white?

There may also be other requirements that you'll want to specify. For example, you might have a confidentiality requirement on releasing the data or the analysis results. You may also need to have the statistician prepare for providing testimony in court, making a presentation to your company's management, or presenting to the public. You should specify whether you expect the statistician to respond to comments after a peer review. Finally, if you have any plans for publishing the results, you should make the statistician aware so that you can discuss proper formatting and attribution.

Expectations

If you don't set goals, you can't regret not reaching them.
Yogi Berra, professional baseball player and coach

If you use a professional statistician, you can expect to get an appropriate statistical analysis and interpretation. However, this does not guarantee that you'll get the answer you want. The condition or phenomenon you are investigating may be too complex to model or your dataset might not be adequate to capture the vagaries of the phenomenon. Furthermore there's no guarantee that reviewers will agree entirely with the approach taken or the interpretation of the results. There may be a variety of methods that can be used to analyze a given dataset, and every statistician may have his or her own preferences as to which method is better. There are also quite a few topics in statistics that are controversial. Frequentist versus Bayesian approaches, parametric versus nonparametric techniques treatment of outliers and missing values, and use of transformations are but a few.

10 Manage to Get It Right

If you are a student pursuing a degree in statistics, you might find this chapter to be the most important in this book. You'll rarely hear much about managing your data analysis projects in school, but once you get a job in the real world, it'll be your biggest headache. You'll find numbers to be quite a bit friendlier than some of the people you'll encounter. Better to know what to expect now than learn from personal encounters with office gremlins.

If you are already an office dweller looking to do a bit of number crunching, you might have already run into some of these issues. Probably the biggest difference between data analysis projects and most other types of projects is knowing when you're done. If you're building something or doing some well-defined task, you usually have a pretty clear end point. Data analysis is like searching the Internet. One thing leads to another until you've gone on seventeen tangents and have forgotten what you were looking for in the first place. Try to stay focused.

If you already have project management experience, you can skim or even skip this chapter. You probably have already heard or even experienced it. However, if you enjoy those *epic fail* videos on YouTube, you might be amused by the stories of a statistician trying to be a PM (i.e., Project Manager).

P-R-O-J-E-C-T Find Out What It Means to Me

The statistician who expects that his contribution to the planning will involve some technical matter in statistical theory finds repeatedly that he makes a much more valuable contribution simply by getting the investigator to explain clearly why he is doing the experiment, to justify the experimental treatments whose effects he proposes to compare, and to defend his claim that the completed experiment will enable its objectives to be realized. ... These comments are offered with diffidence, because they concern questions on which the statistician has, or should have, no special authority, and because some of the advice is so trite that it would be unnecessary if it were not so often overlooked.

W. G. Cochran and G. M. Cox,, *Experimental Designs, 2nd Edition,* 1957, p. 10.

Sometimes all you have to worry about is getting a data analysis completed successfully. More often than not, though, you have to consider how your analysis fits with your client's directed scope, schedule, budget, and other requirements. Ask any project manager; it's not that easy. You have to understand who the stakeholders are, what the risks are, where you fit in, and how everything interacts.

Conducting a statistical analysis can be like traveling to a foreign country that you've never been to before. You had better have a map and some idea of what you want to do there or you might end up wasting a lot of time, or worse, getting totally lost. That's what this chapter is about—how to plan a statistical analysis so that you don't get mugged and end up in the gutter.

The Client

There are two essential rules to management. One, the customer is always right; and two, they must be punished for their arrogance.

Dogbert, in the Dilbert comic strip on
February 28, 2001, by Scott Adams

Hi. I'm your client. Nice to meet you. Now get me some tuna.

You will always have a "client." He or she is the person who initiates, guides, pays for, signs off on, and uses your work. If you're lucky, your client won't be a five-headed dragon. So the first thing you have to ask about is "who's the client"? This may sound like something that should be obvious, but believe me, it's not always the case.

Say you are working as a statistical consultant to an engineering firm that, in turn, is working for a law firm that is representing some "ultimate" client. The ultimate client is the person or company that pays for your work, even if the money has to trickle through several other companies before what's left gets to you. Sometimes, even the "ultimate client" may have more than one head, most commonly a technical manager and a business manager. Aside from the problem of getting paid (which can be a *big* concern), not working directly for the ultimate client can create several problems.

First, you may never get to talk to the ultimate client. Their words are filtered through several organizations, perhaps to maintain confidentiality or perhaps to prevent "short-circuits."[1] What you are told the client wants may not be what they said. For example, a client might tell his lawyer they don't want to spend more than $100,000 on sample

1 Companies that hire you to work on projects for one of their clients often don't want you to communicate directly with their client. If their client gets to know you, they may decide to hire you directly and bypass or "short-circuit" the company that originally hired you. That company would not

analysis. The lawyer tells the engineer that the client has already collected samples and doesn't want to collect a lot more. The engineer tells you the client doesn't want to collect samples, so you should just use the available data. You spend a month wringing some meaning out of a patchwork dataset only to hear the client say that the old data are crap and you should have recommended collecting new samples. Guess who looks like the idiot? So do you say it's the engineer's fault? Not if you ever want to work for them again.

Think of the client as the person you have to satisfy with your analysis. It may be your boss or some other manager in your own organization. It may be a government regulator or other reviewer. It may be your dissertation advisor. Whoever the client is, he or she may not have any say in your compensation for your statistical work but will have the power to influence what you do. Your dissertation advisor will tell you to do things with no thought to how much work it will be for you. Your boss will tell you your analysis isn't what he wanted, as he tosses months of your work in the trash. Your client will assign a new manager to your project who will want you to do everything over, *his way*. The reviewer who barely scraped through a statistics course a decade earlier will suddenly be qualified to disparage your most sophisticated analysis.

Second, more often than not, you'll have at least two clients—the *real* client who pays the bills and your boss. Pleasing many masters isn't easy. The client will want all kinds of status reports, meetings, preliminary findings, conference papers, advice, and other freebies. Your boss will also want status reports, meetings, and preliminary findings as well as your free time to organize the office Super Bowl party. At the same time, he'll be nagging you to finish this project because it's already over budget and he needs you to get started on the next project.

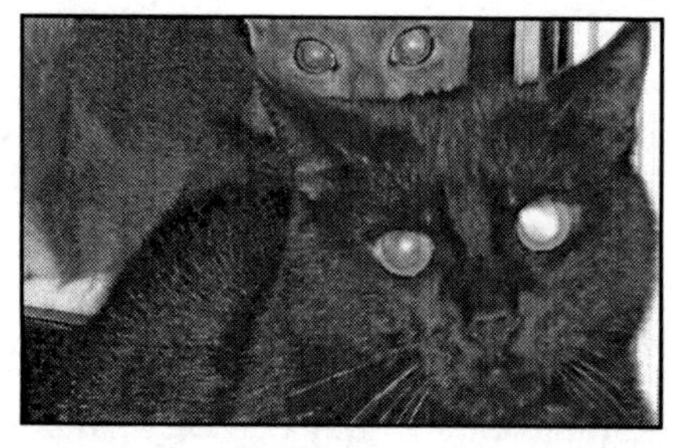

That's all I need you to do but my business partner has a few requests.

If you aren't dealing directly with the client, you can also expect several rounds of reviews of your report. The engineer reviewer will tell you to add the mathematical details and references. The legal reviewer will have you take out everything that isn't a fact or won't help his or her case. The client will ask you to rewrite it so that they can understand it and present it to their board of directors. The regulatory reviewer will have you reformat it so that it looks like the last report he or she read. Sometimes rewriting reports can cost more and take longer than writing the original report. You usually know you're finished with revisions when the rewritten report looks pretty much like your original report.

Third, avoid being your own evil twin. Be sure you know who the ultimate client is so that you don't end up in a conflict of interest. On two occasions in the past I've been asked if I would review a report for a potential new client. After asking a few questions, I figured out it was my own work that they wanted me to review.[2] Sometimes, there may be

only lose whatever management fees they make on you but may also have to take directions from you instead of giving them to you.

2 This is not as weird as it sounds. You would think that the second party would just identify the author of the report and avoid the conflict. Some companies, however, do not as a policy identify who

potentially embarrassing situations that clients need to be made aware of. For example, I've worked on the same project for three different offices of one government client, decades apart. I've even worked on the same project for different private clients with their knowledge and consent. But if you don't catch these conflicts before you start, someone else will before you're done.

Project Goals

Sometimes the questions are complicated and the answers are simple.
Theodor Seuss Geisel, a.k.a. Dr. Seuss, U.S. writer and illustrator

As with all projects, you need to be sure you're clear on the analysis scope, schedule, budget, and deliverables if you don't want any unpleasant surprises along the way. Begin by identifying the project goals. These define where you're starting and where you want to end up.

Project goals are usually set out by the client but may be based on regulatory requirements or guidance. Goals may involve:

- **Conducting a Specific Analysis**—Some clients want to interpret their own data but lack the expertise or resources to conduct the analysis. All you might be asked to do is run the software. These assignments are common. Search monster.com for "SAS programmer" and you'll see what I mean. Sometimes limiting the scope in this manner is a way to simplify a large and complex project. Sometimes it is used as a way to provide security because no one data analyst would see all the results. Sometimes it is a way to evade having to share credit with a colleague. Sometimes it's just what your dissertation advisor wants you to do. Make sure you understand what you're getting into before you commit.
- **Answering a Specific Question**—Some clients only want one specific thing. Is their new product better than the old product or their competitor's product? They don't usually care about what you do so long as you answer the question. Sometimes a client will know what needs to be done, like improve a manufacturing process, but not know where or how to look for solutions. These projects are usually fairly straightforward especially if the requirements are spelled out in some government regulation or guidance document. Just be sure that that's all you really need to do so that you don't leave your client in the lurch if they aren't as well acquainted with the requirements as you are.

---

wrote a particular report. Some companies even incorporate contributions of subcontractor into their reports without attribution. These companies assert that it is the company not the individual author that assumes liability for the report's contents, so the company's name is the only one that should appear. Or there might be so many authors, editors, reviewers, and support staff involved in preparing a report that attribution of a few individuals would be unfair to the rest of the individuals on the team. Many professionals, however, consider this practice to be questionable if not downright unethical because it is also used to hide contributions from inexperienced or unqualified individuals. In either case, the practice is not uncommon. On second thought, this is as weird as it sounds.

- **Addressing a General Need**—Some clients have a general notion about what they want but can't distill it into a specific question or requirement. These cases can be a bit more challenging because you not only have to ascertain what a client thinks they want but also what you believe they need. Projects with general goals often involve model building. You have to establish whether they need a single forecast, map or model, or a tool that can be used again in the future. If the client is looking for a tool, be sure you are clear on the limits of the model's applicability so that there are no misunderstandings or misapplications.[3]
- **Exploring the Unknown**—Every once in a while, a client will have nothing specific in mind but will want to know whatever can be determined from the dataset. Usually, these projects involve examining large datasets that have been compiled, sometimes over long periods, but never analyzed in total. These projects can be two-edged swords. You can really delve into a dataset and try some of the more esoteric techniques without too much fear of backlash from hostile reviewers. On the other hand, there is usually quite a bit of pressure to come up with *something* no matter how messy the dataset is. There is also the danger of not being clear on budgets and schedules. Is this a job for statistical modeling, data mining, or just descriptive statistics?

Whatever you do, make sure the goals are *SMART*—specific, measurable, attainable, relevant, timely—and agreed to by all parties directly involved in the project. There are three situations that can muddy the waters of your objective:

- **Changing goals** are common when the client has only a general goal to begin with. As you find meaning in a dataset during your analysis, the goals may shift or crystallize into something specific. That's fine. It's the reason for doing the analysis. Just watch the budget if the redefined objective takes you way beyond your original scope of work.
- **Multiple goals** aren't uncommon, either. Sometimes, a client might clearly instruct you to consider two or more goals. No problem. Beware of clients looking to get two for the price of one, though.

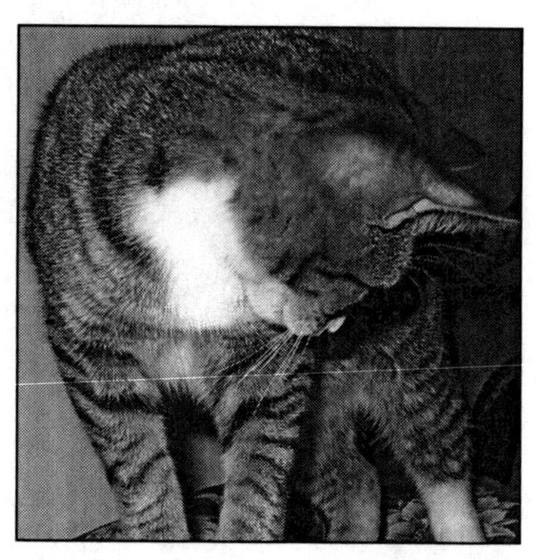

I think we need to go in a new direction. You can do it with the same budget. Right?

3 I once prepared a spreadsheet to analyze a client's statistics that included the use of a statistical procedure not commonly available in spreadsheet software. The client had specifically said they didn't have any statistical expertise and didn't want to do any calculations. However, I later found out that the client reused my spreadsheet for some of their other projects. Unfortunately, they didn't realize the statistical calculations were based on a specific sample size. The calculations I programmed were invalid when the sample size changed. Had the client made it clear that they wanted a tool and not just a specific answer, I would have programmed the spreadsheet differently. That would have cost more, though.

A simple sounding objective might be saddled with additional effort not in your budget; a *freebie* perhaps only mentioned informally (Oh, by the way, can you . . .), that can substantially affect your performance. A typical example might be something like, "Conduct this analysis for us, and by the way, can you give us the spreadsheet when you're done." You might not even have used a spreadsheet if they hadn't asked. And it's one thing if the client just wants the spreadsheet for documentation, but quite another if they plan on using it on a different dataset. Doing the calculation might be easy, but setting up a spreadsheet to handle the different kinds and amounts of data the client might have in the future would be a much larger effort. You also have to consider what professional liability you may have in such an instance.

- **Proposals** are the third special situation to watch for. Some clients will ask for detailed proposals then say they decided not to do the work. In fact, they just needed a plan for doing the work themselves. They get their cake and eat yours too. You can't obsess about this. If a client is going to do this, your only option is to decline the work, which consultants rarely do unless they *know* something is afoot. At the same time, you don't have to give detailed procedures and references for every analysis you might plan to conduct.

If the client isn't entirely clear about their objectives, it may be that they are unable to articulate their goals in your language of quantitative analysis. So start at the very end. Try asking them what decisions they will need to make based on the results of your analysis. They'll understand and be able to articulate those decision points. Then you can translate the decisions into the statistical hypotheses you'll need to evaluate, identify the data you'll need, and select the appropriate statistical methods.

Project Background

Yu [a disciple], shall I teach you about knowledge? What you know, you know, what you don't know, you don't know. This is true wisdom.

Confucius, from *The Analects* 2:17

To know that we know what we know, and that we do not know what we do not know, that is true knowledge.

Henry David Thoreau, U.S. writer

There are known knowns. These are things we know that we know. There are known unknowns. That is to say, there are things that we now know we don't know. But there are also unknown unknowns. These are things we do not know we don't know.

Donald Rumsfeld, former U.S. secretary of defense

What the bleep do we know!?

William Arntz and Betsy Chasse, screenwriters

Once you have an idea of what you'll be doing and who you'll be doing it for, you need to get into the details of the project. This step is like a doctor giving a patient a general physical before diagnosing some malady. First you get a general history of the case, then you ask some generally applicable questions, then you delve into the specifics.

Usually, you might start by learning about the general history of the phenomenon, the study area, or the project. This information is ordinarily descriptive but is essential for putting your analysis into a real-world context. Get as much of the story as you can. Take a tour of the study area or project location if you can, especially if data generation is underway. The information may not seem important at the time, but it may help your interpretation of the statistics later, especially when evaluating the representativeness of outliers.

Once you get an overview of the project, there are several standard questions you might ask (if they haven't already been answered):

- **Client's Vision of Work**—Does the client have any expectations of how the work will turn out? Do they plan to follow some particular course of action based on the work? What work products are they looking for? Obtaining this information isn't to "get the right answer." It's so you can issue preliminary warnings to the client if something falls too far outside of their expectations. For example, say a client expects to curtail an environmental monitoring program based on an expectation of the site passing a statistical comparison to a regulatory limit. You might need to warn the client to budget funds for future monitoring if you see "problems" in the data that might make the oversight agency reluctant to approve the elimination of monitoring.

- **Existing Data and Reports**—Get whatever background materials are available, especially electronic data. The information might help your analysis or you may have to refute some prior misconceptions. In either case, you need to know what's there. If electronic data will be made available, try to get an idea of the number of cases and what variables were recorded so you can think about what to do with the data while you're waiting for it to be delivered.

- **Project Boundaries**—Be sure to ask if there are any limits that are relevant to the project. Does the client want you to limit your analysis to certain processes, or part of his property, or a certain period of historical data, or number of samples?

- **Project Participants**—Find out who else has worked on or is working on the project, including other consultants, oversight agency personnel, and lawyers. You may find that you have your own "history" with another project participant. You may have to deal with unanticipated personal conflicts of interest or awkward personality clashes on prior encounters. The longer you work in a profession, the more likely this becomes. Also find out if you are free to make contact with the client, project participants, and others outside the project. There may be legal, security, or confidentiality restrictions.

- **Interested Parties** —Ascertain what community or other outside interests or involvements there may be. Many community projects are fairly sensitive, especially those impacting land values or human health. There may be competing interests you have to be aware of, such as business competitors. You may need to safeguard your data and draft reports, for instance. Some projects involve personal information that presents privacy concerns. Media interest is also a possibility, although statisticians usually only make the news when they commit capital crimes.

Using this general background information, you will be in a better position to ask specific questions to support your analysis.

Project Scope

After painstaking and careful analysis of a sample, you are always told that it is the wrong sample and doesn't apply to the problem.

Arthur Bloch, "Fourth Law of Revision" in *Murphy's Law and Other Reasons Why Things Go Wrong!*, 1980.

The project scope consists of all the work items you are obligated to complete. Contractual descriptions of scopes are usually brief and results-oriented. This is because most clients neither understand nor care what you actually do so long as they get the information they are looking for.[3]

It can also be easier to administer a contract with a general description of the scope than a detailed version because there are fewer checkpoints to approve. So, for example, you might receive a scope description like:

> Consultant is to conduct a statistical analysis of data provided by client as directed by client's representative.

The client's representative is often a senior manager who you meet at the kickoff meeting and never see again. Your day-to-day contact is a clueless young staff member who can't give you any direction without talking to the senior manager first. If anything goes wrong, the senior manager can claim you didn't do what he directed you to do. On the other hand, such a general scope gives you great flexibility to do whatever you need to do to get the job done. You just have to figure what it is you need to do.

So how do you figure out what you need to do on a project? Well if you've done it before, you can just close your eyes and visualize actually doing each step of the project.

3 Then there's the special case of some statistical analyses being done for a graduate degree or a professional publication in which the scope of the analysis grows in search of significance.

If you've never used visualization in this way before, be patient and practice. You'll get the hang of it before long.

If you've never done the type of statistical analysis the project requires, you have a couple of options. You could ask someone else who has done it, although this would be problematical if that someone could be a competitor. You could do Internet searches to try to find some guidance. You'll find many websites that have quite a bit of statistical guidance, some of which is actually useful. And then there's that old standby, the library.[5] One way or another, you're going to have to learn about the methods if you've never used them before. You might as well do it before you commit to the budget and schedule. And be sure you have some appropriate software, without which, all that knowledge is useless (unless you want to spend your personal time writing a program that will do the analysis).

For the general scope described above, you might come up with activities like:

1. Attend kickoff meeting; obtain background information and data.
2. Prepare database, conduct exploratory data analysis, and resolve outliers and other issues.
3. Conduct statistical testing and evaluate assumptions and power.
4. Prepare report.

Then consider checkpoints or milestones. Are there places in the work effort when you think you'll need some feedback from the client or you'll want to provide the client with a status report on your progress? Build these into your plan. The motto to work by is:

No Surprises!

Hey! I said no surprises!

You know how much you hate it when your auto mechanic hands you a bill that's

4 Assuming they still exist by the time I finish writing this book.

$200 more than his original estimate? Don't do the same kind of thing to your client. If you don't think you will be able to deliver the agreed upon product on time and within budget, give your client some advanced warning.

So that might be all the detail you need to start. If there are some special activities you have to remember to do, be sure to write them down so that you can factor them into the budget and schedule. The same goes for assumptions you might have made to develop the scope.

Deliverables

In products of the human mind, simplicity marks the end of a process of refining, while complexity marks a primitive stage.

Eric Hoffer, 1954 notebook entry quoted in *Eric Hoffer and the Art of the Notebook*, by Tom Bethell, 2005

Deliverables are the products of your work. They are usually specified in your scope of work. Most clients won't understand the details of your statistical analysis, but they will have an impression of what they expect your deliverable to look like. Do they want a simple letter report with just the answers? Do they want a comprehensive report complete with all the statistical printouts and the data burned on a CD? Do they want specialty items, such as graphics for court or public presentations or conference papers? If you're smart, you'll find out what their vision is before you start writing.

Other Requirements

Cogito ergo pensio.
(I think, therefore I get paid.)

Be sure to ask the client if there are any other requirements you should plan for. You can count on there being meetings, but you'll need to find out how many and if they'll require nonlocal travel. Will any of the meetings involve the oversight agency or the public? Will you be required to make a presentation or answer questions? Similarly with conference calls, should you budget for regular status calls? Are written progress reports necessary?

Security and confidentiality are often a concern, especially if there is public or media interest. Be sure the client is comfortable with faxes and emails as means of transferring information. Some clients may want all information transfers to be face to face or in the presence of their lawyers to protect the paper trail.

Don't be hesitant to ask about getting paid. You'll need to find out who to send invoices to and if the client has any special requirements. Most clients will accept a letter with a few major cost items like labor, travel, and supplies. A few, however, will want detailed backup, such as copies of employee timesheets and travel expense reports. Get a notion of the complexity of the invoice approval process. How many people have to sign-off on the invoice before you get paid? How long does it usually take? Can part of your invoice be paid if there are only a few items in question? Will the client retain a percentage of the invoiced dollars pending the project's completion? Remember, it's hard to concentrate on statistics if you don't know where your next meal is coming from.

Budget

You know, if I had the salaries they pay those idiot blowhard politicians to put into my budget... I'm sorry, did I say that out loud?

Major General Hank Landry (Beau Bridges) on *Stargate SG-1* episode "Flesh and Blood (#10.1)," 2006

By now, you should have a pretty clear idea of what the client wants, what the client needs, and what you have to do to make it all happen. Be sure you also have a clear idea of how much the client wants to spend. Are they looking for a top-end Lexus or a budget-friendly Kia?[6] You can probably complete the job at any of several levels of effort. The analysis can be quick-and-dirty or detailed, peer-reviewed, and documented thoroughly. Don't offer a Lexus scope with a Kia budget. The client might just take you up on your offer, and you'll be stuck making up the difference.

Remember, you're getting paid to solve a problem. Don't get caught up in the statistics or other project activities and lose sight of your purpose. Budgets are never as much as we would like, but they're usually enough to get the job done.

So here are a few hints for how to budget a statistical analysis project if you've never done it before:

1. Start by breaking down your statistical analysis into at least these five parts:

    1. Database development (or procurement) and scrubbing
    2. Exploratory data analysis (EDA)
    3. Statistical analysis
    4. Report preparation and revisions
    5. Project coordination, reviews, support, and other activities

2. Write down all the activities you'll have to complete during each part of the project. Then close your eyes and visualize doing each activity. Estimate how many working hours each activity will take to complete and write down that number next to each activity. As a rule of thumb, 50–80 percent of the total hours should be for dataset development and scrubbing; 5–20 percent of the hours should be for the EDA and statistical analysis; and 10–40 percent of the hours should be for reporting and other activities. There will be overlap. During the EDA or even the statistical analysis itself, you may have to go back and sort

---

5 Twenty years ago, the question was whether they wanted a Cadillac or a Chevy. Times have changed.

out some data problems you uncover. Likewise, during report preparation, you'll probably have to run a few more descriptive statistics or create new charts to tell the whole story.

I'm visualizing the project.

3. Apply the "newbie" adjustment factor. If you're a newly graduated statistician, multiply all the hours by two. If you've never done a statistical analysis before, multiply by three. If you think I'm kidding about this, multiply by four. What's the rationale for this step, you ask? Well, if you're right out of school, you might not think about all the daily happenings that conspire to ruin your efficiency. Things like getting coffee, going to the restroom, answering your email, going to meetings, discussing last weekend's game with your office mates, and a thousand other interruptions too insignificant to think about but too time-consuming to ignore. If you've never done a statistical analysis before, you might not think about the scores of little decisions you'll have to make scrubbing your dataset and evaluating the model, not to mention the hardware and software problems that can stop an analysis in its tracks. Finally, if you think I'm kidding, you're overconfident. You'll probably make some fundamental mistake early in the project that will require you to redo your work.
4. Obtain the direct billing rates for the people you plan to have working on the project (usually their annual salary divided by 2,080 hours per year). Multiply the labor dollars per hour from step 3 by the number of hours estimated in step 2 to estimate direct labor costs.
5. Add in your other direct costs like travel, copying and printing, telephone and computer time (if you charge for them), and whatever materials and supplies you'll need. If you've never done this before, ask someone for help.
6. Apply the corporate "taxes." If you've never heard of these before, talk to your boss. Corporate fees usually include multipliers for overhead, fringe benefits, general and administrative charges, and sometimes materials and subcontractor charges. There may also be standard allocations of hours for project manage-

ment, QA/QC, and other corporate niceties. And above all, don't forget the next-to-the-bottom line—profit. If you've never done this before, have your boss look over your estimates. He or she may not understand the statistical work (in which case, your estimates need to be pretty good), but at least the corporate accountants won't roast you for forgetting something.

That's the easy part. The last thing you have to do is look at the whole package and make sure it looks like a reasonable offer. If you have information from the client about how much they want to spend, you can adjust your scope and budget to match. That's why it's so important to try to get this kind of information during your interview. But if you don't know *their* budget, you'll have to decide what *feels* right. This takes a bit of experience and confidence since you'll be bidding against your own insecurities as well as your competitors.

Deciding how much to bid on a competitive project is an agonizing decision. If you overbid, you won't get the work. That result may be catastrophic if you need the work or just a minor disappointment if you're busy. So if you really need or want the work, and you think your competitor is planning to outsource the work to China, cut back the number of hours as much as you can. *Don't cut the budget without cutting the scope.* If you underbid the work, it won't be your boss who has to finish the project on weekends during football season.

You won't necessarily be home free once you win a competitive bid. The client will usually want to negotiate some elements of what you proposed. On the surface, the negotiations will appear to involve scope, schedule, or another constraint, but usually they come down to money.

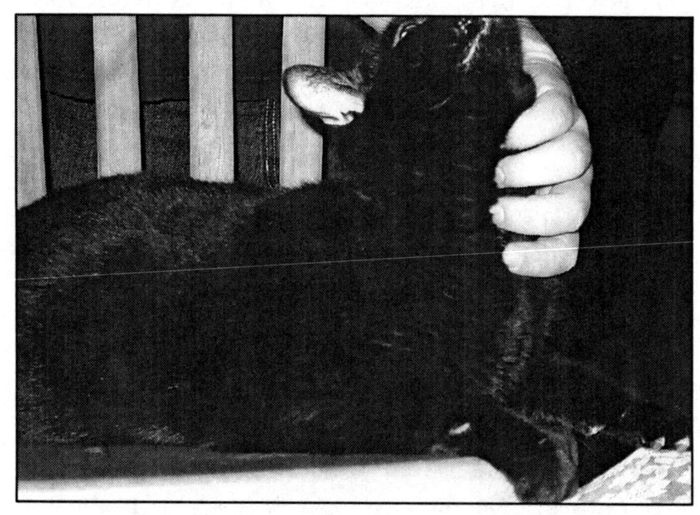

OK. OK. You can have an extra $10K, just don't stop.

As a rule, the budget you'll want will always be more than what the client wants to pay. Compromises will be necessary. So when it comes time to trim your proposed budget, don't be dollar wise and data foolish. Simplify the analysis and the other work you planned to do, but don't undermine your planned (and possible future) uses of the data. Don't create missing data by not making expensive measurements on some samples. You won't be able to use either the sample or the variable in later analyses. The same goes for metadata and other background information about the samples. Be sure you will have enough samples to accomplish your objective. If you won't, simplify your objective. Don't skip regular sampling rounds in a time-series study, for example, you won't be able to use autoregressive modeling techniques. And don't discount the importance of data quality. QA/QC is expensive but indispensable. Variability doesn't go away just by ignoring it.

Schedule

Hofstadter's Law: It always takes longer than you expect, even when you take into account Hofstadter's Law.

Douglas R. Hofstadter, *Gödel Escher Bach: An Eternal Golden Braid 20th anniversary ed.*, 1999, p. 152

Nothing is impossible for the person who doesn't have to do it. In this case, unfortunately, you *are* the one who has to do it, so make sure you don't set impossible deadlines.

Schedules are often prepared interactively with budgets, but there are exceptions. Some managers prefer to define the scope completely, which then dictates the budget and the schedule. If the client has limited funds to spend, though, the budget dictates the scope, which dictates schedule. If there are activities that have some component of timing, like seasonal sampling or summer fieldwork, the scope dictates the schedule, which dictates the budget. If there are set deadlines, like regulatory reports that are due a certain number of days after sampling is complete, the schedule dictates the scope, which dictates the budget. If the client has to spend his funds by the end of the fiscal year, the schedule dictates the budget, which dictates the scope.

If your project has a large number of interrelated activities,[7] consider creating a critical-path schedule. Critical-path scheduling was developed in the late 1950s to facilitate the management of complex government and industrial construction projects. The method is used to distinguish between critical activities (i.e., tasks that will delay the entire project if they are delayed) and noncritical activities (i.e., tasks that will not delay the entire project if they are delayed). This allows extra attention to be paid to the critical activities to ensure they do not slip. The timing of noncritical activities can also be rearranged so that resources are used most efficiently. Critical-path scheduling isn't difficult, but it becomes confusing if there are more than a few activities, so most professional schedules use software to complete all the laborious calculations and prepare the graph-

6 What's a large number of activities in a statistical analysis? It depends on how finely you slice the activities. If all your activities are less than eight working hours long, then fewer than twenty activities isn't a lot but fifty activities is. Complexity also depends on the timeframe and on how narrowly you define your activities. Twenty complex or long-duration activities are a lot. Even twenty simple activities will be a lot if they are scheduled to be completed in a week's time. Most statistical analyses won't consist of as many as twenty activities. But if you are also responsible for the sample collection and other aspects of an investigation, you'll find that one-hundred-activity projects are not unusual.

ics. Most scheduling software can also be used to manage resources, like labor hours. Critical-path scheduling is addictive. Once you use it to plan and manage a project, you'll never want to do it any other way.

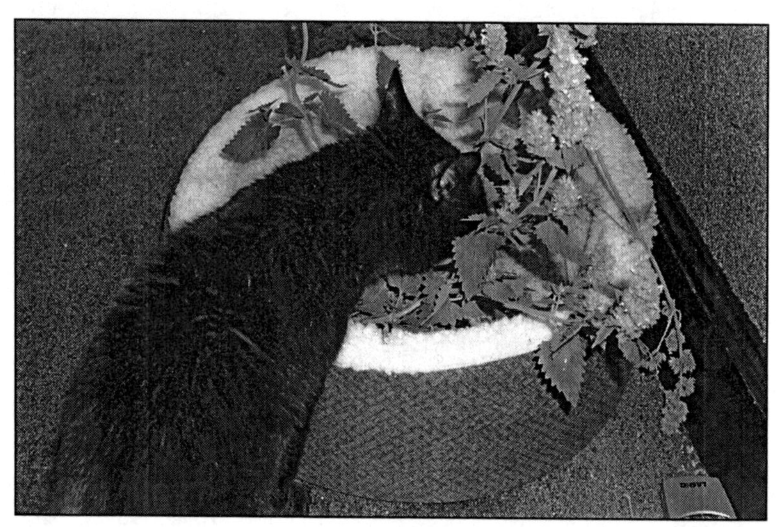

Meet the Critical Path? I thought you said eat the catnip stash. Nom, nom, nom.

To construct a critical-path schedule using software such as Microsoft Project or Oracle P5, identify:

- Project activities you want to track.
- Relationships between the activities. There are three types of relationships and two relationship modifiers. FS relationships mean one activity must finish before the next activity can start. SS relationships mean two activities must start at the same time. FF relationships mean two activities must finish at the same time. These relationships can be offset by leading (moving up the start of an activity) or lagging (delaying the start of an activity) one of the activities. The framework of the activities and the relationships between them are called the project's logic.
- Anticipated durations (i.e., the number of working hours, days, or other consistent work-time unit needed to complete the activity).
- Starting date and a calendar that specifies working days (excluding nonworking times such as weekends, holidays, and vacations).
- Milestones (zero-duration activities; milestones that have a fixed date once the baseline schedule is set are called deadlines).

Then you enter all the information into your scheduling software. The software will use the project logic and calendar to calculate for each activity the earliest and latest starting times (EST and LST) and the earliest and latest finishing times (EFT and LFT). If the schedule is realistic and likely to be acceptable to the client, you're home free. If it's not, you can adjust activity relationships and durations until you have a schedule that you're satisfied with.

The difference between the calculated earliest and latest starting times (or finishing times) for an activity is the activity's float. The float is the amount of time you can delay completing the activity without delaying the entire project. Any activity with zero float is a critical activity. The sequence of linked critical activities through the project is called the critical path. Any delays along this path will delay the overall project. So if you concentrate on completing the critical activities as scheduled, you'll probably finish the entire project on time. Some critical-path scheduling software also allows you to associate resources (e.g., staff, equipment, other direct costs) with each activity. These capabilities can simplify the management of a complex project.

Whether you create a critical-path schedule or not, there are a few things to consider before finalizing your schedule:

- **Build in thinking time**. A cow may only be hooked up to a milking machine for an hour a day, but it has to eat grass for twelve hours to produce the milk. It may only take you a day to calculate all the statistics you planned, but it could take a week to think about what the results mean. Then you might want to go back and redo some of the analyses. You usually don't get paid for all of your thinking time. Some of it happens while you sleep or when you're in the shower. But you can't make yourself think faster any more than that cow can make milk faster. Some things just have to proceed at their own rates. So the schedule for a statistical analysis should always be longer than the hours budgeted for the project staff.

- **Add in milestones**. Most schedules contain some milestones that tie a project's implementation to its schedule. Usually, milestones represent important project events, often involving client involvement such as meetings and deliverable submissions. Milestones occur when some scheduled activities are completed, so they can move in the schedule. Fixed dates, usually specified by the client, are called deadlines. Deadlines are especially important if they are specified in contract documents. Also consider adding checkpoints, key review points within an activity, and tollgates, review points that require a client decision to proceed. Examples of checkpoints might include completion of 30 percent, 60 percent, and 90 percent of data gathering or analysis based on the budget, schedule, or earned value. An example of a tollgate could be approval of a project work plan.

- **Have a data validation strategy.** If you will use laboratory data, you may need a plan to have the data validated.[8] Find someone who can do it, and ask them how long it will take. You may have to decide whether to analyze the unvalidated data to meet a deadline.

- **Remember the reviewer's time**. Cut your internal reviewers (e.g., your boss or colleagues) a break and give them some time to look at your work before sending it to the client. Similarly build in time for the external (i.e., client and oversight agency) reviewers. It may take an external reviewer six months to comment on a

8 Data validation is a process for ensuring that data and metadata were generated and documented in a known manner so that their quality can be assessed.

statistical analysis you completed in a month. If any activities depend on receiving the reviewer's approval, be sure to build the reviewer's time into the schedule. In general, any handoff[9] in a project is likely to cause delays because the receiving parties often don't know when to expect the handoff and they have other commitments to complete. If you can't build in extra time for the transitions, consider adding checkpoints to warn the parties of the impending handoff.

- **Check your calendar**. Most schedules are first developed in relative terms (i.e., working days) pending client approval. Scheduled dates aren't real until a starting date is agreed to. As a consequence, a well-thought out schedule can become a disaster if there are conflicts in real time. For example, you may propose a project work plan in May that a client approves in August but can't commit funding for until the next fiscal year in January. The soil sampling you planned for upstate New York in August now falls in February, when you'll have to chisel out frozen chunks of soil for analysis. Vacations and holidays can also disrupt schedule implementation. Beware of summer vacations and the Thanksgiving through New Year's holiday season.

If you're done, I have to check my calendar to see what I have planned for tonight.

- **Check for budget-schedule inconsistencies**. Calendar time should be greater than the number of working hours budgeted (i.e., you can't charge 160 hours of an individual's time for an activity that is scheduled to be completed in three weeks unless you're a lawyer).

---

9 A handoff on a project is when one worker or group of workers completes their assignment and passes the work product on to the next worker or group of workers. This is easy to observe on an assembly line, but it also happens on data analysis projects. The staff responsible for sample collection gives the samples to staff at the laboratory. The staff at the lab sends their results to the data analysts, and so on.

Contracts

You don't know what it's like out there. I've worked in the private sector. They expect results.

Dr. Raymond Stantz in *Ghostbusters*

Hopefully you'll have legal support so that you won't have to worry too much about the embarrassment of contractile dysfunction. Here are a few things to look for that might have a bearing on your statistical analysis:

- **Description of Services**—Make sure the scope statement is accurate. It also shouldn't be so specific that you cannot make midanalysis changes without violating the contract. For this reason, it is not a good idea to incorporate the detailed approach that you wrote in the proposal to win the work into the contract.
- **Provided Data and Reports**—There should be a statement about the client's responsibilities to provide data and reports (if they are). In particular, make sure there's a statement about the databases they provide being error-free and in a specific (don't say acceptable, it's vague) electronic format. If you have to reenter the data or redo the analysis because of dataset errors, you'll have grounds for negotiating a budget increase and schedule extension.
- **Best Effort**—The nature of statistical analyses does not necessitate that a particular result will be reached.[10] You can only promise that you will conduct analyses aimed at fulfilling the objective and put forth your best professional effort to provide the services requested. Some lawyers hate the phrase *best effort* because it is vague. After all, how do you measure effort? However you phrase it, the point is that you should not guarantee the results of your work.
- **Means, Methods, and Procedures**—The contract should state that you will be responsible for selecting your own means (equipment and other things necessary for getting the work done), methods (strategy for doing the work), and procedures (set of instructions for specific tasks). If you don't, you may find that the simple t-test you were going to do using Excel on your home-office PC now is an ANOVA that has to be done using SAS on the client's workstation in Smellbad, New Jersey.
- **Use of Deliverables**—You should specify that the deliverables you provide are not intended to be used at other times or under other conditions or for different purposes. Why wouldn't you want a client to take a report you did for them and reuse it? Consider this. You set up a spreadsheet to display the results for some statistical procedure. To save time, you just input the results of some of the steps instead of programming them. If the client changes just the data and re-

10 Unless you are a grad student, in which case you must continue working until you find significance ... or a job.

I've got your contract claws right here.

submits it to the regulatory agency the next quarter, it might turn out that the doctored report will tarnish *your* reputation. If the client wants a tool and not just an analysis, they need to tell you up front.

There are, of course, many other clauses that you would want in the contract, such as payments, Force Majeure, and indemnification. Be sure your lawyer looks over any contract before you commit to it.

11 Weapons of Math Production

In theory, you may be able to calculate your statistics using nothing more than a pencil, paper, and some references. Of course, it's a lot easier having a statistician's version of Norm Abram's workshop to fall back on. Here are a few statistical tools you may find helpful.

Software Solutions

Where the ENIAC is equipped with 18,000 vacuum tubes and weighs 30 tons, computers in the future may have only 1,000 vacuum tubes and weigh only 1½ tons.
Popular Mechanics, March 1949

Whether you're planning a career in statistics or just looking to analyze your current dataset, you're going to need software to do your calculations. Yes, there are some people who still calculate descriptive statistics manually, but this practice is so prone to errors that it's only applied to very small datasets. And yes, there are some people who develop their own statistical routines usually with **R**, a programming language for statistics, or with matrix manipulation software like MATLAB, MAPLE and MATHEMATICA. Unless you're a mathematical statistician developing a new statistical technique, though, you won't need to take this approach. There's plenty of software available. All you need to know is the kind of statistical analyses you're likely to need and your price range.

Software for General Statistics

With a few exceptions (discussed in the next section), almost all of the statistical software you'll find is geared to the most common types of statistical analysis. Software used for statistical analysis can be sorted into five categories:

- **Web-based Calculators**—Web sites that perform simple statistical calculations can be found at statpages.org/. This is the low end of cost, but also usability. You usually have to enter your data and edit it manually, so it's not really suitable for production.
- **Spreadsheets**—You probably already have a copy of Microsoft Excel or some

other spreadsheet software on your computer. If you are a beginner at data analysis, you'll find that you can accomplish most of what you want to do using spreadsheet software. Be advised, though, that some statisticians warn against the use of spreadsheet software.[1]

- **Basic Statistical Software**—This category includes software that is used mainly for less sophisticated types of statistical analysis. Most can be purchased for less than about $500. Key examples include StatsDirect, In Stat, Analyze It, and Assistat.
- **Intermediate Statistical Software**—This category includes software that can be used for many types of statistical analysis except some of the more sophisticated techniques like multivariate analysis. Most but not all are a single module and cost less than about $1,000. Examples include NCSS, Statistix, Costat, Origin, Prostat, Soritec, MVSP, and Simstat.
- **Major Statistical Packages**—This category includes software that can be used for a variety of purposes. Most have a base module and a variety of optional add-on modules. They are usually purchased through annual licenses specifying a number of users and cost more than about $1,000 (in some cases, *way* over). Key examples include SPSS, Statistica, S-Plus, Stata, Systat, Minitab, Statgraphics, JMP, and SAS.

Some of the major packages like SAS and SPSS have been around since the mainframe days of the 1960s while others like Statistica are products of the 1980s development of personal computers. All have graphical user interfaces and many also allow you to write your own code for specialized applications. Almost all have downloadable demos, usually fully functional (at least for basic statistics) for thirty days. Nevertheless, there is still a great bit of difference between the packages.

Figure 3 contains screen captures from two of the major statistical software packages. The programs look straightforward. Both have spreadsheet screens for data. Both have utilities for data management and graphing. These are typical features of all the major statistical packages. There can be large differences in the statistical procedures (discussed in Chapter 22) available in the statistical packages, however, because most of the packages are modular. You get the basic package and the optional modules you pay for.

---

1 Four reasons are cited in the literature for why Microsoft Excel should not be used for statistical analysis. First, Excel doesn't do some calculations that statistical packages do. Well, of course it doesn't. It's a spreadsheet program that sells for less than $200 (by itself, not part of Office) compared to statistical packages that cost ten times as much. Second, the graphics are limited. True, only common statistical graphics are included, but most clients don't understand anything they haven't already seen in the *Wall Street Journal*. And if you know how, Excel graphics can be annotated as well as graphics produced using much more expensive software. That being said, there are two real issues. Excel's calculated probabilities are reportedly incorrect in the third decimal place, but this shouldn't affect decisions. If you would base a decision solely on whether a probability is 0.051 instead of 0.049, you're taking statistical tests way too seriously. Also, Excel's random number generators are not of research quality, so if you planned to do Monte Carlo simulations with Excel… don't.

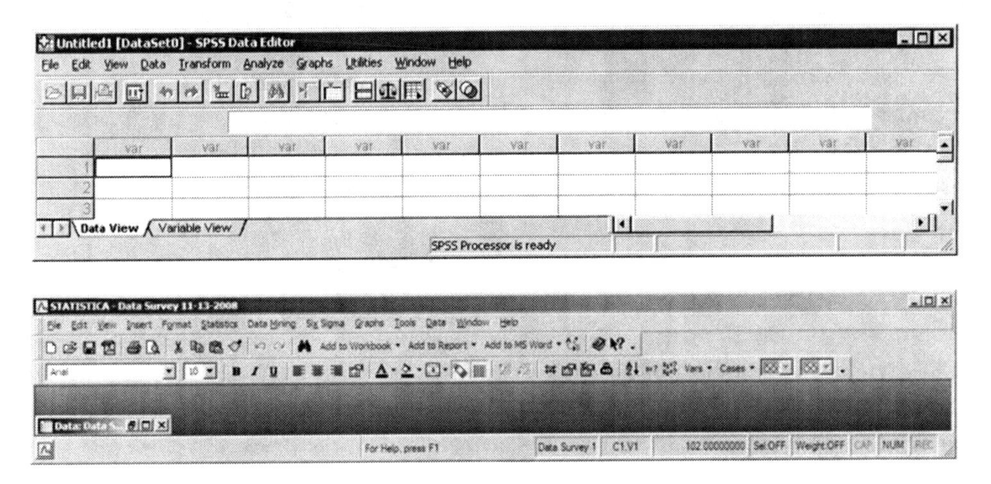

Figure 3. Screen Captures from Two Major Statistical Packages.

To access the statistical procedures, click on the top menu selection (Analyze for SPSS or Statistics for Statistica), and a dropdown menu appears listing the available statistical procedures. Select the procedure you want to conduct from the drop down menu, and then a special menu pops up with all the specifications and options for the procedure. So you can do a lot of statistical analyses with just a few mouse clicks, but you really have to understand what all those specifications and options are about.

Which software package is best? They all have their fans. SPSS was created in the 1960s by graduates of Stanford who continued development at the University of Chicago. It used to be called Statistical Package for the Social Sciences, which is probably why it is very popular in the social sciences. SPSS was bought by IBM in 2009. SAS, formerly called the Statistical Analysis System, was developed in the early 1970s by professors at North Carolina State University. S-Plus started out as a programming language developed by Bell Laboratories in the 1980s. Minitab was created by professors at the Pennsylvania University in the 1970s from statistical spreadsheet software developed at the National Institute of Standards and Technology (NIST).

There is no real best statistical software. They're all pretty good, dollar for dollar. A lot of what determines a user's preference is what software is (was) available at their college or place they work. For example, if you go (went) to Penn State, you probably think Minitab is the best. If you work at a pharmaceutical company, you probably use SAS because that's what the entire pharmaceutical industry uses. Social scientists like to use SPSS. If you like programming your own procedures, look into the **R** programming language for statistics. It's free!

Assuming you don't have access to software through your school or work, you can evaluate your software needs by answering three questions:

- How sophisticated are the statistical techniques you need to use?
- How often would you likely need to use the software?
- How much do you have to spend for the software?

If you are planning on doing only one analysis, see if you can use what you have.

You may be able to do all your calculations in a spreadsheet program or use free software or web-based software. If you are going to do full-time statistical consulting, and you can't afford a license for a major package, try a basic or an intermediate package and move up as you can afford to. If you're going to be an occasional user, any of the statistical packages will be better than using a spreadsheet (except perhaps for dataset scrubbing), so purchase whatever you can afford.

If you aren't acquainted with statistical software, conduct a web search or start at en.wikipedia.org/wiki/List_of_statistical_packages._Explore websites you find to be sure the software has the statistical procedures you think you will be using. Almost all of the sites have free downloads, such as brochures, white papers, and demonstration software. But don't download the demo software until you're ready to make a decision. Most demos are good for only thirty days after which the software won't work even if you download a new copy.

Software for Specialized Applications

There are a few kinds of analysis you might run into that will require specialized software. For example, have you ever seen an icon plot using sparklines or Chernoff faces? How about a ternary diagram or a piper plot? Some day you may have to produce one of these specialized graphics. Software you could look into would include: Sigmaplot, Origin, AquaChem, GraphPad, EasyPlot, Delta Graph, and Grapher.

If you ever have to do time-series analysis, you could start with some of the high-end statistical packages. Or, you could look into specialized software including Autobox, Eviews, ForecastX, and RATS. If you have to produce maps, find a GIS expert to help you. But if you only need to do some contouring, try Surfer. If you're not into meteorology or geology, you probably don't run into orientation data very often. If you ever do, get Oriana. For critical-path scheduling, try Microsoft Project or P5, an update to Primavera Project Planner, now a product of Oracle. There's also software for resampling statistics, control charts, ANOVA, neural networks, nonparametric statistics, power analysis, Bayesian statistics, data mining, and many other specialties.

The software market changes rapidly. The big packages keep getting bigger, spawning optional modules from procedures that used to be part of the basic package. At the same time, new statistical software appears, usually for specialized application. Spreadsheet software is also becoming more sophisticated. Introductory statistics classes are now taught with spreadsheet software; even the calculator is a thing of the past. Do some research and get what's best for your situation.

Information Support

If the Internet teaches us anything, it is that great value comes from leaving core resources in a commons, where they're free for people to build upon as they see fit.

Lawrence Lessig, "May the Source Be With You," 2001 *Wired* magazine article adapted from *The Future of Ideas: The Fate of the Commons in a Connected World*

If you plan to do your own analysis, you'll need information on the statistical procedures you plan to use. This is true whether you are an experienced statistician or a novice. When I was starting my career, there was no Internet, at least in the publicly available form that exists today. Now it's the first place I look for statistical information, not because it's the best information, but because the information is virtually always free and readily available. These features allow users to conduct many web searches to find independent and corroborating information. Sometimes the searches culminate in the purchase of a book or a personal conversation with an expert, but usually the information is available for instant download. What could be more simple?

Keep essential resources close by.

Here are some places you can get information for cheap.

Search Engines, Wiki, and Libraries

Where else would someone start an Internet search but a search engine? Everyone seems to have their favorite search engine. Some prefer the big general engines like Google, Bing, and Yahoo. Others prefer metasearch engines, search engines that search other search engines, like Dogpile, Webcrawler, Excite, and Clusty. It doesn't seem to matter much which engine you use for all but the most esoteric statistical topics. Your problem will more likely be too many hits rather than too few.

Though it's probably always best to start with a search engine, there are a few other categories of general information sources worth exploring. Wiki are sites with content provided by users. The information is usually pretty good so long as you stay away from politics and religion. An Internet library is what you might expect, a collection of a wide variety of types of information. Most libraries are associated with a brick-and-mortar library, university, or organization. The Library of Congress (loc.gov) and the New York Public Library (nypl.org) are two examples.

Government Websites

Government websites are particularly valuable as resources for statisticians because they provide both data and educational information. For federal government data, start with www.fedstats.gov/. There are probably a hundred or so federal agencies that supply statistical information. Once you find the type of data you want at fedstats.gov, you can explore the individual agency's website further.

Many federal agencies also provide educational information and guidance on the use of statistical methods. The National Institute of Standards and Technology provides its *NIST/SEMATECH e-Handbook of Statistical Methods* at www.itl.nist.gov/div898/handbook/index.htm. Many agencies provide references on statistics for the type of data and situations they commonly deal with. Several agencies also provide software for implementing selected statistical procedures.

As you might expect, you can also find numerous statistical websites sponsored at the state and local levels of government. Links to many of these sites are provided at www.usa.gov/index.shtml. There are also many international sites that provide statistical information.

Private Websites

Academic institutions provide a vast amount of information for statistical analysis. If you do a search for some specific statistical information, maybe half of your top ten hits will be from academic institutions. The more esoteric the information requested, the more likely you'll end up at an .edu site. Most statistical information provided on academic websites is attributed to its author. Where this is especially valuable is if you want to identify an individual, with which you can correspond. Websites in government and industry may have contact information, but it's usually a manager, webmaster, or salesperson and not the person who provided the content. The University of Florida, Department of Statistics also provides links to many university statistics departments, software companies, statistical associations and journals, and other relevant sites at www.stat.ufl.edu/vlib/statistics.html.

Many professional organizations often have pages devoted to statistics. Examples include the American Statistical Association (amstat.org) and the American Society for Quality (asq.org). For links to hundreds of sites with software, tutorials, and related statistical resources, visit statpages.org/. Excellent descriptions of many statistical techniques are provided by the StatSoft Corporation, makers of Statistica, at www.statsoft.com/textbook/.

Mailing Lists and Discussion Groups

Statistical information can also be received by email by subscribing to a mailing list. Mailing lists can be found by searching the Internet or visiting www.lsoft.com/lists/listref.html. Mailing lists usually focus on a specific topic, such as a statistical technique or piece of software. After subscribing by providing a valid email address, emails on the lists topic are automatically sent by other subscribers through the list. Subscribers can ask questions of the list and provide answers to others' questions. Some information on mailing lists is very good, and some is very bad, so never accept one poster's information without independent corroboration. Mailing lists are usually moderated, monitored by an administrator, to minimize off-topic posts.

Discussion groups tend to be less formal than mailing lists in that anyone can enter or leave the discussion at any time. There are hundreds of discussion groups on statistical topics that may provide support. Many groups can be found by searching Google or Yahoo groups. Discussion groups focus on a variety of interests, including types of statistics and data, software, and applications. Two popular moderated groups are *sci.stat.consult* and *sci.stat.math*. Some of the groups are moderated to provide more control over the topics that are discussed. Unmoderated groups are often infiltrated by spammers who can bury the useful information with offers for Viagra, credit repair, and weight loss.

Textbooks

Textbooks will never become obsolete, especially ones that have pictures of cats. They are convenient to browse and a nice break from staring at a computer monitor all day. But reference books can be very expensive, so unless you have a big budget for publications, make sure you know what's in them before you buy. If you're more applied than theoretical, you don't want to pay several hundred dollars for a mathematical treatise or a compilation of conference papers you'll never use. You might be able to peruse reference books at a local university library, but a better bet might be to look for books at barnesandnoble.com and www.amazon.com. If one of these businesses has the book, there's a good chance that they'll also have portions of the book available online for review. You can often purchase electronic versions of the book so you don't have to carry it from home to office to job site.

Journals and Conference Proceedings

You can find fairly complete lists of journals that publish statistical articles by following links from www.statsci.org/journals.html. Statistical journals are quite challenging to read, so this isn't the place to launch your exploration of the world of number crunching.

I have my books / And my poultry to affect me / I am shedding my excess fur / Hiding 'neath my chair / safe within my lair / I scratch no one and no one scratches me / I am a cat / I am a feeeeeeeeeline. (Sing to the tune of "I Am A Rock" by Paul Simon and Art Garfunkel.)

12 Tales of the Unprojected

No plan survives implementation. That's why there are whole organizations with extensive libraries dedicated to project management. Search the Internet for information on the intricacies of managing scope, schedule, and budget. You'll probably turn up links to the Project Management Institute (www.pmi.org), perhaps the leading organization in the discipline. If you want to be a project manager, get to know the PMBOK, the Project Management Body of Knowledge. If you're just interested in data analysis, here are a few situations that you should be aware of.

Staffing

It is commonly believed that anyone who tabulates numbers is a statistician. This is like believing that anyone who owns a scalpel is a surgeon.

Robert Hooke, *How to Tell the Liars from the Statisticians*, 1983, p. 1.

The first step in quality is staffing. Chapter 9 details the issues involved in deciding who should conduct a statistical analysis. If you're looking for a statistician to do an analysis for you, Chapter 9 is a good place to start. If you plan to assign someone who works for you to the project, be sensitive to the competing concerns of ability versus availability. Those who are able are often not available; those who are available are often not able. Keeping somebody busy is only worthwhile if you don't have to redo the work. If you're a solitary consultant, the issue is whether you have the knowledge, experience, and, most importantly, time to do the work yourself. Don't let your ego get the best of you. Turn down work you can't really do. If you can't do the work, you certainly won't be able to do it over.

Communications and Relationships

Remember not only to say the right thing in the right place, but far more difficult still, to leave unsaid the wrong thing at the tempting moment.

Benjamin Franklin, U.S. author and diplomat

People are people, so it shouldn't be surprising that two of the key ingredients in work plan implementation are how individuals on the project behave and communicate. There are no more common reasons for project difficulties than these.

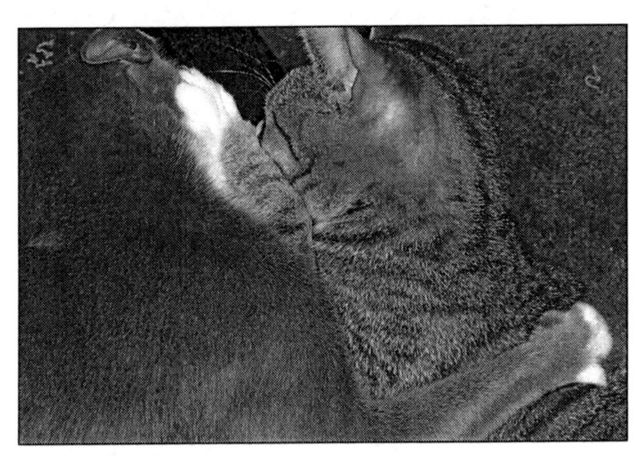

Communications and relationships always seem to be the biggest problems.

Relationships between project participants that can derail a project include:

- **Client's Organization**—No matter who your client contact is, he or she works for someone else who in turn works for someone else and so on. Within their organization, then, there may be a variety of competing interests. Even your contact may not be aware of some of the office politics. Management may want a quick answer. Accounting may want documentation of your work before paying you. The legal department may want you to guarantee your results or have your report phrased in certain ways. The plant manager may resent the intrusion of the home office who you work for. You may be the one who has to accommodate these interests in getting your work done successfully.
- **Client-Stakeholders**—You and your analysis may never be seen by anyone outside the client's organization. Your client, on the other hand, may have to make a decision based on your work that is of great interest to shareholders, employees, customers, neighbors, local action groups, the media, and even the public. Consequently, you have to be sensitive to the client's thinking about how your results will be perceived by the stakeholders. He or she may present your results in simplistic terms that may not be technically correct.
- **Client-Reviewer**—There may be reviewers for your work who are not part of the client's organization. Some reviewers may be linked to the client, such as a legal firm hired by the client for advice. Other reviewers may be independent or even antagonistic to the client, such as regulatory or law enforcement agencies. Sometimes clients dig in their heels and refuse reviewer requests. This can cause delays that can wreak havoc with your schedule and staffing. Sometimes clients tell you to just give the reviewer whatever he or she wants. This can involve out-of-scope work that might impact your budget. The strangest client-reviewer dynamic I have ever seen involved a client-reviewer relationship that was alter-

nately cooperative and adversarial. When the client was obliging, the reviewer, who represented a regulatory agency, was demanding. When the reviewer was acquiescent, the client was obstinate. I was told to stop, then go, then stop, then go. As it turned out, the regulatory agency was trying to extract a larger settlement from the client, who as a large multinational corporation, was perceived to have "deep pockets" (i.e., a lot of money). What the reviewer (and I) didn't know was that the client was on the verge of declaring bankruptcy and didn't want any settlement to complicate their filing. In the end, there was no settlement, the multinational corporation was liquidated, and the regulatory agency had to start over with the successors and settled for a small fraction of what the original client had first offered.

- **Client-Statistician**—This is the relationship that you as the statistician have the greatest chance to manage. Usually the relationship is a good one or else you wouldn't have been selected to do the work. During the project, be sure you are clear on any differences between what the client wants, what the client asks for, and what the client needs. Be sure you are clear on how the client plans to use the results. You don't want the results misrepresented in a way that will affect your reputation. There are many examples of clients repackaging results in ways you might not expect. I had one client use a report I prepared as the basis for a conference presentation. Although he knew nothing about statistics, the karaoke PowerPoint got him management approval to travel on the company's tab. Fortunately for me, conference attendees tend to zone out when you put numbers on the screen, so it wasn't a big deal.

- **Statistician's Organization**—You would think that communications within your own organization wouldn't be a big issue. Well, people are people. I once did a project for a manager who said he had an urgent deadline. But first he delayed a week in providing the data. Then he demanded a partial draft report well ahead of the scheduled review date. He tried to use the hurriedly prepared report to convince his superiors that poor quality work by the staff was making the client dissatisfied. As it turned out, his superiors had already figured out that it was his own incompetence and rude behavior that was upsetting the client. I was lucky; he wasn't. He was fired shortly after the project was completed successfully.

- **Statistician-Reviewer**—Don't assume that the reviewer knows as much about statistics as you do. He or she may just have been the only person available to review your work. (It's not such a tough job; see Chapter 24, Grasping at Flaws). Even so, most of the time these relationships are fairly straightforward. There may be differences of opinion over an approach or the number or origin of samples, but usually this relationship is handled professionally by both sides. There are times, though, when inflated egos and hidden agendas cause conflict. One reviewer I worked with agreed to an analysis plan that called for a specific statistical procedure. After the data were collected and the analysis was completed, the reviewer refused to approve the report because "the analysis didn't work out the way [he thought] it should." After trying two other statistical procedures with the same result, he relented. On another project that involved a statistical comparison to a control group, the reviewer was surprised that the difference was not significant, even though he had participated in the selection of the control group. He demanded and got a new analysis on new samples from a new control group. The results were the same, and he backed down. Yet another reviewer refused to approve an analysis unless published references were provided to show a precedent for the analytical procedure. When the references were provided, the reviewer refused to approve the analysis unless additional statistical studies were done to support the analysis. When the statistical studies supported the analysis, even the reviewer's support staff encouraged her to approve the analysis. She refused because she "didn't understand it."

There's always somebody looking over your shoulder. Be prepared.

Reviewer's Organization—As the last example shows, you usually can't do much to change interactions in the reviewer's organization. I've had cases in which the reviewer was told to reject the report before it was even submitted. One reviewer I worked with, a university professor contracted with a regulatory agency, provided unusual comments on a statistical analysis. Each part of the review consisted of one paragraph containing an incomprehensible polemic followed by several paragraphs of eloquent prose describing some statistical issue related to the analysis. On a hunch, I scanned the textbook the professor was using in his graduate courses. The well-written comments provided by the reviewer were taken verbatim from the textbook. When I informed the agency the reviewer worked for about the plagiarism, they withdrew the comments but elected not to take any action against the professor.

Dealing with Dilemmas

I love deadlines. I especially like the whooshing sound they make as they go flying by.
Douglas Adams, quoted in M. J. Simpson, *Hitchhiker: A Biography of Douglas Adams*, 2003, p. 236.

A decade or so ago, I always feared and was the frequent victim of hardware and software problems. It was a logical consequence of a statistician routinely pushing his tools up to and beyond the limits of their capabilities. But the software is far better now, and the hardware is cheap enough to allow extraordinary redundancy. It isn't often that a problem goes away so completely with so little fanfare. Nevertheless, there are a host of technical problems that can cause major project dilemmas. Three of the most common are:

- **Inadequate Data**—This problem seems to occur on every project in which the client is responsible for providing previously collected data. Data delivery might be late, incomplete, or in the wrong format. More times than I can count, clients have given me spreadsheets they used as a data table in a report—with footnotes, blank rows, and columns, and all kinds of extraneous formatting—all the time thinking that the table was ready for statistical analysis. Those things happen, and in fact, should be anticipated. The real problem is when the client provides incorrect data. Worst of all, is when the errors aren't noticed until after you've already started, and in some cases, completed the analysis. Your dilemma is telling your client, in a nice way, that they screwed up and there are consequences. If you haven't done much work, you can grit your teeth, get the correct data, and start again. But if you're past the dataset scrubbing and into the statistical calculations, you've passed the point of easy return. You have to explain to the client that they have two options: let you finish the analysis with the data you have or pay for you to redo the analysis. Changing only a couple of numbers might not change their decision based on the results, but it will change all the numbers presented in the report. So if the client plans to release the results to adversarial reviewers, they need to understand their alternatives.

- **Unwelcome Results**—Most of the time, your analysis will confirm what you and your client already suspect. No problem. Occasionally, you'll reach some unexpected finding. Most clients don't even mind this. They feel they got something new for their money. But there are two other kinds of findings that are problematical—complex and inconclusive results. Exceedingly complex findings are difficult to communicate, especially to a non-technical audience. If the client doesn't understand your findings, and especially, the *value* of your findings, your work will never see the light of day. Even more troubling are inconclusive results. It's difficult explaining to a client that you finished the work, spent all the money, but didn't reach any conclusive find-

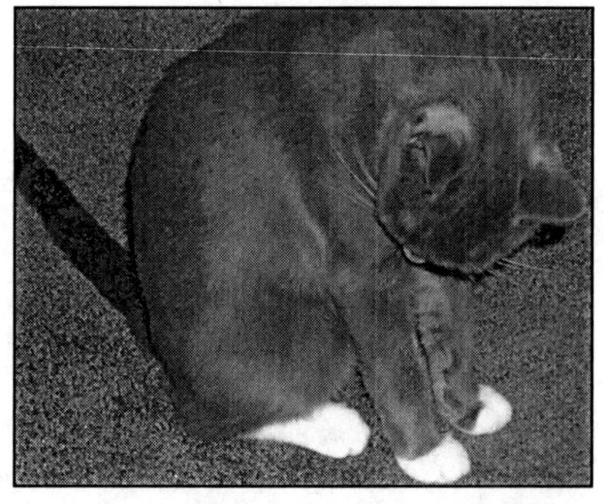

I'm sorry. It didn't turn out the way you wanted.

ings. Imagine how you might feel if your mechanic were to tell you he couldn't find or fix the problem with your car, but then charge you $500.

- **Unavailability of Key Staff**—This happens on all projects not just data analysis projects. Sometimes people get sick or resign and take new jobs. Sometimes, management reassigns your staff during lulls in the work, never to return to your project. There's not much you can do to prevent these dilemmas. You just have to react quickly when the problem arises.

There's an old saying in consulting:

Faster, better, cheaper. Pick two. Get one.

Consultants always want to do a better job than their competitors, complete the job sooner, and charge less for their work. It never happens that way though. Some consultants always do superior work, but they may take longer to achieve their vision of perfection. Some consultants pride themselves in being the lowest cost, but often their work is mediocre. Other consultants specialize in quick response, no matter what it takes.

It's like college. Most students have to do "academic triage"—pick the courses they will excel in and coast through the rest. Nobody is good at everything, but that's what clients want and expect. Besides, you probably said in your proposal that you were faster, better, and cheaper. Now it's time to deliver.

So is it best to be faster? Should you try to be better? Is being cheaper what clients want most? Consider this analogy. Say you hire a painter to paint the outside of your house. You tell him what you want done and agree to a price and a schedule. Then something goes wrong. Maybe you have to leave town, or the painter can't get the paint you want, or it rains for two weeks straight. Suddenly the whole agreement is in upheaval. Now fast-forward a few years. Do you remember that the job took a month longer because of the rain or cost more because the paint had to be special ordered? Maybe, but chances are, you don't think about it nearly as often as you think about the appearance of the chipping, bubbling paint caused by the poor application.

In general, the memory of poor quality lasts far longer than memories of missed schedules or overrun budgets. On the other hand, it's easy to tell when budgets and schedules are missed. Quality, however, is a matter of opinion. So you have to try to balance all three. But if you find you can't be faster, better, *and* cheaper, you'll have to do "management triage." If there's no money left in the budget, you may have to put in some free time even if it results in a delay. If you have an immovable deadline, get help even if you have to eat some costs. If you have no budget or schedule flexibility, stop where you are and package the deliverable with recommendations for the work you wanted to do but couldn't finish.

Picking between faster, better, and cheaper is both a technical and a business decision that is never pleasant. But if you decide not to pick quality, beware of the long-term consequences. Whatever you decide to do, don't wait to inform the client. Clients hate surprises. Confirmed bad news delivered late in a project is much worse than potential bad news delivered early in the project.

PART 3

Is That a Dataset in Your Pocket?

I'm glad to see you made it this far because Part III gets into the real nitty-gritty of selecting samples and variables and putting them into a dataset. If this is your first time designing a dataset for statistical analysis, take heart. It gets easier each time you do it, at least until it is so routine that it becomes tedious. Here's what you'll find in Part III:

Chapter 13—In Search of . . . Variables discusses considerations in selecting the attributes of the samples you might measure and what measurement scales you might use.

Chapter 14—Not-So-Simple Samples will tell you how you can decide how many samples you need and which samples they should be.

Chapter 15—The Heart and Soul of Variance Control describes three strategies for minimizing variance when measuring the attributes of your samples.

Chapter 16—Functional File Formats provides a few tips for putting your data into a matrix that will be well suited for a statistical analysis.

Don't worry about screwing up when you start building datasets for analysis, just so long as you are committed to learning from your errors. You'll find that you'll be making adjustments to your dataset throughout the analysis, such as adding, deleting, and reordering variables. If you're not doing the analysis yourself (i.e., you're passing the dataset on to someone else), they'll be making changes anyway, so any mistakes you make are likely to be caught at some point. With that reassurance, here's what you need to know to create a dataset.

13 In Search of ... Variables

Now that you know a little about variables, samples, and data, it time to get into the nitty-gritty of what variables you might use in your analysis. At this point, concentrate only on variables that you plan to measure. Later you can take these measurements and other existing data and expand their usefulness.

What a Phenomenal Concept

> *Intuition and concepts constitute ... the elements of all our knowledge so that neither concepts without an intuition in some way corresponding to them, nor intuition without concepts can yield knowledge.*
>
> Immanuel Kant, *The Critique of Pure Reason*, 1781

The first step in assembling a set of variables for your analysis is to identify the concepts or aspects of the phenomenon you want to investigate. By concepts, I mean to include hypotheses and theories as well as ideas, suppositions, beliefs, assertions, and premises, which may be less definitive or accepted. These concepts will come from the relationships known and supposed about the phenomenon. The reasons for doing this are that concepts can be multifaceted and linked to other concepts creating a framework of relationships underlying the phenomena. In traditional research, this is what a literature search is for. Literature searches, though, are considered by some to be an academic activity not applicable to analyses done on the job. Not true. The process of thinking through what you want to measure is necessary.

Consider this example about a study of workload. Ask any employee how busy they are, and they won't hesitate to tell you. Whether they have too much or not enough work, they'll have a good idea what their workload is relative to their work capacity. Workload is a simple concept, but it is not always easy to define and measure. Nevertheless, businesses live and die by managing or mismanaging their workloads.

The ideal way to characterize workload would be to measure it directly, but workload is not always clearly defined. Workload can be characterized in terms of:

- **Inputs**—the work that needs to be done

- **Activities**—the number of actions or hours taken to do the work
- **Outputs**—the results of work activities

But work is more complicated than these simple amounts. Say an employee is assigned two projects to work on. One project has a very high dollar value and is being conducted for a very demanding client. The other project is for a moderate amount money and has a fairly routine scope. Clearly, these two projects have different risks and rewards, and so, different business motivations for completing them. The routine project would lead to easier profits and wouldn't demand the use of specialized or limited staff resources. The high-value project might have a bigger payoff and a greater opportunity for future work. Staff experience is also important. The high-value project couldn't be done by inexperienced staff and even the routine project would require more effort than for a more senior staffer. Even two equally experienced staff members will take different efforts to complete a work assignment. It's just human nature (natural variance, perhaps).

Work. Work. Work. Work. Work. Work. Work. Pay attention to ME.

The issue with managing workload, of course, is that it is usually the basis for staffing. But staffing can also affect workload. If an organization is understaffed, it must either hire more staff or it won't be able to complete as much work as it has and it accumulates a backlog. If an organization is overstaffed, it can try to obtain new work to accommodate the staff level or lay off some employees. So the goal is to achieve an equilibrium between staffing and workload. Even so, it is possible to be either understaffed or overstaffed from year to year, quarter to quarter, day to day, and even within a day (e.g., waiting in checkout lines because there aren't enough cashiers).

So, the workload concepts you might want to explore include work inputs, activities, outputs, project size, complexity, risk, profit, scope, staff size, experience, productivity, hiring, layoffs, backlog, and scheduling, as well as new business development concepts related to workload like marketing, advertising, and so on. Depending on what you wanted to know about workload, there are quite a few concepts you could incorporate into your study. From that list, you expand and enhance your concepts making them more definite and measurable. But, of course, you can't measure any of these concepts until you decide at what organizational level you plan to collect the data—individuals, offices, departments, or whole companies—and so on.

Once you have specific ideas you want to explore, identify ways they could be measured. Start with conventional measures, the ones everyone would recognize and know what you did to determine. Then consider whether there are any other ways to measure the concept directly. From there, establish whether there are any indirect measures or surrogates that could be used in lieu of a direct measurement. Finally, if there are no other options, explore whether it would be feasible to develop a new measure based on theory. Keep in mind that developing a new measure or a new scale of measurement is more difficult for the experimenter and less understandable for reviewers than using an established measure.

For example, say you wanted to assess the quality of the water used in the study of coffee shops. You might use standard laboratory analysis procedures to test water samples for specific ions known to affect taste, like iron and sulfate. These would be direct measures of water quality. An example of an indirect measure would be total dissolved solids, a general measure of water quality that responds to many dissolved ions besides iron and sulfate. An example of a surrogate measure would be the water's electrical conductivity, which is positively correlated to the quantity of dissolved ions in the water. Developing a new measure based on theory might involve using professional taste testers to judge the qualities of the waters.

On a Scale of ½ to VIII

> *Anything that is in the world when you're born is normal and ordinary and is just a natural part of the way the world works. Anything that's invented between when you're fifteen and thirty-five is new and exciting and revolutionary, and you can probably get a career in it. Anything invented after you're thirty-five is against the natural order of things.*
>
> Douglas Adams, *The Salmon of Doubt*, 2002

Of the possible measures you identify, select scales of measurement and consider how difficult it would be to generate the data. For example:

- **Qualities** are usually more difficult to measure accurately and consistently than quantities because there is more complex judgments involved.
- **Counts** are straightforward when they involve simple judgments as to what to count. Some judgments, such as species counts, can be relatively complex because you have to be able to identify the species before you can count it. Counts have no decimals and no negative numbers.
- **Amounts** are usually more difficult to measure than counts because the judgment process is more complex. Amounts have decimals but no negative numbers unless losses are admissible.
- **Ratio measures**, such as concentrations, rates, and percentages, are usually more difficult to measure than amounts because they involve two or more amounts. Ratio measures have both decimals and negative numbers.

Once you know what you might measure, evaluate the sources of measurement variability (benchmark, process, and judgment described in Chapter 5) in each measure. Finally, take into account your objective and the ultimate use of your statistics. For example, if you want to predict some dependent variable, quantitative independent variables would usually be preferable to qualitative variables because they would provide more scale resolution. Furthermore, you could dumb down a quantitative variable you measured to a less finely divided scale or even a qualitative scale. You usually can't go in the other direction. If you want your prediction model to be simple and inexpensive to use, don't select predictors that are expensive and time-consuming to measure.

Consider building some redundancy into your variables if there is more than one way to measure a concept. Sometimes one variable will display a higher correlation with your model's dependant variable or help explain analogous measurements in a related measure. For example, redundant measures are often included in opinion surveys by using differently worded questions to solicit the same information. One question might ask "Did you like [something]?" and then a later question ask "Would you recommend [something] to your friends?" or "Would you use [something] again in the future?" to assess consistency in a respondent's opinion about a product.

The Santa Claus Strategy

I'm working all out
Deadline is near
Model's in doubt
Dooming my career.
Sta-tis-tics will chill my meltdown.

I'm adding new vars
Testing them twice
Trying to find out which ones'll suffice
Sta-tis-tics will give the lowdown.

I see the best predictors.
I know what steps come next
I clean up my dataset and
Regress my y on my x.

Ohhhhh!
My work is all through
My deadline was met
My client paid up
Now I'm out of debt.
Sta-tis-tics helped thwart my shutdown.

Sing to the tune of "Santa Claus Is Coming to Town"

Make a list. Check it twice. Sage advice from an old fat guy with a beard.

Figure 4 is a checklist you can use to help you think about your variables. Complete a checklist for each variable you plan to record. This may seem like a formidable amount of work, but it's worth the effort. The checklist will help you think about your measurements, visualize how they will be generated, and ultimately produce results with less bias and variability. The checklists also provide concise documentation that can be added to a report appendix or project file. Furthermore, if you work with the same data often, you'll find that completing such a checklist becomes much easier once you have thought through the process the first time. If this checklist doesn't meet your needs, use it as a starting point to create your own. The important point is to think about what you plan to do.

Checklist for Variables

Variable Name: ______________ **Variable Label:** ______________

Variable Description: ______________

Relationship to Phenomenon and Analysis

Direct/Indirect Measure: ______________ Hard/Soft Information ______________

☐ Dependent Variable: ☐ Grouping Variable ☐ Continuous Variable

☐ Independent Variable: ☐ Predictor ☐ Strata ☐ Covariate ☐ Other: ______________

Variable Origin

☐ Existing Source: ______________

☐ Newly Generated: ______________

Measurement Variance

Reference: ______________

Process: ______________

Judgment: ______________

Scale

			Special Scales	
☐ Quantitative	☐ Continuous	☐ Ratio	☐ Time-Dependent	☐ Counts
		☐ Interval	☐ Location-Dependent	☐ Cyclical
☐ Qualitative	☐ Discrete	☐ Ordinal	☐ Concatenations	☐ Repeat Units
		☐ Nominal	☐ Restricted Range	☐ Orientations

Variance Control

☐ Procedural: ______________

☐ Quality Samples/Measures:

☐ Replicates: ______________

☐ Comparisons: ______________

☐ Tests: ______________

☐ Sampling Control

☐ Random ☐ Stratified ☐ Systematic ☐ Cluster

☐ Other: ______________

☐ Experimental Control: ______________

☐ Statistical Control: ______________

Metadata: ______________

Figure 4. Sample Checklist for Creating Variables.

This isn't a once-and-done process. You'll revisit your thought process periodically throughout your analysis.

High-end statistical software senses some of the characteristics about your data (i.e., metadata) and uses it to assist you in setting up analyses. For example, Figure 5 is a screen capture of SPSS' variable specifications. The specifications include:

- **Type**—whether the variable contains text, dollars, percentages, or numeric values.
- **Values**—valid values, used for nominal and ordinal scales.
- **Missing**—a dummy value used to indicate that there is no actual value.
- **Measure**—the variable scale as either nominal, ordinal, or ratio.

Within limits, you can change some of the metadata, such as the format and even the scale, to try different statistical procedures. For example, you could change what SPSS thinks is an ordinal scale to a ratio scale so you could calculate correlations between variables.

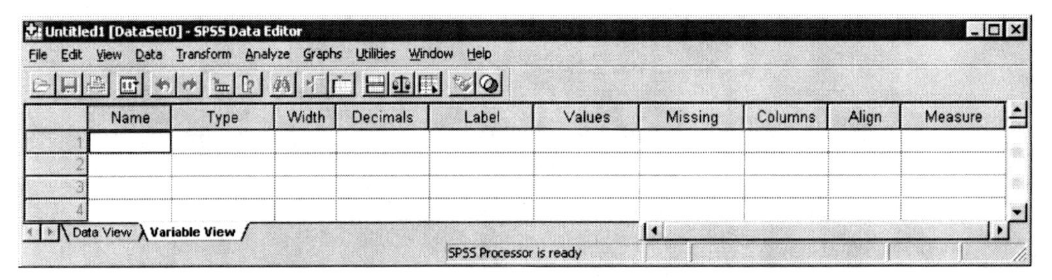

Figure 5. Screen Capture of SPSS Variable Specifications.

Now that you know what variables you're going to use, you need to define the population you want to investigate and determine the best way to sample it.

14 Not-So-Simple Samples

You can't get through Statistics 101 successfully without thinking *population* whenever you see the word *sample*. So if you're going to select samples for an analysis, you first have to ask yourself, "What is the population I want to analyze?" Your samples need to be true representatives of that population. If you are analyzing drug effects in the population of people who might take the drug, your focus is on specific patients who are taking the drugs. If you are analyzing the distribution of pollutants in the soil at a hazardous waste site, your focus is on samples of that soil. So far, so good. Right?

Focus of the Analysis

Below every tangled hierarchy lies an inviolate level.

Douglas Hofstadter, *Gödel, Escher, Bach: An Eternal Golden Braid,* 1979, p. 683

As you think about what data you should collect, you need to identify what the focus of the analysis will be. What's the issue? In a nutshell—hierarchical organizations.

There are many examples of data that are hierarchical, some overt and some arcane. Students attend a specific school in a specific district. Birds build nests in specific types of trees in a specific habitat. Cells reside in a specific organ in a specific body. In all of these cases, there are several levels that could be analyzed, for example, students, or schools, or districts, and so on. That leads to two problems. The first is that sources of natural variability may be different at each level. For example, the reasons for variation in an individual student's performance would be very different from the reasons for variation in the performance of the student's class, school, district, county, or state. More importantly, the levels may not be independent of each other, a key assumption for statistical procedures. For example, the schools in a district may have very different demographics or funding limitations than schools in other districts. So hierarchical data structures call for extra thought in designing a dataset.[1]

1 As you might expect, there are specialized statistical techniques for handling hierarchical data structures. If you think you have hierarchical data, search the Internet for "hierarchical linear modeling." You'll need specialized software, and probably, some professional help if you plan to go this route.

Say a thousand-person company hires you to analyze their workload. This is a common issue especially in service industries because they can't store excess inventory. The company employs one hundred people in their central office and three hundred people in each of three divisions. Each division consists of ten regional offices of thirty people, as shown in Figure 6.

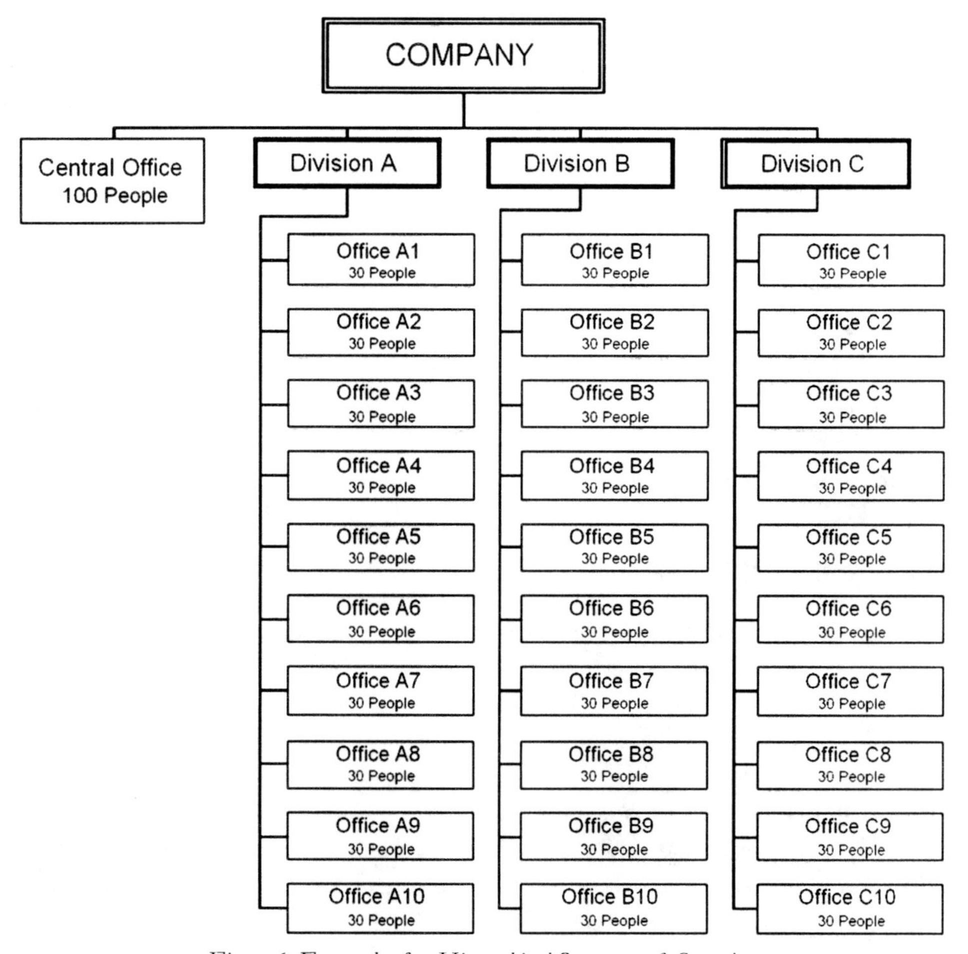

Figure 6. Example of a Hierarchical Structure of Samples.

Should you measure the workload of individuals, or offices, or divisions? There are a couple of considerations. First and foremost, what does the company want to do with the results? Do they want to see how they compare, as a company, with the rest of their industry? Are they looking to forecast future work based on historical patterns? Are they looking to reorganize or transfer work between offices or divisions? Are they looking for individuals who are overworked or underutilized? So your first question should be about their objective, perhaps followed by "what do you mean by *workload*?" Is their aim to determine how well the company is doing, how well each office is doing, or how well each individual is doing? In general, your client's reason for doing the analysis and how they collect data will tell you what your focus should be. If you have any doubt as to what the proper focus should be, collect your data at the lowest level of a hierarchy. It will probably be more expensive and time-consuming to do, but you'll have more options for what you can analyze. You could always aggregate lower-level data up to a higher level, but you can't split higher-level data into lower level data.

Once you know what your focus will be, you should consider how many samples you would have to analyze. If you focus on the divisions in Figure 6, you only have four sources of information, the three divisions and the central office. That's not enough samples for a statistical analysis. You could analyze the workload of the divisions over time, say months, quarters, or years, which could provide enough samples. Since workload would be expected to fluctuate over time, this approach would allow temporal variation to be evaluated. The downside is that you would probably need some specialized procedures to handle the temporally autocorrelated data. If you decided to focus on the thirty offices, you would barely have enough samples and you might need to compensate for different local economies or demographics. If you focused on the thousand employees, you would have plenty of data. You might have to analyze parts of the company separately, for instance, the hundred central office staff, if their jobs were different enough that workload could not be measured in the same way.

How you measure the dependent variable in a statistical model (in this example, workload) will depend on the focus of your analysis. If you plan to focus on a business unit—divisions or offices—there is probably some data the company routinely and consistently collects that you can use. For example, you might consider using number of staff, sales, work in progress, or other performance statistics. If you plan to focus on individuals, there may or may not be data that you can use. In some industries, employees complete forms describing what projects they worked on. In other industries, the type of work isn't recorded but the hours worked are, such as by punching a time clock. Sometimes the information is recorded only informally or not at all. Even if it is recorded, it may not be available electronically. Imagine having to enter time card data for a thousand employees for a year. Is this what you need?

Individuals

> *It is of considerable importance, then, to look for large groups of individuals bound together not by temporal ties of tradition or political artificiality, but by tested ties of common interest so worldwide, indeed so universal, as to be recognized by any individual.*
>
> John H. Manley, "Science in Crisis," *Bulletin of the Atomic Scientists,* 1959, Vol. 15, No. 3, p. 114.

So before you pick the individuals that will be your samples—the rows of your data matrix—you have to know what your population focus is and be confident you'll have enough samples to do a statistical analysis with. Now it's time to select your subjects.[2]

---

2 If you decide to select all the individuals in the population, it's called a *census*. You don't have to worry about representativeness in a census, so long as you can actually access all the individuals in the population. Every decade, the Census Bureau conducts a census in which they count (or at least try to

In an ideal world, you as the experimenter would have total authority to select any of a limitless number of candidate samples for your study. In this reality, though, you are hampered by two constraints:

- The number of candidate samples is sometimes very limited.
- Some candidate samples can challenge their selection.

Consider the examples shown in Figure 7. Some types of candidate samples are almost totally under the experimenter's control, such as environmental samples, manufactured products, and lab animals. If an experimenter is going to sample soil or another environmental medium, he or she can select any of the candidate locations (provided the landowners agree). Likewise with a study of manufactured products, the experimenter can select any candidate although the number of candidate products wouldn't be infinite. Lab animals might be in shorter supply than manufactured products, but they also would have no power to refuse their selection. In contrast, feral animals do have some choice in that they can avoid traps, within the limits of their natural instinct. Any experimenter who has tried to collect fish or other wildlife for a statistical study can tell you that no amount of persuasion will convince a feral animal to participate if they don't like the bait.

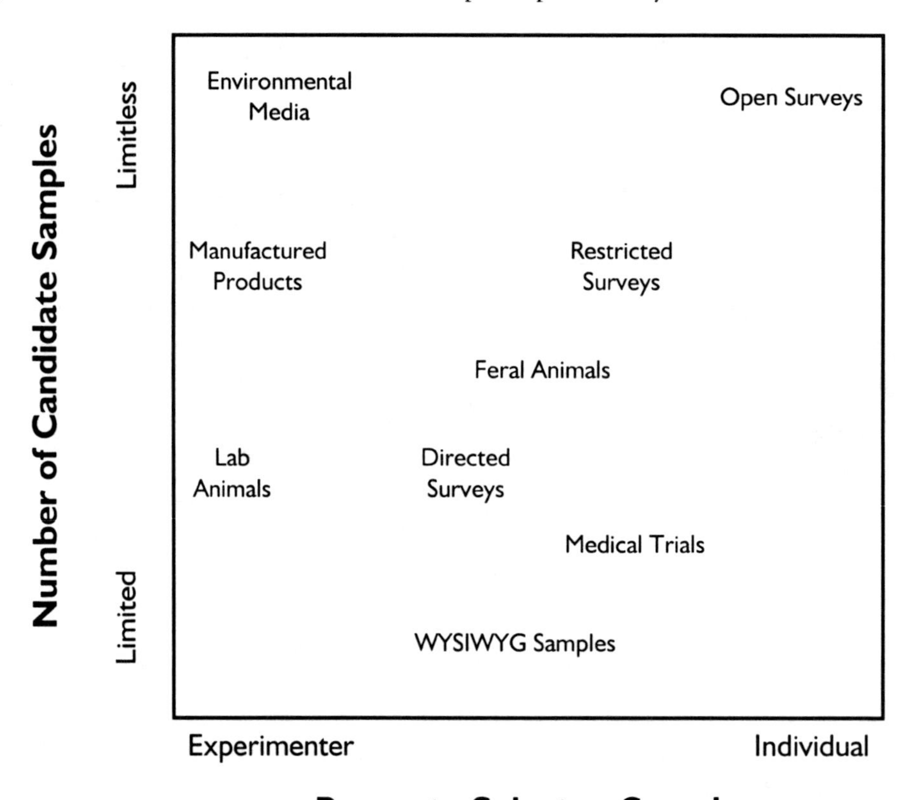

Figure 7. Constraints on Sample Selection.

count) every U.S. citizen. The bureau also selects individuals to serve as a sample of the population. This sample of individuals is asked for more detailed information, which is extended to characterize the entire population defined by the census.

Humans tend to fall at the far end of the control spectrum. In medical trials, for instance, all the experimenter can do is exclude candidates. If individuals do not want to participate, they can't be forced, at least legally. As with medical trials, candidates for opinion surveys can choose not to participate, but if the initial pool of candidates is comparably large, the experimenter has more flexibility in restricting who is invited to respond. In directed surveys, experimenters may place restrictions on who can participate, send surveys only to those individuals, and hope enough of the invitees choose to respond. In open-invitation surveys, experimenters give up all control over selection and allow anyone who notices the survey to participate. These results are usually reported on talk radio.

WYSIWYG (what-you-see-is-what-you-get) samples are in such limited supply that experimenters have no choice but to take whatever is available. Time-series data fall into this category if the data are available only by discrete periods, such as days, months, or years. Organization data, in which there are a limited number of organizational divisions, is another example. Business and economic studies are commonly limited to WYSIWYG data.

Why is this important? In a word—bias. Bias can come from either the experimenter *or the subject*. The experimenter is the anchor of impartiality, at least in the ideal world. If there are enough candidate samples to pick from, an experimenter is supposed to select an unbiased sample. It even happens that way sometimes. Individuals, on the other hand, will do whatever suits them given the chance. An open invitation to a web survey is almost guaranteed to be a biased sample, especially if it involves politics, religion, or marijuana. In general, the more control your subjects have in determining their participation, the more you as the experimenter have to do to minimize sampling bias and variability.

So the first principle you need to follow in selecting candidate samples is to avoid bias, preconceived notions, or any other expectation you might have in your selections. The other principle you should adhere to is to use a statistical plan that will minimize variability attributable to your sample selection.

Statistical sampling seeks to characterize a population, but sometimes you can't identify the individual members in the population so that you can select them. You need an alternative, which is called a *frame*. For example, if you can't identify every potential user of a company's new consumer product you plan to study, you could use the company's registration lists from similar products as your frame. The frame has to correspond to the focus of the study. From the population or the frame, you use a statistical sampling scheme that will provide a reasonable assurance that the samples you select are representative of the population.

How's this for a frame?

There are a variety of statistical sampling schemes, the most commonly used of which are:

- **Random Sampling**—Random sampling involves picking samples from the frame purely on the basis of some randomization tool, such as a table or software algorithm. Random sampling is incorporated in some form in most of the other statistical sampling schemes.
- **Stratified Sampling**—Stratified sampling involves subdividing the population into parts or groups, called *strata*, and then randomly selecting samples from each stratum in proportion to the number of candidate samples in the strata. You need to know a lot about your population to use stratified sampling effectively. For example, if you were going to conduct a one-hundred-person survey of members of a gym having six hundred male members and four hundred female members, you could divide candidates for the survey by sex, and then randomly select sixty male gym members and forty female gym members. Stratified sampling is also called proportional sampling because the number of samples selected from each stratum is proportional to the size of the strata.
- **Cluster Sampling**—In cluster sampling, you select locations where you will sample and then randomly select individual samples at those locations. In the gym example, you might randomly select the weight room and the pool area from the ten exercise areas at the gym, and then randomly select fifty participants from each of those two areas. Cluster sampling is also called two-stage sampling because there are two random selection steps, first clusters and then individuals. For comparison, strata are an inherent property of the population whereas clusters are coincidental or experimenter-imposed groupings of candidate samples. Composite sampling, also called three-stage sampling, involves taking the samples from a cluster location and compositing them by physically mixing the samples or mathematically averaging the measurements on the samples. For example. You could collect three water samples and mix them together and conduct one chemical analysis or you could analyze all three samples and average the results of the chemical tests. Compositing samples is cheaper but averaging results provides more flexibility in analyzing the data.
- **Systematic Sampling**—Systematic sampling is based on the assumption that the underlying population is randomly organized so that samples can be selected in a fixed pattern (i.e., systematically) because further randomization is not necessary. In the gym example, you might systematically select every third person to enter the gym until you had surveyed one hundred members. For this to provide a representative sample, the arrivals must be random and independent of each other, which is not very likely. Grid sampling is a form of systematic sampling used extensively in environmental studies in which a grid is placed on a site being investigated and samples are selected in each grid cell. If the samples are selected at the same relative location in each cell, the scheme is called a systematic-grid sample. If the samples are selected randomly within each cell, the scheme is called a systematic-random sample. If the grid size is designed to identify enti-

ties or properties that cannot be seen based on a probability of occurrence, the scheme is called search or probability sampling.

- **Surrogate Sampling**—Surrogate sampling involves using an easy-to-select sample as a substitute for a difficult-to-select sample. Surrogate sampling requires the experimenter to demonstrate that the responses from the two types of samples are equivalent. If you wanted to survey young children about their preferences in playground equipment in the gym example, it might be more effective for you to survey their parents as surrogates for the children. Parents can usually answer questions better than their children and there would be less of a chance of you being arrested for looking like a pervert. Surrogate sampling is also called double sampling.

These sampling schemes can be combined in a variety of ways to address complexities of the population being sampled. So which sampling scheme do you select? It takes some experience, as you might expect, but here are three steps you can try:

Step 1. Decide if there is a way (or ways) to subdivide your population that would make sense in light of your objective. If so, use a stratified sampling scheme. Stratified samples are more work to analyze, but if the strata are well selected, the reduction in variance will be worth the effort.

Step 2. For each of the strata from step 1, decide if the population attribute you plan to measure is randomly organized in the population. If so, use a systematic sampling scheme, as it is easier to lay out grids or select samples on a schedule than it is to make random selections. If the attribute is not random in the population, decide if uniform coverage of the sample space is important. If uniform coverage is not important, use a random sampling scheme. If uniform coverage is important, use a systematic-random sampling scheme.

Step 3. For each of the strata, decide if access to the samples is restricted or difficult or if the sample locations are clustered. If so, use a cluster sampling scheme.

These three steps will lead you to one of five schemes: stratified, systematic, random, systematic-random, and cluster sampling. Although these are the most commonly used, there are many, many others. Table 8 provides examples of how these sampling schemes might be used. There are scores of books and websites on statistical sampling that provide additional detail on these and other sampling schemes.

Table 8. Examples of Commonly Used Sampling Schemes

Sampling Scheme	Example of a Spatial Array of 25 Samples	Example of a Telephone Survey of 25 People	Example of a Manufacturing Study of 25 Gizmos
Systematic		Select every 10th name from a company's five-page telephone list (frame) of 250 customers (100 past customers and 150 current customers).	Select every 10th gizmo coming off the assembly line until 25 gizmos are selected.
Random		Number all 250 names on the list and use a random number generator to select 25 unique numbers corresponding to customer names.	Use a random number generator to decide how many minutes to wait between selecting gizmos.
Systematic-Random		Randomly select 5 names from each of the 5 pages of the telephone list.	Select 1 gizmo from the production line every hour by using a random number generator to pick at which minute of the hour the gizmo should be selected.

Sampling Scheme	Example of a Spatial Array of 25 Samples	Example of a Telephone Survey of 25 People	Example of a Manufacturing Study of 25 Gizmos
Cluster	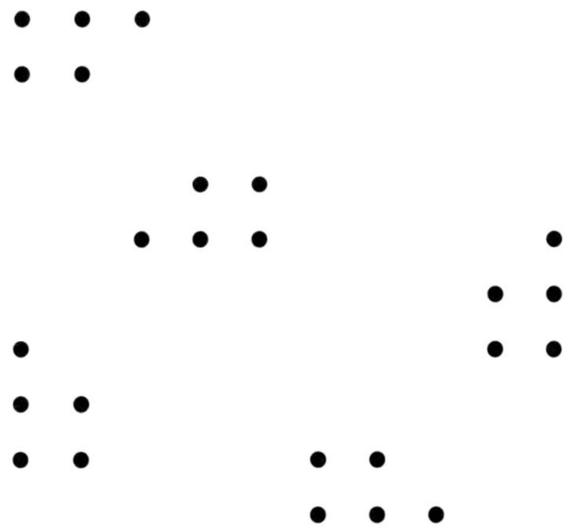	Sort the phone numbers by area code, then select a roughly equal number from each area code.	Select 8 gizmos from each of the 3 shifts of workers
Stratified-Random (40%) and Stratified-Systematic (60%)	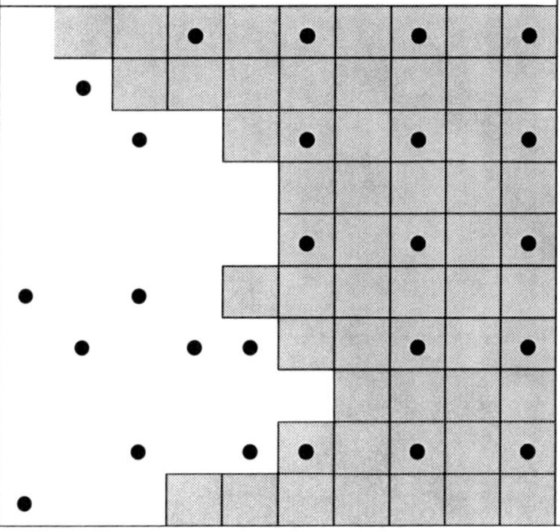	Randomly select 10 names from the 100 past customers and systematically select 15 names from the 150 current customers.	Randomly select 10 gizmos produced by experienced staff and systematically select 15 gizmos produced by newly-hired staff

Samples and Potato Chips

Samples are like potato chips. You're never satisfied with just one. Every one you take makes you want more. And you're never sure you've had enough until you've had way too many.

Charles Kufs, *Making Sense of Statistical Models*, 1989

No matter what their area of expertise, statisticians are asked certain questions with such predictability that it borders on the deterministic. No question is asked more often than:

How many samples do I need?

Most statisticians wish they could answer the sample size question definitively instead of mumbling about effect sizes and whatnot. It's just not that simple. But if you've ever taken any graduate courses in disciplines that use statistics for data analysis, you were probably exposed to the mystique of thirty samples. Countless times I've heard statistician do-it-yourselfers tell me that you need thirty samples for statistical significance. This statement is so misguided, it merits more than a little clarification. For now, suffice it to say that if there were any way to make the answer that simple, you would find it in every textbook on statistics, not to mention TV quiz shows and fortune cookies. Still, if you do an Internet search for "thirty samples," you'll get thousands of hits.

Chips all gone. Get more.

Thirty Samples—More Lure Than Law

Like many legends, there is some truth behind the myth. The thirty-sample rule of thumb may have originated with William Gosset, a statistician and head brewer for Guinness. In a 1908 article published under the pseudonym Student,[3] he compared the variation associated with 750 correlation coefficients calculated from sets of four and eight data pairs, and 100 correlation coefficients calculated from sets of 30 data pairs, all drawn from a dataset of 3,000 data pairs. Why did he pick thirty samples? He never said but he concluded, "With samples of 30 . . . the mean value [of the correlation coefficient] approaches the real value [of the population] comparatively rapidly," (p. 309). That seems to have been enough to get the notion brewing.

Since then, there have been two primary arguments put forward to support the belief that you need thirty samples for a statistical analysis. The first argument is that the t-distribution[4] becomes a close fit for the Normal distribution when the number of samples reaches thirty. That's a matter of perspective. Figure 8 shows the difference

3 Student. 1908. Probable error of a correlation coefficient. Biometrika 6, 2-3, 302–310.

4 The t-distribution, sometimes referred to as Student's distribution, is also attributable to W. S.

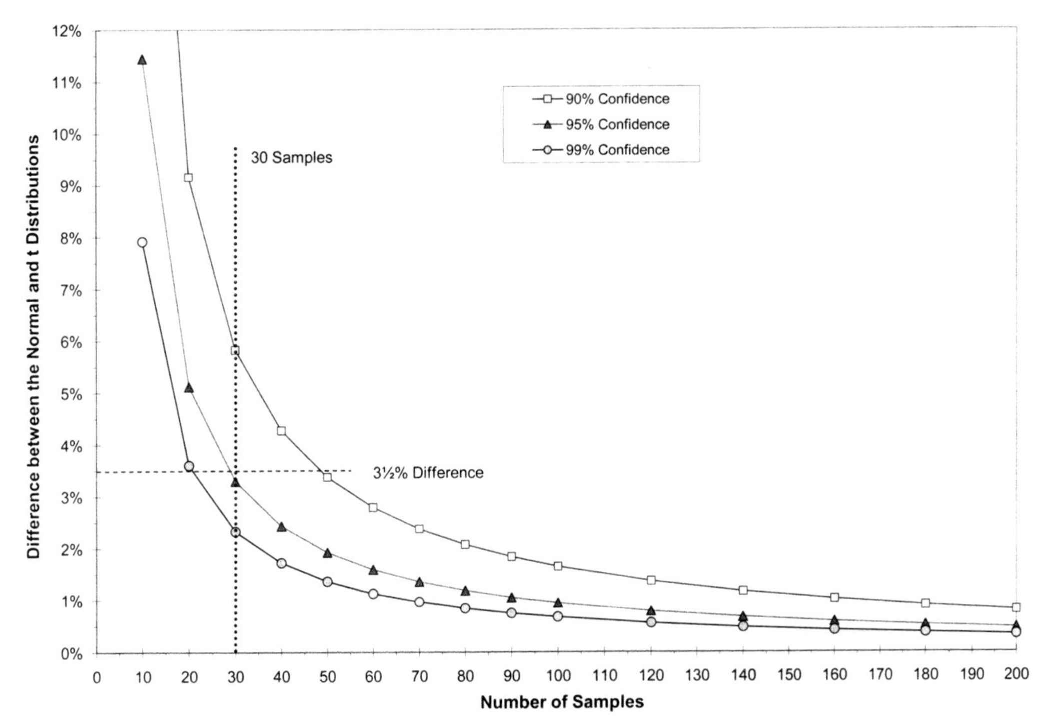

Figure 8. Differences between the Normal and t-distributions.

between the Normal distribution and the t-distribution for 10 to 200 samples. The differences between the distributions are quite large at 10 samples but decrease rapidly as the number of samples increases. The rate of the decrease, however, also diminishes as the number of samples increases. At 30 samples, the difference between the Normal distribution and the t-distribution (at 95 percent of the upper tail, used for statistical testing with 95 percent confidence) is about 3½ percent. At 60 samples, the difference is about 1½ percent. At 120 samples, the difference is less than 1 percent. So from this perspective, using 30 samples is better than 20 samples but not as good as 40 samples. Clearly, there is no one magic number of samples that you should use based on this argument.

The second argument is based on the Law of Large Numbers, which in essence says that the more samples you use, the closer your estimate of population parameters, like the mean and standard deviation, will be to their true population values. This sounds a bit like what Gosset said in 1908, and in fact, the Law of Large Numbers was 200 years old by that time. Figure 9 shows how differences between means estimated from samples of different sizes compare to the population mean.[5] The small inset graph shows the largest and smallest means calculated for datasets of each sample size. The large graph shows the difference between the largest mean and the smallest mean calculated for each sample

Gosset. The t-distribution is used to estimate the mean of a normally distributed population from a limited number of samples from the population.

5 These data were generated by creating a normally distributed population of 10,000 values, then drawing at random 100 sets of values for each number of samples from 2 to 100 (i.e., 100 datasets containing 2 samples, 100 datasets containing 3 samples, and so on up to 100 datasets containing 100 samples). Then, the mean of the datasets was calculated for each number of samples. You can try this at home.

size. These graphs show that estimates of the mean from a sampled population will become more precise as the sample size increases (i.e., the Law of Large Numbers). The important thing to note is that the precision of the estimated means increases very rapidly up to about 10 samples then continues to increase, albeit at a decreasing rate. Even with more than 70 or 80 samples, the spread of the estimates continues to decrease. So again, there's nothing extraordinary about using 30 samples. So what's the point? Keep reading.

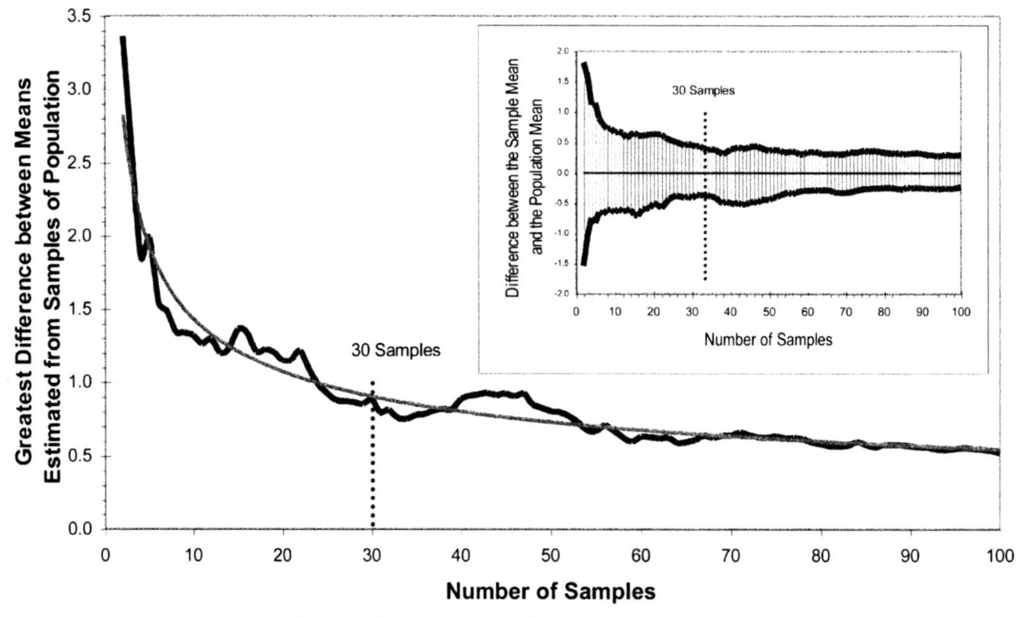

Figure 9. Effect of Sample Size on Estimates of a Population Mean.

Purrfect Resolution

One way to look at how many samples you need for an analysis is in terms of how much information you need to answer your question. How many notes do you need to name that tune? How many letters do you need to guess the secret phrase? How many shots do you have to take to sink someone's battleship? How many samples do you need to reach some objective?

Another way of asking how much information do you need is how much *resolution* do you need. Think of the resolving power of a telescope or a microscope or the number of pixels in a computer image. The greater the resolution, the more detail you'll see. Consider the pictures in Figure 10. You couldn't make out the image with a resolution of 9 pixels per inch and maybe not even with a resolution of 18 pixels per inch. At 36 pixels per inch, you can tell it's an image of a kitten, even if it's a bit fuzzy. At 72 pixels per inch, the image is sharp, and you can tell that the kitten is Kerpow. Doubling the resolution again adds little to your perception of the image. Likewise with statistics, the greater the number of samples, the more precise your results will be. But beyond a certain point, adding samples adds little to your understanding. The trick is to collect the fewest samples that will achieve your objective.

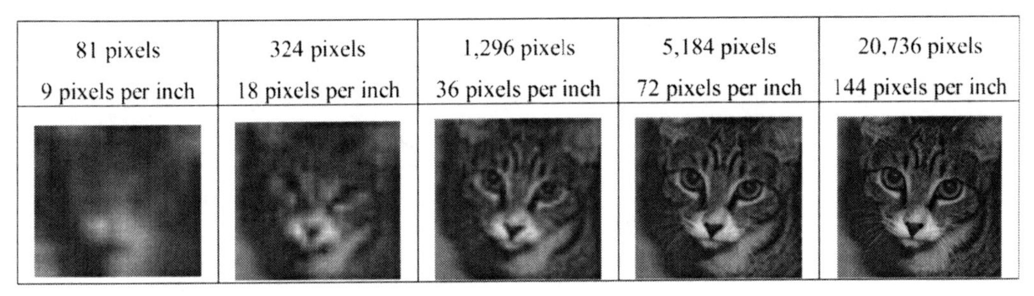

Figure 10. Resolution—Samples are to Statistics as Pixels are to Images.

So deciding how many samples you'll need starts with deciding how certain your answer needs to be given your objective. Now here's the bad news. There's no way to know exactly how many samples you'll need before you conduct your study. There are, however, formulas for estimating what an appropriate number of samples *might* be. In the situations in which the formulas don't apply, there are rules of thumb or other ways to come up with a number. Unfortunately, it seems that no matter how many samples you estimate you'll need, the number is always a lot more than your client wants to collect. After all, they are the ones who have to pay for collecting and analyzing the samples. So any answer you come up with ends up being a subject of negotiations. But here are a few places to start.

How Many Samples for Describing Data?

Say all you want to do is to collect enough samples to calculate some descriptive statistics. Maybe you want to characterize some condition, like the average weight of a litter of kittens or the average age of your favorite professional sports team. How many samples do you need? Well if your population is small enough, like five kittens or twenty-five baseball players, you simply use all the members of the population, a census.

But what if you want to calculate descriptive statistics to characterize a large population? The number of samples you'll need to describe it will depend on the precision you want, not the accuracy.[6] The greater the number of samples, the more precise your estimate will be. More specifically, the precision will be proportional to the variance divided by the square root of the number of samples. So maximize the number of samples if you can (by a lot, remember, precision is proportional to the *square root* of the number of samples), but if you can't, try to control the variance. Chapter 15 describes how you might go about doing that.

How Many Samples for Detecting Differences?

Often, the point of calculating statistics is to make an inference from a sample to a population. You can estimate how many samples you might need to conduct a statistical test of one or more populations by using the equation for the test you plan to use.

---

6 Look back at the inset graphic in Figure 9. The calculated estimates of the mean are more variable, less precise, for small samples but still all centered on the population mean. You might recognize this as the central limit theorem at work.

Rearrange the equation and solve for the number of samples instead of the test statistic. Take a t-test as an example:

$$\text{t statistic} = \frac{\text{difference you want to detect}}{\text{(standard deviation) / (square root of the sample size)}}$$

You can rearrange this equation into:

$$\text{sample size} = (\text{t statistic})^2 * \text{variance} / (\text{difference you want to detect})^2$$

Forgetting the t-statistic for a moment, there are two other things you need to know—the difference you want to detect and the population variance.

You should have some idea of the size of the difference you want to detect; otherwise you shouldn't be doing the analysis. Think about what might be a meaningful difference. Say you want to compare how long it takes you to commute to work via two different routes. Differences of a few seconds probably aren't meaningful, but differences of a few minutes probably are. If you work as a NASCAR driver, go with seconds. You get the idea? The smaller the difference you want to detect, the more samples you'll need.

Then there's the population variance, the Catch-22 of statistics. You can't calculate the number of samples you'll need without knowing the population variance, and you can't estimate the population variance without already having samples from the population. See the problem? Now, there are maybe a half dozen ways to try to get around this problem, but they all require you to know or guess at some aspect of the population. So this is a tough way to go. The approach is used, however, after a preliminary study (called a pilot study) is done to estimate the population variance. There's more to it than that, so if you're interested, search the Internet for "power analysis."

How Many Samples for Opinion Surveys

If you're going to survey a small population, like your colleagues at work, send surveys to everybody and hope you get a representative sample from the people who do respond. If the size of your population is large compared to the number of samples you might take, a quick way to estimate the sample size is:

$$\text{sample size} = 1 / (\text{approximate percent error you want})^2$$

So if you want a ±5 percent error with 95 percent confidence, you would need about 400 samples (i.e., $1/0.05^2$). Political pollsters use this approach all the time. With 625 samples, they get about a 4 percent error. With 1,000 samples, the error drops to about 3 percent. But to get to 2 percent error, they would have to collect 2,500 samples. That's why most political polls are usually based on no more than 1,200 samples.[7] It's

7 Exit polls after voters have cast ballots usually collect far more samples for two reasons. First, the respondents are readily available upon leaving the polling place. Second, a much smaller error rate is essential because exit polls are used to forecast results for the mainstream media. Election monitoring

more complicated than this of course. If your sample will be a sizable proportion of your population or if the opinions aren't evenly divided, the short-cut formula will over-estimate how many samples you need. Also, if you carve up your sample to explore demographic groupings, the variances of the groupings will be larger, usually much larger, than for the overall survey.

How Many Samples for Evaluating Trends?

I'm being a linear trend.

Say you plan to do a regression analysis to evaluate the relationship between two sets of measurements. How many samples do you need? There are two answers to this question: a difficult answer and an easy answer. The difficult answer is that you can calculate it the same way as you would if you were looking to detect differences. This approach requires a sophisticated understanding of statistical tests and the populations being tested. It is most often used in experimental situations.

The simpler approach is to base the number of samples on a rule of thumb comparing the number of samples to the number of independent variables. The idea is that the more independent variables there are, the more samples are needed to define their relationship to a dependent variable. A number of researchers[8] have conducted simulation studies (called Monte Carlo studies) to estimate how many samples might be needed. The guidelines are not hard and fast, but boil down to these:

- **10 times the number of predictor variables**—The bias may be large but there are often enough samples to estimate simple linear relationships with adequate precision.
- **50 times the number of predictor variables**—The bias is relatively small, linear relationships can be estimated with good precision, and there are usually enough samples to determine the form of more complex relationships.
- **100 times the number of predictor variables**—The bias is insignificant, linear relationships are estimated precisely, and complex nonlinear relationships can be estimated adequately.

---

groups also use exit polls to detect fraud in elections. In fact, exit polls are often the only way to detect election fraud and are used extensively for that purpose. For example, discrepancies between exit polls and purported vote counts during the 2004 presidential election in the Ukraine led to nationwide protests that caused the reported results to be reversed. For some unknown reason, similar discrepancies in the 2004 presidential elections in the United States were not taken as evidence of election fraud.

8 Including Bob Barcikowski, my advisor in grad school. His article on this topic is Barcikowski, R., & Stevens, J. P. (1975). A Monte Carlo study of the stability of canonical correlations, canonical weights, and canonical variate-variable correlations. Multivariate Behavioral Research, 10, 353–364.

- **250+ times the number of predictor variables**—The bias is insignificant, and most complex relationships can be estimated precisely.

As with the mean, estimated correlation coefficients for a simple linear relationship between one dependent variable and one predictor variable will increase in precision as the number of samples increases. With ten samples, the spread in estimates will probably be large, perhaps on the order of 10 percent. With one hundred samples, the spread would be less than 1 percent.

How Many Samples for Forecasting Time Series?

Deciding how many samples to use for analyzing a time series can be a challenge. Here are two popular rules of thumb:

- Collect samples at regular intervals from at least three or four consecutive cycles or units of any pattern in which you might be interested. For example, if you are interested in seasonal patterns (i.e., a pattern lasting a year) collect data for at least three or four years.
- Collect samples at time units much smaller than the duration of the pattern in which you might be interested. For example, if you are interested in seasonal patterns, collect data weekly, biweekly, or at least, monthly.

How Many Samples for Identifying Targets?

Sometimes the goal of sampling is to find one or more targets. For example, in World War II, destroyer captains needed to know how many depth charges to drop to be reasonably certain of destroying an enemy submarine. Likewise, adventurers looking for sunken ships, like the *Monitor* and the *Titanic*, use statistical sampling to find their targets. In the environmental field, sampling is often done to look for "hot spots" of contamination in soil. Imagine, for instance, that your employer wants to buy an old industrial property that has supposedly been cleaned up (i.e., a brownfield). Your pointy-haired boss tells you to "collect a few samples and tell me if there's any pollution out there." How many samples do you collect?

You found me!

There are two ways this type of problem is typically handled: judgment sampling and search sampling. The strategy behind judgment sampling is that an expert collects samples he or she believes are most likely to reveal a target. In the brownfield example, the expert might walk around the property looking for possible signs of pollution (e.g., stained soil, dead vegetation, organic odors) and collect samples at those points. If no contamination is detected in the areas the expert believes to be most likely to be contaminated, then the rest of the site is probably also not contaminated. With this approach, it is assumed that the expert's observations are adequate for identifying areas most likely to be contaminated.[9] Judgment sampling (a.k.a. judgmental sampling, biased sampling, haphazard sampling, directed sampling, professional judgment) has the advantage of involving far fewer samples than other approaches. The disadvantage is that there is no way to quantify the uncertainty of the result. Sometimes, judgment sampling allows decision-maker ostriches to bury their heads in the sands of ignorance. They believe if uncertainty isn't quantified, it does not exist. So what would you tell your pointy-haired boss if none of the four samples the expert collected from the twenty-acre site were contaminated? Would you pronounce the site clean and ready for redevelopment?

The statistical approach to finding targets is called search sampling. Search sampling involves sampling on a regular grid so that it is possible to estimate the probability of finding randomly located targets. In essence, the probability of finding a target depends on the size and shape of the target and the size and shape of the cells of the sampling grid. In search sampling, sample locations are planned in advance on the basis of some expectation of the size and shape of the target. The downside of this sampling approach is that it usually involves many more samples than the judgment sampling approach and the results do not always sound very reassuring. In the brownfield example, say the property was 23 acres (1,000,000 square feet) and you decided you wanted to have an 80 percent probability of finding a circular hotspot 100 feet in diameter. You would need over 10,000 samples taken on a 100-foot grid. OK, maybe you don't want to present that plan to your pointy-haired boss. Instead, you could propose collecting 100 samples on a 10,000 grid and have a 50 percent chance of finding a 7,980-foot hotspot. That doesn't sound too good either, but it's the price you pay for being able to quantify uncertainty. If you understand the uncertainty, you are one giant step closer to controlling adverse risks. Now can you imagine what the probability might be of the expert finding contamination with just 4 samples?

Too Much of a Good Thing

Can you eat too many potato chips? Of course you can. Likewise, there are some disadvantages to having *too many* samples. Here are five.

Information Overload—Statistical software tends to be very efficient, but when you have tens of thousands of samples, you start to see performance slow

---

9 This approach would be a form of surrogate sampling if it could be shown that there was a close correlation between the expert's opinion and the laboratory analysis. In prospecting for oil, for instance, knowledge of large-scale geologic structures allows geologists to predict locations and depths where wells may encounter producing zones.

a bit. What's more important, though, is the inefficiency you run into when you scrub your dataset, especially if you use a lot of spreadsheet formulas.

Chasing Tails—In any dataset, you may have 5 percent influential observations not to mention the outliers and errors that you'll have to check to determine if they should be corrected or removed from the dataset. This is a very time-consuming process. With a small dataset, you may have to investigate just a few samples. With a 1,000-sample dataset, you may have to investigate 50 samples.

Data Intimacy—When you're working with only a few dozen samples, you get to know each data point. You can look at plots and tables and see how individual details fit into a bigger picture. You can't do that with a thousand data points.

Graphic Mud—It's tough to see patterns with only a few samples, but plotting thousands of samples can be just as perplexing. You won't be able to use any small plots like matrix plots. Even with full-scale plots, it will be difficult to see subtle differences in data point markers, like size, shape, and even color. Points will overwrite each other, so you won't be able to tell it there is one point at a graph location or a hundred points stacked on top of each other. And even the best statistical software will choke when trying to print graphs with thousands of data points.

Meaningless Differences—Sometimes you can have too much resolution in a statistical test. If the test can detect a difference smaller than would be of interest in the real world, you used too many samples.

After the Estimate

So say you've estimated that you need 1,000 samples to do a statistical analysis but your client doesn't have enough money to collect that many. What can you do? Here are a few suggestions:

Change the Study—Lower your confidence and power[10] or look for bigger effects (e.g., differences between means, size of targets, and so on). Subdivide or stratify the sample or use other methods to reduce variability. Limit the area, level of detail, or scope of the study.

Take Smaller Bites—Take as many samples as you can and use the information to decide what to do next. This is sometimes called a pilot study. You can use the samples collected during a pilot study to estimate how many more samples you'll need to get the statistical resolution you want. You might also be able to collect samples in phases to accommodate your client's budget cycle.

Use Supporting Data—If all else fails, there may be historical data available that

10 Confidence is equal to 1 minus the probability of a Type I (false positive) error. Power is equal to 1 minus the probability of a Type II (false negative) error.

you can use to reassess the number of samples you'll need and even augment the samples you plan to collect (i.e., provided the quality of the historical data is appropriate). You can also consider surrogate sampling, in which you correlate the results of many inexpensive samples to the few expensive samples your client can afford.

So that's a long answer to the question "How many samples do I need?" And though this section just scratched the surface of what you'll find in advanced statistical textbooks, the number of samples you'll need for a statistical analysis really all comes down to resolution. How weak a signal do you want to be able to detect from your noisy population? Needless to say, it's a very unsatisfying answer compared to … *thirty samples.*

15 The Heart and Soul of Variance Control

In an ideal universe, your dataset would contain no bias and only the natural variability you want to analyze. It never happens that way. In fact, most of the "disappointing" statistical analyses you'll see are more likely to suffer from too much variability rather than too little accuracy. There are even statistical methods that attempt to trade off accuracy for precision.[1] So to get a good result, whether in marksmanship or in data analysis, you have to control variation.

This chapter discusses how you can plan to control and minimize variability. Extraneous variability can conceal data patterns. Bias in particular can imply misleading information. Understanding the sources of variability in measurements is fundamental to understanding data and producing a valid statistical analysis.

The Shivaree Dance of Data

You can't understand the data without controlling variance.
You can't control variance without understanding the data.
Charles Kufs, *Making Sense of Statistical Models*, 1989

Out, damn'd variance.

To control variability, you have to understand it. But that's not enough. Data and variance are thoroughly intertwined. Data is the sweet and variance is the peppery taste of an ambrosia and parsley salad. You can try to focus on one taste, but the other is always there refusing to be ignored. Variance doesn't go away by ignoring it. It's been tried; it doesn't work. You must be proactive in planning your data collection efforts to control as much of the extrane-

1 For more information, search the Internet for "Ridge Regression" or "bias-variance trade-off."

ous variability as possible so that only the components of natural variability that you are interested in remain.

The classification scheme for variance described in Chapter 5 involves the cause of the variability. Two other ways to look at variability involve how it affects data:

- **Control**—the extent to which variability can be controlled so that data aren't affected.
- **Influence**—the proportion of data points that are affected by uncontrolled variability.

Consider Figure 11. Sampling and measurement variability usually tend to be under your control. Sometimes you can control environmental variability, and sometimes you can't. These types of variability tend to affect all or most of the data. Natural variability, on the other hand, can't be controlled, and it affects all data. Biases affect all or most of a dataset and usually can be controlled if they are identifiable and unintentional. Intentional bias of only selected data is exploitation. Mistakes and errors may or may not be controllable, and they tend to affect only a few data points. Shocks are uncontrollable short-duration conditions or events that can influence a few or even most of the data in a dataset. Examples of shocks include heavy rainfall upsetting a sewage treatment plant, missing a financial processing deadline so one month has no entry and the next has two, having a meter lose calibration because of electrical interference, mailing surveys without realizing that some have missing pages, assembly line stoppages in an industrial process, and so on.

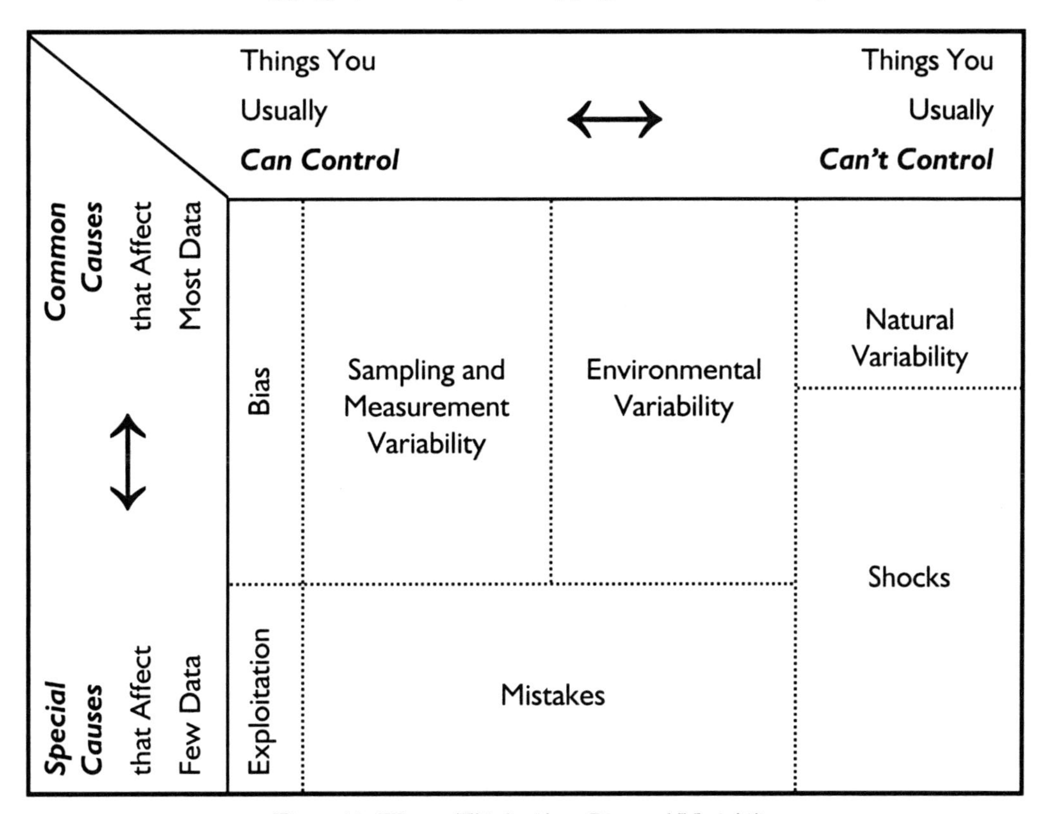

Figure 11. Ways to Think About Bias and Variability.

No one scheme for classifying variance will be best for all applications. Think of variance in terms of the data and the particular analyses you plan to do. You'll know it's right because it will help you visualize where the extraneous variability is in your analysis and what you might do to control it.

Three Rs to Remember

Data maze, error haze
Keeping the scatter small pays
Reference and Repeat and Randomize
Ought to reduce the dispersion's lies.
Error control is apropos
For testing data 'cause you know
That statistics can tell
You yes or no
When you don't feel like anecdotes.

Sing to the tune of "School Days" by Will Cobb and Gus Edwards

The fundamentals of education that we all learned in elementary school are **R**eading, '**R**iting, and '**R**ithmetic (obviously, not spellin'). With these concepts mastered, we are able to learn more sophisticated subjects like rocket science, brain surgery, and tax return preparation. Similarly, if you plan to conduct a statistical analysis, you'll need to understand the three fundamental **R**s of variance control—**R**eference, **R**eplication, and **R**andomization.

Reference

The concept behind using a reference in data generation is that there is some ideal, background, baseline, norm, benchmark, or at least, generally accepted standard that can be compared to all similar data operations or results. References can be applied both before and after data collection. Probably the most basic application of using a reference to control variation attributable to data collection methods is the use of standard operating procedures (SOPs), written descriptions of how data generation processes should be done. Equipment calibration is another well-known way to use a reference before data collection to control extraneous variability.

References are also used after data collection to assess sampling variability. This use of a reference involves comparing generated data with benchmark data. The comparison doesn't control variability, but allows an assessment of how substantial the extraneous variability is. A more sophisticated use of a reference is to measure highly correlated but differently measured properties on the same sample, such as total dissolved solids and specific conductance in water. Deviations from the pre-established relationship may be signs of some sampling anomaly. Further, data collected on some aspect of a phenomenon under investigation can be used to control for the variability associated with the measure. Variables used solely to control or adjust for some aspect of extraneous variability are called covariates.

Perhaps the most well-known application of a reference is the use of control groups. Control groups are samples of the population being analyzed to which no treatments are applied. For example, in a test of a pharmaceutical, the test and control groups would be identical (on relevant factors such as age, weight, and so on) except that patients in the test group would receive the pharmaceutical and the patients in the control group would not.

Replication

If you can't establish a reference point to help control variability, it may be possible to use replication, repeating some aspect of a study, as a form of internal reference.

Replication is used in a variety of ways to assess or control variability. Replicate sampling or measurements are one example. You might collect two samples of some medium and send both samples to a lab for analysis. Differences in the results would be indicative of measurement variability (assuming the sample of the medium is homogeneous).

In addition to the data source (i.e., sample, observation, or row of the data matrix) being replicated, the type of data information (i.e., attribute, variable, or column of the data matrix) can also be replicated. For example:

- Asking survey questions in different ways to elicit the same or very similar information, such as, "Did you like this . . . ," "Did this meet your expectations . . . ," and "Would you recommend this . . . ."
- Measuring the same property on a sample using different methods, such as pH in the field with a meter and again in the lab by titration.

Replicated samples or variables require a little extra thought during the analysis. If you are looking for a fair representation of the population, a replicated sample would constitute an over-representation. Typically, replicated samples are first compared to identify any anomalies, then, if they are similar, they are averaged. Sometimes, either the first sample or the second sample is selected instead. Never select a sample to use in the analysis on the basis of its value. For replicated variables, first compare the variables to identify any anomalies, then select only one of the variables to use in the analysis. Highly correlated variables will cause problems (called multicollinearity) with many types of statistical analysis.

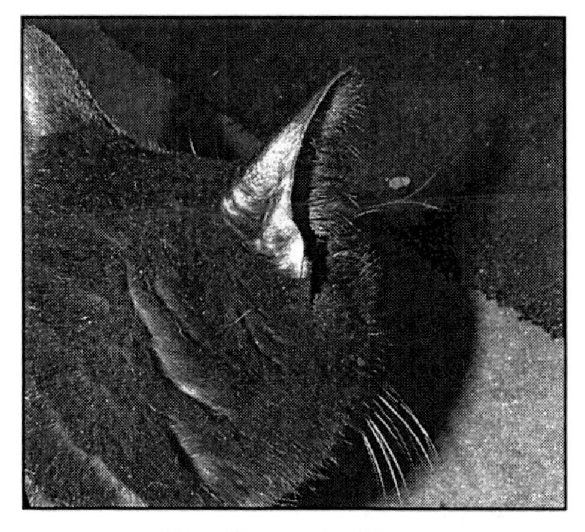

Wow. That looks like me.

The concept of replication is also applied to entire studies. It is common in many of the sciences to repeat studies, from data collection through statistical analysis, to verify previously determined results.

Randomization

Randomization refers to any action taken to introduce chance into a data generation effort.[2] Randomization is desirable in statistical studies because it minimizes (but not necessarily eliminates) the possibility of having biased samples or measurements. As a consequence, randomization also minimizes extraneous variability that might be attributable to inadvertent inconsistencies in data generation. It is an irony of nature that introducing irregularities (randomization) into a data generation process can reduce irregularities (variability) in the resulting data.

As with replication, randomization can be applied to both samples and variables. Samples or study participants can be chosen at random or following a scheme that capitalizes on their existing randomness. Values for variables that are not inherent to a sample can be assigned randomly. This is done routinely in experimental statistics when study participants are assigned randomly to the treatments. Random assignments are simple to make using random number tables or algorithms.

It's All in the Technique

[Japanese businesses] realized that the gains that you get by statistical methods are gains that you get without new machinery, without new people.

W. Edwards Deming, *If Japan Can… Why Can't We?*, 1980

Based on the concepts of reference, replication, and randomization, extraneous variability can be minimized in five ways:

- **Procedural Controls**—used primarily to prevent measurement variability by reference.
- **Quality Samples and Measurements**—used primarily to identify sources of measurement and environmental variability by reference.
- **Sampling Controls**—used primarily to correct sampling variability by reference and randomization.
- **Experimental Controls**—used primarily to correct measurement and environmental variability by reference and randomization.
- **Statistical Controls—**used primarily to correct environmental variability by reference.

2 Statisticians use the term *randomization* to refer specifically to the random assignment of treatments in an experimental design. Here, the term is used in its common sense, the arrangement of objects to simulate chance.

These techniques are summarized in Table 9.

Table 9. Techniques for Controlling Extraneous Variability

		Variance Control Technique						
			Quality Samples and Measurements					
		Procedural	Replicates	Tests	Blanks	Sampling	Experimental	Statistical
Objective of Variance Control Technique	Prevent	✱	●	●	●	✦	●	●
	Identify	●	✱	✱	✱	✦	✦	✦
	Correct	●	✦	●	●	✱	✱	✱
Variance Control Strategy	Reference	✱	✦	✱	✱	✱	✱	✱
	Replication	●	✱	✦	●	✦	●	●
	Randomization	●	●	●	●	✱	✱	●
Type of Variance Controlled	Sampling	✦	●	●	●	✱	●	●
	Measurement	✱	✱	✱	✱	✦	✱	●
	Environmental	✦	✦	✦	✦	●	✦	✱

✱ Primary Importance ✦ Secondary Importance ● Not Important

Procedural Controls

Procedural controls include the field, office, and laboratory procedures that specify how activities related to data generation should be carried out. Forms of procedural control include standard operating procedures (SOPs), chain of custody (CoC) procedures, survey scripts and instructions, checklists, calibration procedures, training courses, and so on. Procedural controls are the main way to minimize extraneous variation before it occurs.

Quality Samples and Measurements

When a physical sample is collected, especially for laboratory analysis, there are a variety of samples that can be collected with it to verify that the analyses produce valid results. Quality samples, sometimes called QA/QC[3] samples, used to assess sampling and measurement variability include:

- **Replicate Samples**—multiple samples collected of the same subject to assess variability. Replicate samples may be collected by splitting a large sample into two or more subsamples (called split samples). Alternatively, discrete samples may be collected sequentially (called co-located samples). Most replicates are co-located samples because they are less time-consuming to collect. Co-located samples of mixable media (e.g., fluids like water and air) will usually be very similar compared to co-located samples of nonmixable media (e.g., solids like soil). *Replicate* is the generic term for a multiple sample. More commonly, the terms duplicate (two samples) or triplicate (three samples) are used.
- **Blind Samples**—split samples submitted with unique sample identifiers so they appear to be from different sources. Blind samples are an independent check of laboratory methods.
- **Trip Blanks**—distilled water placed in sample containers and sealed in the lab. Trip blanks are carried by the field sampling team and returned to the lab with the test samples. If any aspect of the sample handling and transport process might have compromised the quality of the samples within their containers, the effect should be detectable in the trip blank. It is very rare for trip blanks to contain contaminants.
- **Rinse Blanks**—distilled water that has been poured over reusable sampling equipment that has been decontaminated. Rinse blanks are sometimes called field blanks, rinsate blanks, equipment blanks, or decontamination blanks because they are a test of the cleanliness of the sampling equipment.
- **Atmosphere Blanks**—distilled water placed in sample containers that are opened during sample collection. Atmosphere blanks are a test of whether any air pollutants or windblown particulates might have contaminated test samples during the process of collecting the samples.
- **Water Blanks**—potable water used for site activities, particularly rock drilling.[4]

---

3 The acronym QA/QC refers to quality assurance and quality control. Quality assurance is a system of activities undertaken to ensure that the data will be of a level of quality appropriate for the intended use. QA components might include staffing, training, standard operating procedures, audits, and documentation. Quality control refers to the tests and other activities undertaken to ensure that the quality specifications are fulfilled. In brief, QA focuses on the data generation process; QC focuses on the resulting data.

4 Perhaps the first telling use of water blanks was during the 1979 groundwater investigation of Love Canal in Niagara Falls, New York. Groundwater samples from bedrock wells contained

Water blanks are also collected from wells when domestic plumbing is being assessed for lead contamination.

- **Preservative Blanks**—preserved samples of distilled water. Preservative blanks are used to assess possible contamination of the acids, bases, and other reagents used to preserve test samples.

Analytical results from these samples are usually checked during data validation and exploratory data analysis. Except for replicates and blind samples, they are not included in datasets to be used for statistical analysis. Other QA/QC samples used to check laboratory procedures include:

- **Laboratory Blanks**—distilled water or purified solid prepared in the lab to assess contamination from reagents, glassware, and analytical hardware. These samples are sometimes called method blanks because they are a test of the laboratory method.
- **Performance Evaluation Samples**—samples containing known concentrations of specific compounds. PE samples are usually used to certify laboratories rather than check quality on a specific data collection task.
- **Calibration Samples**—samples containing known concentrations of an analyte similar in chemical behavior to the analytes of interest. These samples are used to calibrate instruments and assess method bias.
- **Matrix Spikes**—samples of the media being analyzed that have been spiked with known concentrations of a representative analyte. These samples are used to check for interferences between the analytes being tested and the sample matrix. Matrix spike duplicates are routinely analyzed by laboratories to assess measurement variability.

Analytical results from these samples are usually checked during data scrubbing. They are not included in datasets to be used for statistical analysis. Most of these QA/QC samples are used only for chemistry laboratories. Quality samples for other types of laboratory analyses (e.g., geotechnical, radiological, biological) are usually limited to replicates.

These are commonly used quality samples, but there are many others possible. You create the samples to fit the experimental situation. Don't feel limited to laboratory samples. You can use the same approach to create tests or other methods of variance assessment. If you understand the possible sources of variation in your data, you can create relevant quality tests for survey questions, industrial processes, or whatever your study will involve.

---

chloroform and other chlorinated organic compounds, but shallow wells in unconsolidated materials did not. Tests of the potable water used for rock coring (but not the overburden drilling) revealed it to be the source of the chlorinated compounds (probably the result of chlorine disinfection at the municipal treatment plant). This surprising finding had a major impact on how future studies of hazardous waste sites were conducted and led to further research on health impacts of drinking water chlorination.

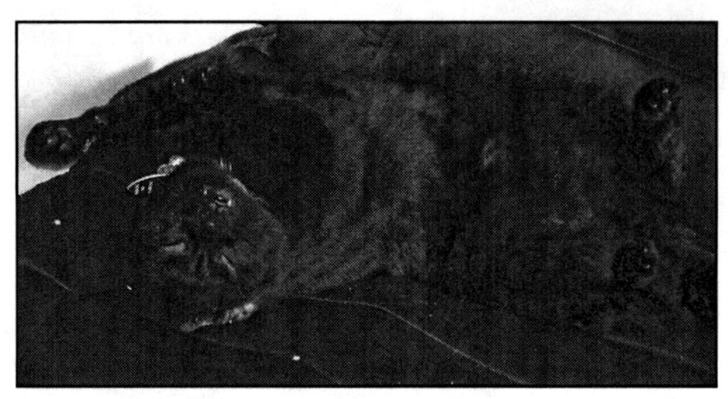

"To err is human, to purr feline." —Robert Byrne, U.S. writer

Sampling Control

Samples have inherent properties that may introduce extraneous variability into data being generated. For instance, differences in sex, age, and social class, may introduce extraneous variability in sociological surveys. Environmentally related examples might include: soil type, geologic strata, species, location and depth, and season (or other time unit). Applying sampling controls involves grouping the data by the control factor and calculating statistical analyses for each homogeneous group. Stratified sampling designs are one form of sampling control.

Sampling controls can be used to prevent, identify, or correct all kinds of extraneous variation. As a consequence, they are used to some extent in most statistical studies.

Experimental Control

Statistical studies can be categorized into two types—observational and experimental. In an observational study, the phenomenon under investigation is a characteristic of the objects being sampled. The concentrations of arsenic in the soils of a waste management facility would be an example. Variability and bias in observational studies can be assessed through the use of control samples. Control samples (like QA/QC samples) are groups of samples that don't have the condition being tested but are otherwise identical to the experimental group. Adding an offsite (or background) area to the study of the waste management facility would be an example of a control group.

In an experimental study, the phenomenon under investigation is assigned to the objects being sampled. Testing a cleanup technique on several plots of contaminated soil would be an example of an experimental study. In such a study, the cleanup techniques would be randomly assigned to the plots in a manner that would help control some of the variation in the plots.

So, the difference between experimental and observational studies is that in an:

- Experimental study, samples are randomly assigned to controlled conditions.
- Observational study, samples are randomly selected from preexisting conditions.

Another way to control variability and bias is through the use of placebos. A placebo is usually thought of as a faux drug because of its association with statistical tests of pharmaceuticals, but it can be any item or action that gives the appearance of being a valid treatment. For example, give two patients blue pills that are the same size and shape, one of which has some active ingredient and the other does not. To test a new pesticide, spray two agricultural test plots, one of which contains the pesticide and the other contains just water. The key is that the subject and the data generator can't tell the difference.[5]

Placebos are a type of blinding. In any experiment, there are subjects or samples, experimenters, data collectors or generators, and data analysts. Sometimes, one individual will fill several roles, such as the experimenter who designs the experiment and then generates and analyzes the data. But whenever there are humans involved, there exists the possibility of intentional or unintentional bias. Blinding is simply the act of denying one or more of the study participants with information that might induce them to behave differently.

I'm not telling who got the placebo. (Cat stays in bag.)

To test the effects of a pharmaceutical, for example, a single-blind study might involve not telling subjects whether they are receiving the active drug or a placebo. A double-blind study might involve not telling either the subjects or the data generators (the nurses who collect physical measurements on the subjects or the lab technicians who analyze blood samples) who received the drug and who received the placebo. A triple-blind study might involve not telling the subjects, the data generators, or the data analysts who received the drug and who received the placebo.[6]

Statistical Control

Special statistics and statistical procedures can be used to partition variability shared by measurements so that extraneous variability can be assessed. For example, *partial correlations* quantify the relationship between two variables while holding the effects of other variables constant. A covariate is a continuous variable that is incorporated into an analysis of variance design to eliminate extraneous variability so that tests of the grouping factors will be more sensitive. Statistical controls are typically used only when variation cannot be controlled adequately through other means.

---

5 In Season 5, episode 14 of the television series *House*, the nurse can tell which patients are being given the placebo during Dr. Foreman's drug trial because the real pharmaceutical has a strong odor. This would not be a good example of an effective placebo.

6 It's ironic that the less informed the participants are, the better the statistics. Too bad government doesn't work that way.

Putting It All Together

Dr. Foreman: Her oxygen saturation is normal.
Dr. House: It's off by one percentage point.
Dr. Foreman: It's within range. It's normal.
Dr. House: If her DNA was off by one percentage point, she'd be a dolphin.
From *House* episode "Autopsy" (2.02)

Say you had to determine if rodents were being impacted by soil contamination at a hazardous waste disposal site.[7] You plan to capture rodents and measure their size and weight and then take a soil sample for lab analysis from the capture location. How would you control extraneous variation?

You might start by developing a QA/QC program that specifies the qualifications needed by members of the field team and the procedures they would use to collect and measure the rodents and soil samples. For QA/QC samples, you might plan to use trip blanks and rinse blanks for the soil, but replicates wouldn't be very helpful for QA/QC because soil is neither homogeneous nor mixable. Replicates would, however, be helpful for identifying variability attributable to differences in individual soils or animals. There are several sampling controls you could apply, for example, to control for species or soil types. However, they wouldn't be helpful for controlling the soil geochemistry or animal age. If you couldn't use replicates, you might be able to control such sample-specific factors using covariates, a statistical control. Because this would be an observational study; experimental controls wouldn't be useful.

If you believe that all other options for controlling variability may not be enough—randomize! In the rodent-impact study, randomization could be used to introduce chance into a data generation effort in several ways. Most important would be the random selection of sample locations so that each point in the study area has an equal chance of being selected. It might also be possible to randomize the assignment of field staff, equipment, and even analytical labs to compensate for any variation that might be inadvertently introduced by these data generation components.

Now consider the following real-world example of a pharmaceutical study:[8]

> In this randomized, double-blind study, 211 patients with [medical condition] who were not achieving [measurement of condition] using [conventional drug] received either placebo or [new drug] in addition to their [conventional drug] for 16 weeks. The [conventional drug] doses were adjusted at regular intervals based on an established algorithm to target [measurement of condition]. [New drug] dose began at 60 micrograms and increased to 120 micrograms as instructed. Patients using [other

7 This can be important because toxins can be bioaccumulated in the food chain.

8 This example has been simplified. The original description of the study can be found at www.medicalnewstoday.com/medicalnews.php?newsid=51399.

prescription drugs] continued with their usual regimen throughout the study. Baseline [measurement of condition] for the study population was 8.5%, and average body weight was approximately 225 pounds.

Table 10 summarizes what the experimenters might have done to try to control extraneous variance and the possible effects. If you were a medical professional, perhaps at a conference or reading the study in a journal, you might question the investigator about steps taken to select subjects and effectively blind the study. These actions are sometimes quite difficult to implement effectively and often have a great bearing on bias and variability.

Table 10. Possible Effects on Study Bias and Variance

Experimenter Action	Effect on Bias and Variance
Advertised for study participants through newspapers and medical colleagues.	This step is problematical; there may be no other way to find subjects for a study. There is always the possibility that an ad may be worded in such a way that biases certain potential participants to preferentially avoid or apply for a study. For example, if the ad promises free health care, people who do not have health care provided through their employment may be more likely to respond than those who do. This response may bias the study to individuals who tend to receive less medical support than average. That bias may or may not have an impact on the results of the study.
Reviewed patient histories and selected 240 participants.	This step is also problematical but necessary. Bias could be introduced if the investigator selects subjects on the basis of anticipated performance. Having an independent third party conduct this step (i.e., blinding the subject selection) may reduce the possibility of bias.
Randomly assigned participants to receive either the new drug or a placebo.	This action would reduce the effects of bias introduced in the previous step so long as the assignments are truly random.
Wrote algorithm for adjusting drug dosages as well as any other standardized procedures to conduct the study.	Having standard procedures can reduce the possibility if introducing measurement bias and variance. It is important that procedures be made available for peer review. After all, a bad procedure can be worse than no procedure at all.
Set up double-blind conditions	This means that neither the experimenter nor the patient knew who received the drug or the placebo. If conducted properly, this action can limit the introduction of measurement bias and variability. Blinding a study is often more difficult than just not providing information to a study participant (e.g., placebo pills must be disguised to look like the test drug) and may not be possible for drugs having substantial effects (e.g., chemotherapy drugs). Also, care must be taken to ensure support personnel do not compromise the blinding.
Established baselines for measurement of condition, weight, and other important factors.	This information can often be used to control for extraneous effects that cannot be controlled efficiently by statistical sampling. For example, weight change might be used as a covariate to assess the drug's performance relative to this side effect.

So there are many ways to start your statistical analysis off on the right foot by minimizing extraneous variability. And that's not all you'll want to do, either. Keep reading.

16 Functional File Formats

Chapter 3 talked about how statistical calculations are all about matrices. This is something that is rarely brought up in Statistics 101 because the next sentence usually contains the words—*matrix algebra.* If you liked algebra in high school, matrix algebra is no worse than eating raw okra. Feel free to explore it on the Internet. The aim of this chapter, though, is to lay down a little groundwork for the matrix you'll be building for your analysis.

The Matrix Recoded

The first matrix I designed was quite naturally perfect. It was a work of art. Flawless. Sublime. A triumph only equaled by its monumental failure.

The Architect, in *The Matrix Reloaded*

Herding cats. Managing data. It's all the same.

If you're going to do a statistical analysis, your dataset ultimately will have to take the form of a matrix in which the rows represent individual samples (e.g., observations, subjects, or samples collected at a specific time from a specific location), the columns represent variables (e.g., information, attributes, properties, responses, or measurements on a sample), and there are no empty cells where there should be data values.

To do a statistical analysis, you have to put your data in an electronic format, preferably one of these formats:

- **Spreadsheet**—The easiest way to set up a matrix for a statistical analysis is to use spreadsheet software. If you have access to this type of software, you'll find that the row and column format facilitates data editing. Built-in functions also enable some editing processes. High-end statistical software read most major spreadsheet formats. Statistical software that does not read your spreadsheet format usually will read delimited text files, which most spreadsheet software can export.

- **Delimited Text**—A delimited-text file is like a spreadsheet in that it consists of rows representing individual samples. Instead of columns, however, the data entries are separated (i.e., delimited) by special keyboard characters (i.e., delimiters). The most commonly used delimiters are commas, spaces, tabs, semicolons, and dashes. This type of file can be created in any word processor and can be read by virtually all statistical software. Delimited-text files are harder to edit than spreadsheets because data usually aren't perfectly aligned and there are no functions to apply as error checks. Furthermore, it's easier to create major data errors by inadvertently entering too few or too many delimiters. This is a frequent problem when there are missing data in the file.
- **Formatted Text**—Formatted text files have variables in fixed-width columns. These are the classical data files you might have used in a programming class in which you had to specify the data formats. For example, a format of I8 specified an integer eight characters long. A format of F10.2 specified a floating point value having two decimal places and seven values to the left of the decimal (the decimal place gets one space). The problem with delimited and formatted text (besides bad delimiters and formats) is that these files can contain more variables than spreadsheet software allows. Formatted files can even spread a single sample over more than one line, which is not allowable in matrices.
- **Relational Database**—Relational databases eliminate the duplication that is inherent in matrix formats so that they are better for storing, querying, and reporting from large files. Relational databases consist of one or more data tables linked by relational keys. So if you expect to have a large data file, a relational database is probably your best choice. What's large? It depends more on the number of variables than the number of samples. Over a hundred variables or a few thousand samples could be handled easily by either format but at that point you would probably start thinking about using a relational database over a spreadsheet.

You'll be able to get to a usable matrix without too much effort from any of these file formats. If a client sends you a text file, like Adobe Acrobat .pdf files, Microsoft Word .doc files, and Internet .html files, you may be able to extract the data using *cut-and-paste special* into Excel or another spreadsheet program. If the data is in a picture file, like a jpg or bmp, you're out of luck. You'll have to scan the image and use OCR (optical character recognition) software to convert the image into text or just reenter the data. Likewise with paper and even audio files,[1] you'll have to reenter all the data.

At this point, you should have a good idea what your samples will be and what information you are going to collect about your samples. You even should know where you're going to put the information, in a spreadsheet, database, or text-delimited file. Don't worry about the order of the samples in the rows. During the course of an analysis

1 I've gotten them all from clients in the past. Fortunately, this doesn't happen much anymore.

you'll reorder them many times. If you think you'll want to return to your original sort order, create a column with the original row numbers. You could also use this variable to explore serial correlation.

So now the question is how you should order your variables.

Paw and Order

In a statistical data file, the data are represented by two separate yet equally important dimensions: the samples, which represent the sources of the information to be analyzed, and the variables, which represent the measures that characterize the samples. These are their stories.

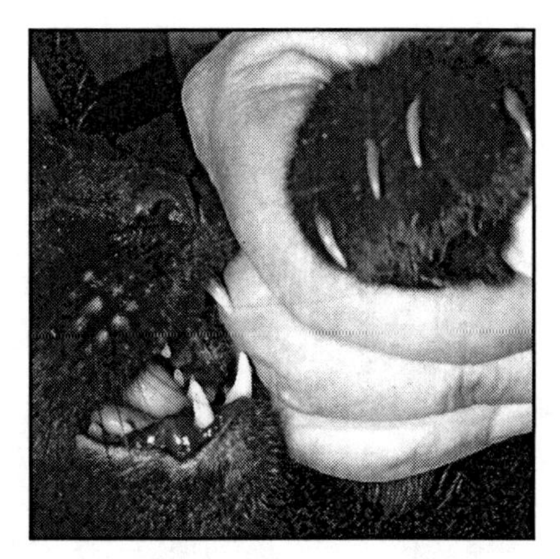

Bad claw. Bad claw. Watcha gonna do? Watcha gonna do when they come for you?

In the distant past, statistical software often required the variables in an analysis to be in a specific order in the columns of your matrix. That is, by and large, no longer true. However, if you have a large number of variables, grouping your variables will make them easier to find and facilitate your analysis. Here are some ideas for putting your dataset together.

Column Names—Variable Identification

The first row in the data matrix should contain unique identifiers for the variables. Preferably these should be no longer than eight characters and contain no spaces or other "illegal" DOS symbols. This is because some data analysis programs still have these limitations. If you know the software you plan to use does not have these limitations, then make the names as descriptive as possible. Parameter names such as "var_1" usually lead to confusion. If your software limits the length of variable names, you may be able to add descriptive labels that are used on output. Using descriptive names or labels greatly facilitates interpreting output.

Row Names—Sample Identification

Sample IDs are more complex than variable names because:

- Like variables, each sample must have a unique ID
- There are almost always many more samples than variables
- Sample IDs sometimes need to contain multiple pieces of metadata
- Sample IDs sometimes need to be short for some statistical procedures

Here are some things to think about. You'll need to have unique identifiers for each sample. This can be the name of the subject, the sample ID used by field staff, or other unique designation. It's best to have a single ID, but this isn't always feasible. Preferably the length of the identifier should be a minimum. This is because long sample-IDs overwrite each other in graphical analyses. You may be able to circumvent this problem by using more than one variable to uniquely identify each sample, or you can add a variable with sample aliases for use in graphing.

Sample IDs should contain some flag to indicate the sample is a replicate or other quality control sample. (These samples are usually excluded from statistical analyses.) If several samples are taken at a specific location (e.g., replicates or co-located samples), the location identifier will no longer be unique. In this case, additional information must be added to the sample ID to make it unique. The easiest way to do this is to use the spreadsheet's concatenate function to combine enough information into a single sample ID. It is a good idea to also include the individual components of the ID in the dataset so that they can be analyzed separately.

If the study is supposed to be blinded to the data users, be sure the IDs do not reveal the identities of the test groups. IDs should be all numeric or all alphanumeric. Mixing the two will create confusion when sorting. This information is best placed either at the beginning (far left) or end (far right) of the data matrix. If your samples are human subjects or other entities requiring privacy, be sure your IDs don't compromise their confidentiality. Don't use social security numbers, bank account numbers, or other private information unless you have to. If you do, the numbers should be encrypted or otherwise encoded so that they are not easy to decipher without the key.

Sorting and Filtering Variables

You'll probably be sorting your dataset many times in many different ways during an analysis. Try to keep your primary sorting variables together, either in the first or last few columns of the dataset. This isn't a software requirement but it's easier to remember where they are if you make a habit of it. It's also sometimes useful to create a variable with the original order of samples if that information is not contained in other variables. That variable with the original order would be useful in testing for serial correlations.

Sample Location Variables

If the locations where samples are collected or measured are different, data about the locations should be included in the main dataset. Often, these data are included in the statistical analysis. Location information may be as simple as a reference to an address or part of a building. Quantitative location information would include:

- **Coordinates**—Locations in the form of degrees-minutes-seconds (such as latitudes and longitudes) generally cannot be analyzed without being converted into northings and eastings in a consistent length unit such as feet or meters. The coordinate system used should be consistent with that used for other maps so that results can be compared. The precision of the coordinates should be reported in the metadata. For analyses involving spatial contouring, put the northings and eastings in the first two columns of your dataset. Modern contouring software will read other data arrangements but this is most efficient if you have to specify the coordinates in many simulations.

- **Elevation (depth/height)**—Elevations (or depths) of specific sampling points should be in the same units as the location coordinates. If the vertical location of a sample is not a point (i.e., the sample is collected over an interval), there are several ways the data can be represented. Usually two variables are used—one for the top elevation or depth and one for the bottom. Alternatively, the interval can be represented by the top elevations and the interval lengths. If all the intervals are the same length, only the midpoint (or starting point or ending point) of the interval need to be provided.
- **Time**—Data generation or sample collection dates should be provided in an absolute date format recognized by the software to be used. Be sure the format is the same for all entries. This information can then be translated into numbers or relative times (e.g., day of week; hour of day; season; month) that can be analyzed.

If the locations are the same for all the samples or are not important to the analysis, they can be handled as metadata apart from the main dataset.

Dependent Variables

Variables that are the object of your analysis, your dependent variable(s), can be listed next. Separate the categorical-scale variables from the continuous-scale variables. If the variables are related in some way (e.g., all chemical analyses, all treatment effects) try to group them together in the dataset. Arrange variables for effects to be tested apart from covariates (variance control variables).

Independent Variables

As with the dependent variables, separate the categorical variables from the continuous-scale independent variables. If the variables are related in some way (e.g., survey questions), group them together in a logical order in the dataset. Again, there is no requirement that the dependent and independent variables be arranged this way. It simply can make model specification easier in some software.

Metadata

Metadata do not have to be on the same sheet as the data unless you plan to use the variables in the analysis. Just make sure the information is somewhere. Examples of metadata that should be documented include measurement data units, coordinate precision, common laboratory methods and detection limits, and the person(s) who generated the data (e.g., field sampler, interviewer, lab). Comments should be provided on any reasons why the sample may not be representative of the population being tested (e.g.,

QA/QC deviation) or whether there is any question on the identity of the sample (e.g., sample has an alias or there is a duplicate location ID). Update the comments or add a separate field to document changes you make to samples during data scrubbing. Do not include comments as footnotes at the end of a data file because they will be mistaken for data by some programs if they are not edited out.

The type of analysis you plan to do and the software being used will dictate the best, and sometimes the only, way you can structure your data. Some statistical procedures require special data designs. For example, datasets for repeated-measures analysis-of-variance must be formatted in specific ways (depending on the software). Some forms of graphical analysis, such as icon plots, may require a different data format, depending on the sophistication of your software. Still other types of analysis require derivative datasets consisting of means, statistical distances, or correlations. These latter datasets are usually created in two steps. The original dataset is used to calculate the statistics, which are then exported as a second dataset.

Cell Contents—The Data

Numbers must be entered as numeric data rather than alphanumeric data. If numbers are entered as alphanumeric data, they will look like numbers but not enter into statistical calculations. It is difficult to tell the difference; sometimes (but not always) alphanumeric data will be flagged or left justified while numeric data will be right justified.

Missing data (blank cells) are usually not a problem in statistical software. Delimited-text files and spreadsheets are another story. Delimited-text files can be corrupted by missing values if too few or too many delimiters are added. Spreadsheet functions can interpret missing values as zeros. If you're not careful, you might write a formula that uses cells you don't expect. Statistical software, on the other hand, generally allows you to assign a value that is not used in the rest of the dataset. For example, missing numbers could be replaced by –999; missing alphanumeric data could be replaced by "NA." An added benefit of this feature is that you can analyze patterns of missing data.

The way you code censored data may also have an impact on how you do your analyses. For example, datasets with analytical chemistry data are coded using at least three columns for each chemical: a concentration, a quality flag, and a detection limit, and sometimes, a unit of measurement and a reference to a procedure number. Some spreadsheet software (e.g., Microsoft Excel) allow you to create a custom format with a "<" or ">" symbol to be incorporated into the concentration format. This approach eliminates the need for the data quality column for each chemical. If you've analyzed a hundred chemicals, that's a lot of columns. However, the specially formatted data are interpreted by the software as if the qualifier were not there, which may not be what you need to do. Using this approach to identify censored data is increasingly inefficient as the size of the dataset increases because each censored value has to be formatted separately.

Some data should not be included in the dataset to be analyzed. Most notably, QA/QC samples should be included in a separate file or spreadsheet.

A Scythe of Relief

The scythe swings, and at once the grass starts to grow back. Cut again and the grass grows faster than ever.

Anonymous

Go ahead. Take a deep breath and let it out along with all the stresses of constructing a dataset. Just don't hold that breath too long. As it turns out, you'll probably have to rebuild the dataset more than once, especially if your analysis is complex. You'll add and discard variables more times than you can count (especially if you don't document it). Plus, different analyses within a single project can require different data structures. Graphics especially are prone to requiring some special format to get the picture the way you want it. That's just a part of the game of data analysis. It's like mowing the lawn. You're only finished *temporarily*.

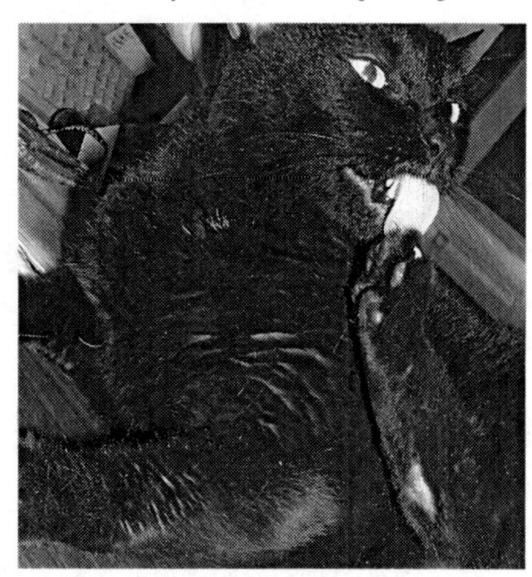

The paws that refreshes.

When you do update a dataset, it usually involves creating new variables but not samples. Data for the existing samples might change, for example, you might average replicates, fill-in values for missing and censored data, or accommodate outliers, but you normally won't add data from new samples. This is because the original samples are your best guess at a representation of the population. Change the samples, and you change what you think the population is like. That might not sound too bad, but from a practical view, there are consequences. If you change samples you have to redo a lot of data scrubbing. The new samples may affect outliers, linear relationships, and so on. It's not just the new data you have to examine, it's also how all the data, new and old, fit together. You have to revisit all those data scrubbing steps.

By now you've probably figured out that statistical analysis projects are pretty much once-and-done endeavors. Statisticians are like builders in the sense that once they get the plans and raw materials, they don't want changes. They have a plan and a goal, and it should be a one-way street to the end. Other professionals, like doctors, want updates. Their next move may be dictated by the most recent data they receive. Their road allows U-turns and all kinds of side streets.

Database managers, unlike data analysts, want routine updates. It's their job to maintain a storehouse of up-to-date information that can be provided to users in reports. But a data report is not the same as a data analysis. It's like with your checking account. Sometimes you just want a quick report of your balance. That information from the bank's database has to be up-to-date and readily available when you need it. Plus, both you and the bank have to be working with exactly the same data. If you want to figure out how much more you're spending on commuting over the past five years, though, you have to conduct an analysis. You'll have to compile the data, scrub out anomalies, like the driving you did on vacation, and look for patterns. That takes time. Sometimes, a lot of time.

You'll find that this point is lost on many clients. Too many times I've asked about possible errors in a database only to get a corrected and updated, totally new data file. That means back to square one for Sisyphus the statistician.

That being said, there are types of data analyses that are conducted continuously, most notably, data mining. Data mining involves a variety of techniques that look for patterns and associations in data, usually for the aim of prediction. Some data mining techniques are statistical in nature, involving concepts of central tendency and dispersion. Some techniques are mathematical, involving multidimensional distances, rates, and areas. Some techniques involve file operations, like filtering, sorting, and merging. Unlike classical statistical analyses, data mining usually involves massive amounts of data, a census of the source if possible. Moreover, the data are often updated periodically, limiting data preparation to routines that can be automated.

Data mining requires specialized software and expertise, so it's not something you'll run into in Statistics 101. Figure 12 shows the data mining techniques provided in Statistica. There are many techniques for mining data. Unlike statistical techniques that have commonly used names, similar data mining techniques may go by different names in different software packages, making software difficult to compare.

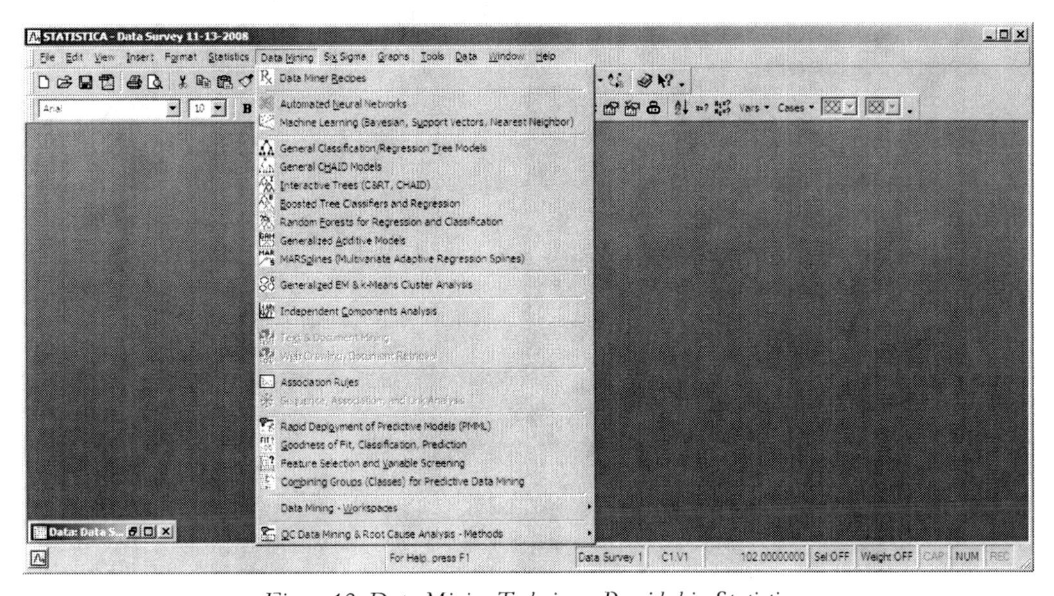

Figure 12. Data Mining Techniques Provided in Statistica.

The use of data mining is growing as fast as the mega-datasets that rely on those analysis techniques. Whether you know it or not, data mining has already influenced your life. It is used by marketing organizations to predict what you might purchase. That's where those recommendations on Amazon.com come from. Financial institutions use data mining to identify good candidates for credit cards. And if you belong to a terrorist organization, you can bet that the CIA has your numbers.

Data mining is a whole different philosophy from statistical modeling even if some of the techniques are the same. But that's part of what makes data analysis so much fun. You can approach data from many different directions and still reach the same knowledge.

And there you have it. That's how you set up a dataset for a full statistical analysis. But that's just the beginning. Next you have to scrub the dataset so you know you have the right numbers expressed in the right way. But first … a story.

The Good, the Sad, and the Faulty

They rebuilt me. Everything works. But they had never seen a human. They had no guide for putting me back together.

Vina, in *Star Trek* episode "The Menagerie"

I thought I had already seen it all when I received a small "analysis-ready" dataset a few years ago. Looking at the hard copy of the data, everything looked fine. Sample IDs were consistent, dates were all in the same format, there were no missing values, there were no Os hiding in place of 0s. But something didn't look right. The font looked funny. Then it struck me. The data columns were ever so slightly misaligned. I couldn't imagine how that might have happened until I opened the spreadsheet. Then it was clear. The spreadsheet consisted of eight columns for the ID, the date, and the six analytes … but only one row. My client had, with great care and a lot of spaces and returns, placed the data for his samples into a single cell for each variable. I can't even imagine how much work that must have been. I felt sorry for him because he had obviously never built a dataset, or for that matter, used a spreadsheet before. As a solitary consultant, he was on his own. He had no guidance for how to create a dataset for a statistical analysis. I figured it would be futile to try to get him to do it right even though that's what the contract called for, so I just reentered the data and did the analysis. I gave him my dataset with the report with the hope that he might look at it and see how a dataset should be constructed. But needless to say, I didn't make any money on that job. I did profit from the experience, though, having included his example in *Stats with Cats.*

So sad but still, so wrong.

PART 4

Statistical Foreplay

So far, you've gone from concepts and measurement scales to variables, populations and individuals to samples, and matrices and file types to a dataset. But that dataset is like a piece of unfinished furniture. Now it's time to take that dataset, rub down the grainy spots, and finish it with a shine that will expose all your data's secrets. Here's what that entails.

Chapter 17—Getting the Numbers Right by making sure all the values in the dataset were generated appropriately and are identical to the values that were originally generated.

Chapter 18—Getting the Right Numbers by deciding what to do about replicates, missing data, censored data, and outliers

Chapter 19—Kicking the Data Tires by looking at descriptive statistics and graphs to decide if the dataset is ready to analyze according to your plan.

Chapter 20—Teaching Old Data New Tricks by adding any of six kinds or variables to enhance the dataset.

There certainly are a lot of ways that a dataset can get screwed up, but there are also a lot of ways to make them better. This part will show you that finding errors is hard, but fixing them is easy, and that finding replicates, missing data, outliers, and censored data is easy, but fixing them is usually hard. Still, you gotta do it. Data scrubbing can take a few hours or a few weeks depending on the size and complexity of the dataset and the objective of the analysis. As you get more experienced at doing it and it becomes routine, it'll go faster even if it feels longer. After reading this part you'll appreciate why preparing a dataset for analysis takes so much effort.

17 Getting the Numbers Right

Open any textbook on statistical analysis and one topic you're not likely to find is how to prepare a dataset for analysis. Is it because it's not important? On the contrary, it's the most fundamental part of *every* statistical analysis. Get this step wrong, and everything that comes after it is garbage. Is it because the time it takes to "scrub" a dataset is small compared to the statistical analysis? No, it's just the opposite. Data preparation is always at least half of the work of a statistical analysis and sometimes over 90 percent. So why don't textbooks devote some space to this essential topic? Because dataset preparation is as much fun as bathing a cat, only without the excitement of canines and claws. Data scrubbing requires a superhuman attention to detail, an ability to maintain focus while running endless repetitions of byzantine validation routines, and a total immunity to mind-numbing tedium. Yes, all of that. If this topic were taught in an introductory course, no one would want to take the second course. There are too many really exciting topics in introductory statistics—like coin flipping and the Central Limit Theorem—to spend time on data preparation.

Measure Once; Check Twice

> *Murphy's Laws of Analysis. (1) In any collection of data, the figures that are obviously correct contain errors. (2) It is customary for a decimal to be misplaced. (3) Any error that can creep into a calculation will. Also, it will always be in the direction that will cause the most damage to the calculation.*
>
> G. C. Deakly, in Paul Dickson, *The Official Rules*, 1978

Garbage in, garbage out is a saying that dates back to the early days of computers but is still true today. If the numbers you use in a statistical analysis are incorrect, so too will be the results. That's why so much effort has to go into getting the numbers right.

The process of checking data values can be divided into two parts—verification and validation. Verification addresses the issue of whether each value in the dataset is identical to the value that was originally generated. Twenty years ago, verification amounted to a check of the data entry process. Did the keypunch operator enter all the data as it appeared on the hard copy from the person who generated the data? Today, it includes

I'm the oversight contractor. Give me all your QA/QC records . . . and a can of tuna.

making sure automated devices and instruments generate data as they were designed to. Sometimes, for example, electrical interference can cause instrumentation to record or output erroneous values. Verification is a simple concept embodied in simple yet time-consuming processes.

Validation addresses the issue of whether the data were generated in accordance with quality assurance specifications. In other words, validation is a process to determine if the data are of a known level of quality that is appropriate for the analyses to be conducted and the decisions that will be made as a result. An example of a validation process is the assignment of data quality flags to each analyte concentration reported by an analytical chemistry laboratory. Validation is a complex concept embodied in a complex time-consuming process.

With some types of data, these two processes may overlap. For example, modern laboratory information management systems (LIMS) have been designed to facilitate both of these processes so that at least some verification and validation checks are conducted before any data is exported. In addition, some types of data may require more of one process than the other. Verification of geologic sample descriptions can often be accomplished with a spell checker whereas validation, when it is done, can amount to redescribing some percentage of the samples. Technological improvements over time are also a factor. Data from electronic recorders, for example, require less verification today than a generation ago when all data were recorded by hand.

Ideally, verification is followed by validation, which is followed by data scrubbing, exploratory data analysis, and finally, the analysis. However, this can create a conflict between a project's budget and its schedule. Take, for example, Superfund[1] site investigations during the 1980s. Lab analysis of samples took four weeks to complete followed by six weeks of data validation and then a few weeks of data scrubbing.[2] The three months during which there were no analytical data to match to the environmental data presented

---

1 *Superfund* is the colloquial term applied to the Comprehensive Environmental Response, Compensation, and Liability Act of 1980, also known as CERCLA. The act taxed generators of hazardous waste and created a fund (a really super-big trust fund) that was used to investigate and clean up abandoned hazardous waste sites. Congress allowed the tax to expire in 1995.

2 These were the target durations for these tasks. Sometimes a lab might take up to three months if the analyses weren't routine. Having a data validation effort six months late was not uncommon. Building the dataset was a crap shoot. Keep the database structure simple and maybe the data could all get entered in a couple of weeks. The numbers might even be mostly correct. But if the IT Department got involved to build a database management system, you might not see the data for a year. A lot has changed since then. Data validation is at least semi-automated. Laboratory Information Management Systems (LIMS) now provide Electronic Data Deliverables (EDDs) that load automatically into Geographic Information System (GIS) or other data management software. But there is still a time

a management conundrum. Early in the program, managers hoping to conserve their budgets reassigned project staff until the data would become available. Unfortunately, by that time, staff members had other commitments and could not return to the project quickly. Staff unavailability led to schedule slippage, which was often addressed by assigning additional staff to the project at extra cost. The result was a late, over-budget project. Having observed this cycle, some managers decided to preserve the schedule by maintaining key staff to analyze unvalidated data directly from the lab. To do this, each discipline, hydrogeologists, engineers, and risk assessors, had to develop their own databases. This, of course, introduced errors and inconsistencies. Having the validated database finally become available didn't help. By that time, the report was complete, and sometimes, approved by the client. Audit any Superfund investigation from the 1980s and early 1990s, and you'll find all kinds of data errors and inconsistencies.

I'm DONE with the analysis. NOW you want to change some of the data points?

Changes in data points create a dilemma for statisticians. Change one data point and you've changed the entire analysis. It probably won't change your conclusions, but all those means and variances, and other statistics not to mention the graphs and maps will be incorrect, or at least inconsistent with the final database. This is the most important warning I issue to my clients when I start a job. Still, on the majority of projects, some data point changes after I am well into the data analysis. It's inevitable.

The Data Scrub

RULES OF THE DATA SCRUB
1st Rule: You should talk about data scrubbing.
2nd Rule: You DO document what you did in data scrubbing.
3rd Rule: If a client says stop or doesn't pay bills, data scrubbing is over.
4th Rule: Only two eyes to a file.
5th Rule: One dataset at a time.
6th Rule: No effort, no clues.
7th Rule: Data scrubbing goes on as long as it has to.
8th Rule: If this is your first try at data scrubbing, you have to look for everything.

Here are some guidelines for scrubbing your data. Only one person should scrub a file at a time. There are a lot of decisions made in scrubbing a dataset that would be needlessly complicated if more than one set of eyes were involved. Similarly, it's more efficient if you only work with one dataset at a time. Data scrubbing begins the process of getting to know your dataset. It takes time and attentiveness, but the effort is always

lag. So, do you gamble and analyze the data or wait? Your decision could have a huge impact on your project's budget and schedule.

worth it, not just because you have to do it to have appropriate error-free data, but also because you will usually get clues about the nature of the data and what you might do with it.

Data scrubbing goes on as long as it has to, the only exceptions being when the client says stop or doesn't pay you for your work. If you've never done data scrubbing before, look for every probable flaw just so you know how. After you've scrubbed a score of datasets, you'll know what needs to be done in a given situation. Always be sure to document what you did in data scrubbing. No matter what you decide, it'll have an impact on your analysis. Also be sure to keep a copy of the original data handy so that you can refer back to it when (not *if*) you need to.

I got them steadily depressin', low-down mind-messin goin' through the cat wash blues.

A Rain of Errors

Data that you may observe
Measured on a ratio scale
If they fit a Normal Curve
Catch the errors in the tail.

Chant to the tune of Eeny, Meeny, Miny, Moe

Data Entry Mistakes

A generation ago, virtually all data were entered manually. Data entry errors were the norm. Consider this, a very small dataset consisting of four variables of two-digit measurements for 30 samples would contain at least 240 keystrokes. A typical typist with an accuracy rate of 90 percent would make 24 errors. Even a superior typist with a 99 percent accuracy rate would make a few errors. Imagine the errors that might have lurked in a dataset consisting of a score of variables with values for hundreds of samples. Needless to say, data entry errors were the bane of any type of computer analysis. Today a large proportion of data is generated automatically by computers, optical scanners, or electronic instrumentation. Nevertheless, there is still a great deal of data that is entered manually at some point in the data generation process.

Data entry mistakes can take a variety of forms. Random mistakes follow no particular pattern. They occur sporadically throughout a dataset. A related type of error is a spelling mistake. Spelling mistakes occur most frequently in words that might be unfamiliar to the typist, such as medical terms, geographic or street names, surnames, sample IDs, and chemical analyte names.

Transposition mistakes involve mixing the order of specified characters. There are a variety of common character transposition errors, including:

- Simple transpositions: typing 56 instead of 65
- Reversed double digits: typing 455 instead of 445
- Added digit: typing 4567 instead of 457
- Dropped digit: typing 467 instead of 4567
- Stutter errors: typing 455 instead of 45

Hand shift mistakes occur when the typist's hand shifts from its proper place on the keyboard or keypad. For example, a downward hand shift on the keypad would produce 12 instead of 45. Sight shift mistakes occur when the typist reads the data from the wrong row or column of the hardcopy. Decimal displacements involve incorrect placement of decimal points, such as typing 44.6 instead of 4.46.

Hand-shift error in 3…2…1.

Finding data entry mistakes is generally a laborious process of manually comparing the original hardcopy with the entered data. Spreadsheet formulas don't help much. A good spell checker might catch incorrect analyte spellings but won't help much for sample IDs. Some mistakes such as decimal displacements might be identified using descriptive statistics, such as minimums and maximums.

An especially frustrating type of entry mistake is alphanumeric substitution, in which a letter is substituted for a number or vice versa. The most notorious example of alphanumeric substitution is the space. You can usually spot them in the middle of data entries, and sometimes, at the beginning of entries if there are only a few so you can detect an offset between entries. But if all the data have spaces at the beginning or any have spaces at the end, they are difficult to find visually. Sometimes, this error is produced electronically rather than by typing when data are exported to other file formats that insert leading or following spaces. Excel users can use the TRIM function to discard extra spaces.

Other examples of alphanumeric substitution include:

- O (upper case o) instead of 0 (zero)
- l (lower case L) instead of 1 (one)
- S (upper case s) instead of 5 (five) or 8 (eight)
- b (lower case B) instead of 6 (six).

These errors can be difficult to identify visually but can usually be detected by spreadsheet formulas. In Excel, for example, the ISNUMBER function can be used to detect nonnumeric entries. Because alphanumeric errors cause statistical software to either ignore the data point or abort a calculation, discrepancies in sample size can indicate the presence of alphanumeric substitution errors. However, many conditions can alter the number of samples used in an analysis, so it is usually a good practice to check all numeric data fields by entering test formulas.

Data Specification Mistakes

Data entry specialists can't be blamed for all the errors in a dataset. Sometimes the errors are in the hardcopy they are given to enter. These are specification errors, and of course, there are a variety of types.

Data mistakes are caused by the person generating or recording result incorrectly. These errors are almost impossible to detect because the error resides on the original hardcopy and there is usually no other source that can be consulted to check for discrepancies. Another type of data mistake involves incorrect calculations or rounding. Data from laboratories and surveyors are most likely to have this type of error. Usually, the data generator will find his or her mistake upon reexamining the calculations but the presence of a possible error must first be identified. If the incorrect data point is similar to correct data, detecting the error is unlikely. Data mistakes can also involve instrumentation. Lack of calibration, voltage irregularities, and recording faults are all types of instrumentation faults that can produce data errors. Sometimes these errors can be detected by quality control procedures or direct observation (e.g., an automated recording device vandalized or interfered with by falling branches).

Another type of specification mistake involves data inconsistencies. Two common types of variations are ID variants and analyte aliases. An ID variant might involve inconsistent spelling or use of separators (i.e., spaces, dashes, underscores). For example, a sample location might be designated SAMPLE 5, SAMPLE005, SAMPLE-5, or SAMP5. Statistical software would interpret these designations as four different locations rather than the same location. Even more subtle are analyte name aliases. For example, trichloroethene can also be trichloroethylene or TCE. You might be able to figure out different sample designations without knowing much about the project, but if you don't know a little chemistry (enough to be able to look up analyte aliases) you'll have problems correcting these mistakes.

Data Format Mistakes

The last type of data error involves how numeric data are formatted. For example, including NAs, NRs, NDs,[3] and other flags in a cell meant for numbers will change a numeric variable into an alphanumeric variable and cause some statistical software to abort a calculation. More insidious, sometimes the software will simply ignore the sample. This may be the cause if the number of samples your statistical software has counted is less than expected. Another format problem occurs if a depth interval is used instead of discrete depths. Because of the dash in the interval, a depth interval is an alphanumeric variable rather than a numeric variable. It is often possible to correct this misspecification without reentering data by using a spreadsheet text-to-columns conversion or functions that extract characters from cells (e.g., RIGHT and LEFT in Excel). Formatting mistakes occur commonly with dates and times because a format supported by the data entry software (e.g., spreadsheet) may not be supported by the statistical software. En-

3 Not Available, Not Recorded, and Not Detected.

tered dates and times may also be ambiguous (e.g., AM or PM, no day, two-digit year, or no year), which can result in unusable data.

Finding Data Problems

It isn't that they can't see the solution, it's that they can't see the problem.
G. K. Chesterton, "The Point of a Pin" in *The Scandal of Father Brown*, 1935

Now that you have an idea of some of the bad things that can happen in a dataset, it's time to start looking for those mistakes and anomalies. Masochists do this by manually cross checking the dataset with all the hardcopies that were used by the data entry specialists. You can take shortcuts, though, such as the following tricks, tips, and tools.

Spreadsheet Tricks

This is the point when using a spreadsheet to assemble your dataset really pays off. Most spreadsheet programs have a variety of capabilities that are well suited to data scrubbing. Here are a few tricks:

- **Marker Rows**—Before you do any scrubbing, create marker rows throughout the dataset. You can do this by coloring the cell fill for the entire row with a unique color. You don't need a lot; just spread them through the dataset. If you make any mistakes in sorting that corrupt the rows, you'll be able to tell. You could do the same thing with columns, but columns usually aren't sorted.

- **Original Order**—Insert a column with the original order of the rows. This will allow you to get back to the original order the data was in if you need to (you'll understand why after reading about lag variables in Chapter 20). You can do this in four steps: (1) enter a 1 or a 0 in the first cell of the column, say cell A1; (2) enter **=A1+1** in the second cell of the column, cell A2; (3) copy cell A2 down the column to the end of the data, that will give you unique numbers for all the rows of data; (4) Click on the column, copy it, then *paste-special* as *values*.[4] This will convert the formulas to numbers so they won't change when you sort.

- **Sorting**—One at a time, sort your dataset by each of your variables. Check the top and the bottom of the column for entries with leading blanks and nonnumeric text. Then check within each column for misspellings, ID variants, analyte aliases, non-numbers, bad classifications, and incorrect dates.

- **Reformatting**—Change fonts to detect character errors, such as O and 0. Change the format on any date, time, currency, or percentages and incorrect entries may pop out. For example, a percentage entered as 50% instead of 0.50 would be a text field that could not be processed by a statistical package. Per-

4 These are the commands for Microsoft Excel. Your spreadsheet may use different commands.

centages may also be entered as whole numbers instead of decimals (e.g., 50 instead of 0.50). This trick works especially well with incorrect dates. Conditional formatting can also be used to find data that fall outside a range of acceptable values. For example, identify percentages greater than 1 by conditionally formatting them with a red font.

Is that a 0 or an O?

- **Formulas**—Write formulas to check proportions, sums, differences, and any other relationships between variables in your dataset. Use cell information functions (e.g., VALUE and ISNUMBER in Excel) to verify that all your values are numbers and not alphanumerics. Use counts and sums to make sure you have the right numbers of samples for each variable. Also check to see if two columns are identical, in which case, one can be deleted. This problem occurs often with datasets that have been merged. A useful formula (in Excel) for this purpose is:

  =IF((comparison of cells in two columns),"-","*****")

  You can create a column in your spreadsheet using the formula to compare two columns. Wherever the ***** appears, there is a difference between the columns. If there are no differences, you can delete one of the columns. This trick is especially useful if you have hundreds of data points to review. You can also use this approach to check for duplicate sample entries by comparing cells in adjacent sorted rows.

Descriptive Statistics

Even before getting involved in the data analysis, you can use descriptive statistics to find errors. Here are a few things to look for.

- **Counts**—Make sure you have the same number of samples for all the variables. Otherwise, you have missing or censored data to contend with. Count the number of data values that are censored for each variable. If all the values are censored for a variable, the variable can be removed from the dataset. Also count the number of samples in all levels of grouping variables to see if you have any misclassifications.

- **Sums**—If some measurements are supposed to sum to a constant, like 100 percent, you can usually find errors pretty easily. Fixing them can be another matter. If it looks like just an addition error, fix the entries by multiplying them by {what the sum should be} divided by {what the incorrect sum is}

For example, if the sum should be 100 percent and the entries add up to 90 percent, multiply all the entries by 1.0/0.9 (1.11) and then they'll all add up to 100 percent. There will be situations though, especially in opinion surveys, when you'll have to try to divine the *intent* of the respondent. If someone entered 1 percent, 30 percent, and 49 percent for responses that are supposed to sum to 100 percent, did he mean 1 percent, 50 percent, and 49 percent? It's like being in Florida during November of 2000. You want to use as much of the data as possible, but you just have to be sure it's the right data.

- **Min/Max**—Look at the minimum and maximum for each variable to make sure there are no anomalous values.
- **Dispersion**—Calculate the variance or standard deviation for each variable. If any are zero, you can delete the variable because it will add nothing to your statistical analysis.
- **Correlations**—Look at a correlation matrix for your variables. Don't worry too much about outliers and nonlinear relationships at this point. Look for correlations between independent variables that are near 1. These variables are statistical duplicates in that they convey the same information even if the numbers are different. You won't need both, so delete one.

There are other calculations that you could do, but it's usually better to defer this work. At this point, your objective in looking at descriptive statistics should be to find errors and eliminate unnecessary variables.

Plotting

Plotting is usually done as part of exploratory data analysis (EDA), which is discussed in Chapter 19. Whatever plotting you do at this point is preliminary. These plots won't make it into your final report, so don't spend a lot of time on them. Here are a few key graphics to look at.

- **Bivariate Plots**—Plot the relationships between independent variables having high correlations to be sure they are not statistically redundant. Don't bother with any subgroups at this point.
- **Time-series Plots**—If you have any data collected over time at the same sampling point (e.g., weather-related measurements like temperature), plot the time-series. Look for incorrect dates and possible outliers in the data series.
- **Maps**—If you collected any spatially dependent samples or measurements, plot the location coordinates on a map. Have field personnel review the map to see if there are any obvious errors. If your surveyor made a mistake, this is where you'll find it.

Auditing

Auditing has a bit more to do with validating data quality than verifying that there are no errors in the dataset. Consequently, these checks involve much more effort than the other steps, but then again, you won't need to audit every dataset. Audits are used mostly for data from analytical chemistry laboratories. Examples of lab audits include:

- **Laboratory Data Validation—**Data validation for analytical laboratories involves checking each analysis on each sample to ensure that the analyses met all the requirements of the lab procedure and contract specifications. Items that are checked in this process include how the sample was collected and preserved, how long the samples were held before being analyzed, and how the lab instruments were calibrated.
- **Inorganic Analyte Balances**—Anion-cation balances can be calculated for samples that are analyzed for a complete suite of inorganic analytes. If the sum of the milliequivalents of the anions (negative ions) in the sample is within 5 percent of the sum of the milliequivalents of the cations (positive ions), the sample is in balance. If the sums are more than 15 percent apart, one of the analyses may be in error or you may not have analyzed for all the important cations and anions.

Audits are used not only to discover errors in data generation but also expose fraud. There have been too many well-publicized instances (and probably way too many more unreported instances) of data being falsified. If you think you may have compromised data, you might be able to test it using Benford's Law.

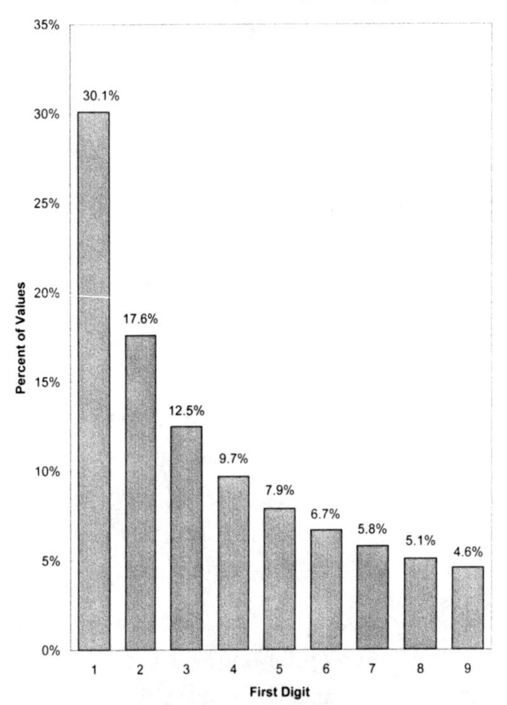

Figure 13. Distribution of First Digits following Benford's Law.

Benford's Law says that the probability of the first digit of a data value being a low number is higher than it being a higher number. In fact, the frequencies of occurrence of the digits 1 through 9 follow a logarithmic function. The odds of obtaining a 2 as the first digit in a number, for example, are much higher than the odds of obtaining a 3 through 9. If a variable follows Benford's Law, the probability that the first digit is a 1 is about 30 percent. Figure 13 shows what the distribution of the first digits in a dataset would look like if the data followed Benford's Law. In contrast, a random distribution of numbers would have the same frequency of occurrence for each digit, 1 through 9.

If you have a measurement that could be any number[5] and it is not random, there's a good chance it follows Benford's Law.

Try this. Take a column of your data and create two new variables, one containing the first digit of your data point, and the other containing the last digit of your data point. Ignore zeros and decimal points. In Microsoft Excel, this is easy to do with the LEFT and RIGHT functions. Then count the number of occurrences of the digits 1 through 9 and create histograms. Compare your results to Table 11. If the results more or less follow Benford's Law, they're probably legitimate. If there are too few 1s and 2s, and too many 5s through 9s, the data may not be legitimate, you may not have enough data, or the variable doesn't follow the law. In any case, it's worth investigating further. This may sound contrived, but believe me, every major financial institution in the world uses Benford's Law to help detect fraud.

Table 11. Interpretation of Benford Tests.

Distribution of Values				**Possible Interpretation**
First Digit(s)		**Last Digit(s)**		
Benford's Law		Equally Likely		Random data from a natural process
		Erratic		Small sample size; rounding effects; possible tampering.
Equally Likely		Equally Likely		Random data from a process measured on a logarithmic scale of from a process that does not follow Benford's Law.
		Erratic		Rounding effects; possible tampering.
Erratic		Equally Likely		Limited range data; process that does not follow Benford's Law.
		Erratic		Non-representative sampling; data tampering.

5 In other words, there are no mathematical or natural reasons for the number to be limited to a range of values. For example, the pH of most natural waters is usually within a range of 6 to 8. Such values do not follow Benford's Law.

Other Dataset Scrubbing

After you are confident that you have caught all the errors in a dataset, it's time to clean out anything that doesn't contribute to the analysis. First, delete all the graphical formatting. Delete all borders, backgrounds, colors, extra lines and spaces, special fonts, titles, footnotes, and long variable names or labels. These are often artifacts left over on spreadsheets that are used for both data reporting and data analysis. Replace the long variable names with eight-character names having no embedded blanks or special characters. Place the new variable names in the first row of the data file. Finally, examine each variable to assess its importance to the analysis. Any variable that has no variance (i.e., all the values are the same) or is a duplicate of another variable, is perfectly correlated to another variable, or contains information that is no longer relevant (e.g., laboratory ID) should be deleted. Then reorder the variables so that IDs, coordinates, grouping variables, and sort keys are on the left side of the file. Consider also whether a program you plan to use has any special formatting requirements. Old DOS-based programs are notorious for requiring specially formatted header lines in their input datasets.

Resurrecting the Unplanned

My data has holes
I gotta fill those
By exhuming data better left alone
I'm resurrecting samples that were long ago withdrawn
Yeah, my data's getting pawned
By filling in holes

Sing to the tune of "I'm Diggin' Up Bones" by Randy Travis

"... the cat is cryptic, and close to strange things which men cannot see."
H. P. Lovecraft, "The Cats of Ulthar"

Even if you took a class in statistics or another form of data analysis, you probably didn't hear about frankendata, a.k.a. data hash. Frankendata is created when data, collected by different people, at different times and locations, analyzed with different procedures and equipment, and reported in different ways, are conglomerated together. In statistics classes, students are provided the same datasets so that they all have at least some chance of getting the same answer. Unlike the saintly sages of statistics who teach those classes, your boss expects you to work with data hash. Often, this is because he sold the client on the idea of cobbling together all the data from the consultants who had the project before your company was hired. The client totally bought into you being able to make sense of the data mishmash.

Too often, data for a statistical analysis are generated without the prior input of a statistician. Sometimes, even the statistical analysis is an afterthought, coming shortly after the investigator realizes that the data defy interpretation by any means known to him or her. In these cases, you have two possible courses of action. You can try to dodge the bullet, perhaps by explaining the problems with the dataset, and then declining the assignment. This never works. Your boss wants the client's money. The sick-child gambit works better and is a lot easier to explain, only you can't use it very often. Most consultants, though, are simply incapable of saying no. This is not just for the money. It's because they become consultants because they like to solve problems. And believe me, doing a statistical analysis using data that were generated without the oversight of a statistician is a problem.

To nonstatisticians, data are data. Concepts like populations and representativeness and randomization and variability aren't relevant. But data generated without statistical oversight are like cookies made by unsupervised kindergarteners. You can't expect that they followed a recipe since they can't read yet. You can't even assume that they know the differences between sugar and salt, or flour and baking powder, or cooking oil and motor oil. You won't t know what you might have until you take a bite. Scary thought, huh!

So what do you do if faced with this situation? You can swallow hard and not take the assignment. Recognize, though, that someone else will. If it's an issue that's important to you, you'll have more control over what gets done if you're involved. You might start by following this recipe:

- **What are the ingredients?**—How were samples picked relative to the population of interest? Were any steps taken to minimize variability and bias? How many good samples do you have? Are the variables appropriate for solving the problem? Are outliers and missing data likely to be issues? Can other information be included to augment the analysis?

- **Is it safe to eat?**—What can you do with the data given the number of samples and variables? If a complete analysis isn't feasible, can an exploratory/pilot study or partial analysis be done?

- **Where's the Maalox?**—What are the limits/caveats/uncertainties of the analysis? Will the results satisfy the client and other reviewers?

If you can think through an approach that will at least get the client to the next step, it's probably a good idea to take the assignment. If you do, be sure the client has a clear idea of what you think you can do

I once had a client who was considering buying some property. They were looking at several parcels in an industrialized area of several square miles. The client wanted to know if the groundwater of the area was contaminated because they did not want to get caught up in a regional problem not of their making. The traditional method for answering this type of question would have been to install and sample wells on each property and then develop contour maps for each pollutant of concern. Because of the size of the area and the large number of chemicals to be analyzed for, such an approach would have been prohibitively expensive.

There were, however, scores of industrial facilities in the area that did have groundwater monitoring data, which was publicly available under a state program. The problem was that each site was a different size, from an acre to hundreds of acres, and had different numbers of wells that were sampled on different schedules for different chemical analytes. Each facility used different chemicals, and so, had different monitoring requirements imposed by the state. No analyte was being tested for in even half of the several hundred wells. In a nutshell, nothing was comparable.

Resurrecting this data involved having groundwater specialists review the data from all of the wells in the area. For each of the wells, the specialists determined whether any of the analytes tested for exceeded the standards established by the state. Wells with groundwater that exceeded a standard were coded as 1; wells that did not exceed a standard were coded as 0. The 0s and 1s were then used to produce a contour map of the probability that the groundwater of the area was contaminated. So the client got the information they needed at a price they could afford.

I guess data scrubbing isn't always that boring after all.

18 Getting the Right Numbers

Now that you're sure you have all your numbers right you have to make sure you have all the right numbers. This step in the data scrubbing process involves addressing replicate samples, missing data, censored data, and outliers.

Replicates

> *Not even the most subtle and skilled analysis can overcome completely the unreliability of basic data.*
>
> R. G. D. Allen, *Statistics for Economists,* 1949

Not all replicate samples are exactly alike.

There are two kinds of replicates you'll encounter in analyzing data—the kind you don't expect and don't want, and the kind you do expect and don't want. The first kind of replicate, the unplanned kind, occurs when the ID variables don't uniquely identify an individual. If you've ever gotten multiple copies of the same catalog from a mail-order house, it's because your name and address are replicated in the mailer's database. Usually this is attributable to variations in the way the name is spelled or abbreviated in the database. For example, Dr. Henry Walton Jones, Jr. could be in a database multiple times as Henry Jones, H. W. Jones, Dr. H. Jones, or even Indiana Jones. Sometimes addresses don't help to resolve conflicts because people use home addresses, office addresses, and mailstops when on travel.[1] If you're lucky, you won't run into any of these replicates because they are usually difficult to resolve.

1 And that doesn't include the people who intentionally change how they provide their name and

The other kind of replicate, the kind you expect, occurs when the sampling plan includes provisions for the collection of replicate samples or measurements as a gauge on quality control. The purpose of these samples is principally to verify the consistency of some data generation process. A typical example would be replicate samples collected to assess the consistency of laboratory analyses. The question is, though, should these replicates be included in a statistical analysis?

The Problem of Representativeness

The issue with including replicate samples in a statistical analysis boils down to the problem of representativeness. You want the samples you collect (or the measurements you take) to be representative of the population from which all possible samples (or measurements) might come. By including replicate samples, you may be biasing your sample in favor of those individuals in the population whose information is replicated. On the other hand, maybe what you think are replicates really aren't. Maybe there really are two Dr. Jones living at a certain address. So how you perceive the samples will direct how you should handle replicates.

For example, replicate samples are often taken in environmental studies for quality control purposes. But what are thought to be replicates may really be co-located samples, samples collected sequentially from essentially the same location. In environmental sampling, this distinction can be thought of as a function of mixability. Fluid media like water and air are easy to mix into a homogenous mass. You can be reasonably sure that replicate samples of a fluid are identical. Two water samples collected sequentially from a stream or pump, for instance, are likely to be identical. Solid media, on the other hand, like soils and feral animals, remain somewhat heterogeneous no matter how well you try to mix them. The only thing you can do with solid media is hope that the proportions of discrete components are about the same in both samples. Take two soil samples, mix them thoroughly and split them, and you may have something approximating duplicate samples. But if you just collect two soil samples side by side, or worse, at consecutive depths, you won't have comparable samples. This makes a big difference when doing quality control checks of a measurement procedure. It's also a good example of why you have to be able to look at the metadata to see how the field technician collected and prepared the sample.

Unfortunately, representativeness of a sample is an undeterminable attribute. If you know the parameters that define your population, you can conduct statistical tests for specific attributes, but there is no general test to assess representativeness. You have to use your judgment and, whenever possible, incorporate some component of randomness into your sampling schemes.

What Should You Do?

Here's a simple strategy for treating replicate samples or measurements. If you collected the samples to assess measurement variability, use all the measurements. Conduct-

---

address so that they can tell if their personal information is being sold to others.

ing rising-head and falling-head slug tests,[2] even when the tests themselves are replicated, are a common example. If your replicate samples come from a very large population of possible samples, consider the purpose you will use the samples for. If your intent is to calculate descriptive statistics or conduct a statistical test involving the population, average the replicate samples. Delete the individual replicates from the dataset and replace them by the averaged measurement. If your intent is to produce a map or conduct a geostatistical analysis,[3] keep all the samples.

The one thing you shouldn't do is pick the measurement that matches your expectations. For example, picking all the lowest replicate analyte concentrations from an area you believe is uncontaminated. Yes it has been done but it should not be considered an acceptable practice because it biases any results derived from the dataset.

Missing Data

God not only plays dice. He also sometimes throws the dice where they cannot be seen.
Stephen Hawking, *Nature,* 1975, p. 257

Ever try to read a book with pages missing? You can usually figure out what's going on so long as there aren't too many pages excluded. But if there are substantial omissions, the story will make little sense. Datasets with missing data, though, are much harder to comprehend than books with missing pages because data patterns are generally much more difficult to recognize than written concepts.

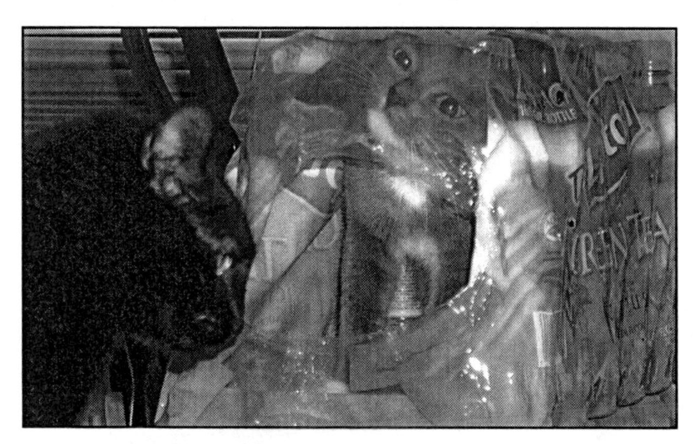

He won't see me in here.

---

2 A slug test is a procedure used in hydrogeology to estimate the ability of soil or rock to transmit groundwater. The term slug refers to a volume of water, not the slimy little critters in my garden. The test involves putting a slug of water into a well (falling-head test) or taking a slug of water out of a well (rising-head test), and determining how long it takes the water in the well to return to it's original level. Aren't you glad you asked?

3 The reason for this is that a geostatistical analysis examines variability between samples located various distances apart. Having replicate samples in which the separation distance is zero allows the evaluation of intrinsic variability, termed the *nugget effect.*

There are several common causes of missing data:

- **Nonresponses**—Missing data usually occurs when an information source provides no response. This occurs commonly in surveys when a subject refuses to answer a question or a patient drops out of a study. Technical problems can also cause nonresponses, especially with physical samples. Occasionally, a sample may be lost, destroyed, or compromised so that it cannot be analyzed. Broken or mislabeled sample bottles are common examples. Alternatively, a meter or device for generating the needed data might be broken or unavailable. Finally, a datum may be rejected because it fails to meet quality standards. Usually this cause of a missing data point is unavoidable but relatively infrequent.
- **Data Not Recorded or Entered in Database**—This cause of missing data occurs most frequently with text descriptions and ancillary measurements, such as field measurements taken in support of sample collection. Sample metadata (data about how the sample was collected, such as sampling device, sampler, weather), are also commonly missing. Often, the information is recorded in the field but not incorporated into a database.
- **Poor Planning**—This cause involves not having clear data requirements based on study objectives and a statistical analysis plan. Sometimes, the problem is related to not specifying the proper types and frequencies for measurements. Other times, the problem is that required information is not entered into a single, appropriately designed database. Most commonly, the problem stems from an attempt to cut costs. While minimizing data collection costs is laudable, it makes no sense to compromise an entire data analysis effort to save a few dollars. If the cost of the data collection is too high, redesign the study.

Whatever the reason, there seems to be missing data in almost every dataset.

You Can't Analyze What's Not There

Statisticians have identified three situations involving missing data. If data are missing completely at random (MCAR), there is no pattern to the occurrence of the missing values. If data are missing at random (MAR), there is a pattern to the occurrence of the missing values for a variable that is related to some other variable in the analysis. For example, an employee satisfaction survey may have more data missing for managers than for other employees. The pattern of missing data is related to employee rank but not their reported satisfaction. If data are missing not at random (MNAR), also called nonignorable missing data, the pattern of the missing values is related to the same variable. For example, there may be more missing data for employees who would have responded with a low level of satisfaction because they don't believe the survey will lead to improved conditions or because they fear reprisals.

So what's the big deal? Well, as it turns out, statistical procedures can't cope with matrices having empty cells. So to do almost any kind of statistical analysis, you'll either have to delete the variable (column) with the missing data or delete the sample (row) with the missing data. If you have several missing values, you might have to delete several

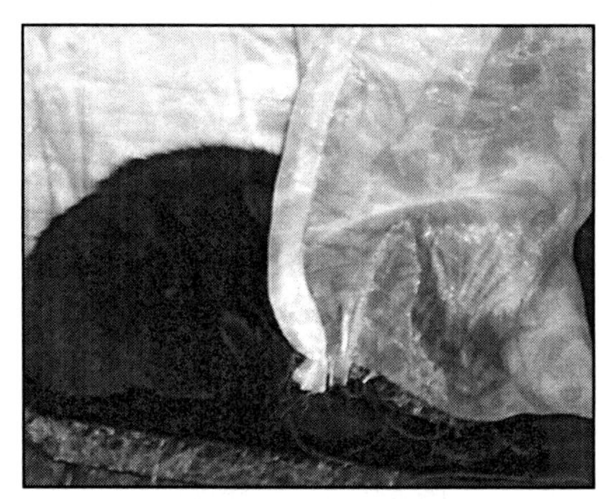

Onyx missing not at random (OMNAR).

samples or variables. This is definitely not good. And it gets worse. If your missing data is MNAR or even MAR, the absent data will bias your statistics

What Can You Do?

The first thing you have to do is look for patterns in your missing data. If you have high-end statistical software, there may be some utilities to do the job for you. If not, a simple approach is to take a *copy* of your dataset and replace missing values by one and real values by zero. Then calculate all the descriptive statistics you would ordinarily do. If there are any patterns, such as means significantly higher in one data grouping than another, then your missing values are MAR or MNAR. In these cases, you might have to go back to your study plan and figure out how to control the cause of the missing data or somehow incorporate the cause into your model.

If there are no patterns in the missing values, they are probably MCAR, so you can let your statistical software cope with the problem. Statistical software usually offers two options for dealing with missing values—casewise deletion and pairwise deletion.

In casewise (or listwise) deletion, all samples (i.e., rows, records, cases) having any missing data are excluded from the analysis even if the missing data is for a variable not included in the analysis. This approach assumes that the missing data is symptomatic of problems with the sample that would make its inclusion in the analysis inadvisable. Pairwise deletion is less draconian, but only a little. In pairwise deletion, all samples (i.e., rows, records, cases) having any missing data are excluded from the analysis *if* the missing data is for a variable included in the analysis. This approach makes no assumption about the quality of the sample but removes the samples that prevent calculations.

When pairwise deletion of missing data is selected, cases will be excluded from any calculations involving variables that have missing data. In the case of correlations, however, calculations will include any sample that has valid data for both variables. Thus, correlations between pairs of variables in a dataset may each be based on different numbers of samples. Some statisticians don't like this.

These deletion approaches are simple, but they present problems. First, it's difficult for clients to understand why you have to exclude any of the samples that they paid a lot of money to collect and analyze. Second, when you delete samples, you may actually be adding bias to your dataset. But you won't necessarily know.

It's not entirely true that you can't analyze what's not there. There are actually a variety of methods that replace the missing value with a value that can be analyzed. Examples of replacement options include:

- **Mean (or Median) Replacement**—Missing values are replaced by the mean or median of the nonmissing values. This approach artificially decreases the calculated variance of the data especially if the number of missing values is appreciable. Correlations may also be substantially effected.

- **Neighbor Replacement**—A "neighboring" value is used in place of the missing value. This approach can take a variety of forms. For time-series data, a missing value can be replaced by the previous non-missing value. This approach is called last observation carried forward (LOCF). Missing spatial data can be replaced by the spatially nearest data value or a composite value calculated by interpolating between several neighboring values.
- **Hot Deck Imputation**—One or more variables are specified to identify samples similar to the sample with the missing data. Missing values are then replaced with values taken from matching samples. The replaced values may be randomly selected, averages of all the candidate values, or predicted from a regression model of the specified variables. This approach was designed to minimize the variance reduction of mean replacement method. However, the process of selecting variables to characterize the missing values may unintentionally introduce bias. It is also computationally intensive. It is a cool term to throw out during conference presentations, though.
- **Multiple Imputation**—A number of new datasets are created by replacing a missing value with values randomly selected from a distribution that represents the uncertainty in the missing value. The new datasets are then analyzed separately using the same techniques. The results are combined into a single decision or conclusion. Research has shown multiple imputation to be able to correct for the variance reduction and biases produced by deleting missing values or using simpler replacement approaches. However, the trade-off is a monumental increase in the amount of data analysis that has to be done.

So if you have missing data, you have to decide what's worse, making up replacement data or deleting real data. You're caught on the prongs of Morton's fork.

What Should You Do?

From this, it should be possible for you to develop a strategy for coping with your missing values. For example, if you don't mind losing a few samples, 5 percent is a good rule of thumb, use the pairwise deletion option in you software. If you need to replace the values,[4] you might use a replacement scheme. Leave imputation techniques to professional statisticians.

There's one other issue you need to consider. Some of your reviewers might consider replacing missing values to be *fraud.* It won't matter that you will be able to cite countless references on why replacement and imputation techniques can be legitimate

---

4 Some statistical procedures have additional data requirements that might not allow samples to be deleted. For example, autoregressive modeling, a type of time-series analysis, requires that samples be collected at equal time intervals. If one or more data points are missing, they must be filled in using some replacement or imputation technique or the analysis cannot be done.

components of a statistical analysis. They will believe you are trying to "pull something." If you run into this attitude, you might as well resort to pairwise deletion. In the long run, such uninformed reviewers will cause much more trouble than just deleting a few samples.

Censored Data

> *It is important to realize that it is not the one measurement, alone, but its relation to the rest of the sequence that is of interest.*
>
> William Edwards Deming, *Statistical Adjustment of Data,* 1943, p. 3

If you took only one course in statistics, you probably never heard of censored data. It's like dirty laundry. You don't air it in public. But like dirty laundry, the reality is that it does occur and it's better to deal with it than try to ignore it.

Why Censored Data Is Important

Consider this example. Chemists call analytical results (i.e., concentrations) below a machine or method detection limit (DL) nondetections. Statisticians call nondetections censored data because it is known that values exist but the true values cannot be determined. Here's a simpler definition—data points qualified as less than (<) or greater than (>) are censored data.

There are actually quite a few examples from the natural sciences beside contaminant concentrations. The thickness of geologic layers, the discharge of streams, and the number of individuals of a certain species in an area can all at times be qualified as less than or greater than. Analytical chemistry data, though, receives the most attention because there is so much of it. And while there are occasional results qualified as greater than, most analytical censored data are those qualified as less than. The focus of the

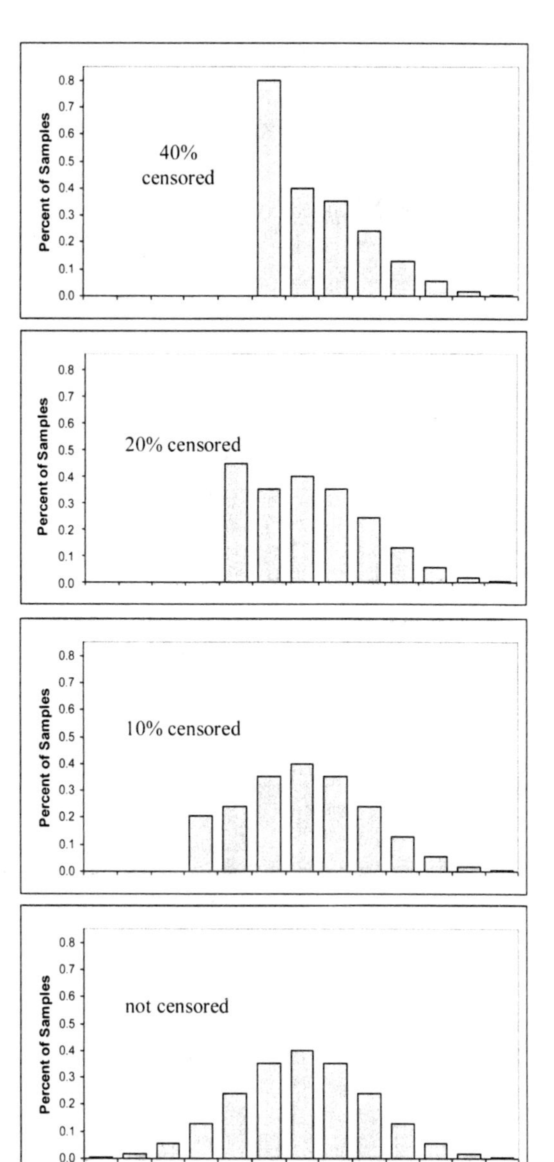

Figure 14. Effect of Censoring on Sample Frequency Distribution.

following sections, therefore, is on less than values, termed left-censored or low-censored data.

Now here's the problem. What do you do with a <10 value? You can't just discard it. It's a real value from the population. You just don't know exactly what the value should be. If you ignore the qualifiers (< and >), you're going to dramatically change the look of your sample distribution (Figure 14), and with it, all the statistics that describe the population. So what do you do?

You Can't Analyze What You Can't See

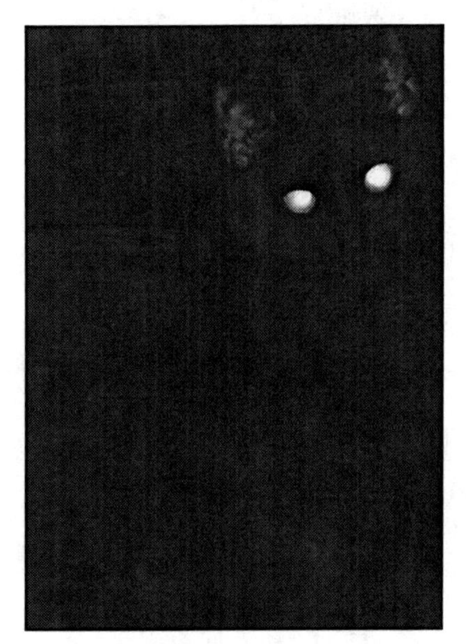

Ninja cat says, you CAN feed what you can't see.

As you may have expected, this is not entirely true. Certainly, if there are ways to analyze missing data, there must be ways to analyze censored data. And there are, in fact, quite a few. One fundamental difference between the methods is whether they are applied to individual data points or to calculated statistics.

Methods Applied to Data

Statistical methods applied to data involve changing each censored data point to a different value that can be used in an analysis. These methods are usually used for mapmaking, graphing, and complex statistical procedures that can't be calculated manually. There are three general approaches—deletion, trimming, and recoding.

- **Deletion** involves removing the censored data from the dataset prior to analysis. This method should only be used for maps and graphics as it radically biases any calculated statistic.
- **Trimming** involves deleting the censored data *and* deleting an equal number of uncensored values from the opposite end of the sample distribution. If you have ever watched how figure skating is judged, you've seen an example of trimming.[5] Trimming can be used for maps, graphics, and statistical calculations of means. It cannot be used for calculating variances or correlations. Trimming is not used often because it involves discarding good data, which most data analysts are very reluctant to do.
- **Recoding methods** involve replacement of censored data by a proportion of the censoring limit. (For analytical nondetections, the censoring limit would be the detection limit). Recoding is perhaps the most commonly used procedure

5 Even if the judges' scores are not censored, it's the same approach. The highest score and the lowest score are discarded before the average score is calculated. Sometimes the two highest and two lowest scores are removed.

for addressing censored data because it is conceptually and computationally simple. It also reclaims censored data points rather than discarding them as deletion and trimming do.

One-half of the censoring limit and one-times the censoring limit are the two most commonly used recoding methods. Recoding as zero or a very small proportion of the censoring limit (e.g., one-tenth or one-hundredth of the censoring limit) are sometimes used for mapping or graphical applications but never for any procedure involving the estimation of a statistic, such as a mean or a standard deviation. That's because recoding methods introduce bias into the dataset. The amount of bias is always unknown because it is a function of the recoding value and the censored values. But for recoding using the detection limit, at least the direction of the bias is known.

Recoding using the detection limit will bias the mean toward higher values and bias the standard deviation toward lower values. The magnitude of the bias is uncertain but, in general, estimates of the standard deviation are far more uncertain than estimates of the mean. Consequently, estimates of a mean based on even substantial amounts of censored data (perhaps up to 30 percent if the coefficient of variation[6] is less than 100 percent) may still be fairly accurate. Statistical tests based on both the mean and standard deviation, however, are likely to be in error for even small percentages of censored data (perhaps as little as 10 percent).

In contrast to recoding using the detection limit, recoding using one-half of the detection limit will also introduce bias, but both the magnitude and direction of the bias will be unknown. Replacement by zero will bias the standard deviation high and the mean low. This is not considered to be an acceptable approach to use for calculating statistics.

Recoding methods have been found to produce acceptable results if the proportion of censored data is below about 15 percent. Above 15 percent censored data, recoding methods are reportedly inferior to the other methods.

Methods Applied to Statistical Calculations

Methods applied to statistical calculations involve calculating a mean or standard deviation and then adjusting the calculated statistic based on the amount of censored data. These methods are generally used only for simple statistical procedures because they are not built into most statistical software. There are three methods of correcting calculated means and variances for censored data—Winsorizing, Aitchison's procedure, and Cohen's procedure.

1. ***Winsorizing*** involves first replacing the censored data with the next largest (or smallest) data values and, like trimming, repeating the action on the other side of the data distribution by replacing the same number of data values with the next smallest (or largest) data values. The mean is then calculated as usual. The calculated standard deviation, though, is increased by a correction factor to adjust for the data that were replaced. The correction factor is the number of

6 The standard deviation divided by the mean, expressed as a percentage.

samples before winsorizing minus one divided by the number of samples not replaced minus one.

2. ***Aitchison's procedure*** is a relatively simple adjustment in which the mean of the uncensored values is multiplied by the proportion of censored data. For example, if the mean of the uncensored values were 10 and the proportion of uncensored values were 70 percent (0.70), the Aitchison-adjusted mean would be 7. The adjustment for the variance is more complex, and is based on both the proportion of censored data and the estimated mean. Aitchison's procedure assumes that both the censored data and the uncensored data come from the same, Normally distributed population. Aitchison's procedure produces results that are reasonably unbiased up to about 50 percent censored data.

3. ***Cohen's procedure*** is a more rigorous procedure for adjusting censored data. It is based on estimates of both the mean and the variance of the uncensored data and the proportion of censored data. From these statistics, tabled values are used to adjust the mean and variance. Cohen's procedure assumes that both censored and uncensored data come from different, Normally distributed populations. Cohen's procedure also produces results that are reasonable up to about 50 percent censored data.

So what should I do?

What Should You Do?

If you know what you are going to do with your data, consider these guidelines. For maps and graphs, replace censored data with the censoring limit. If you then use a weighted average interpolation algorithm, you can establish the limit as a baseline. For calculating descriptive statistics, use Cohen's or Aitchison's methods. These provide the best correction as long as the degree of censoring is less than 50 percent. For more complex statistics, delete the sample or the variable if there are many censored data points. If there are only a few censored data, say less than 5 percent of the dataset, replace them by the censoring limit.

Outlier s[7]

The folly of rejecting an extreme observation was demonstrated when shortly after 7:00 AM on the morning of December 7, 1941, the officer in charge of a Hawaiian radar station ignored data solely because it seemed so incredible.

Anonymous

I'm being an outlier.

Occasionally, a dataset may contain a value that is far greater (or less) than any of the other values. This anomalous value is termed an *influential observation.* If the influential observation is not representative of the population being sampled, it is called an *outlier.*

Causes

Influential observations and outliers occur for a variety of reasons. Some are straightforward data generation or reporting errors. Lab results that are off by a factor of ten are often identified as outliers. Outliers occur in business data for a variety of reasons. Reporting deadlines may be missed, weather or construction may prevent customers from shopping, and there may be one-time corrections for past errors. Some aberrant measurements are caused by instrument error or miscalibration.

One frigid January morning, I was a member of a team conducting a gamma radiation survey of an industrial facility. We calibrated the meter to ignore background radiation and started walking around the site. After a few minutes, we heard a click. A while later, we heard another click, and then another, and another. The clicks kept coming faster and faster no matter where we walked. It wasn't long before the meter sounded like a cicada on speed. We quickly retreated back to our vehicle wondering if we had already received a lethal dose of whatever was out there. Gradually the meter returned to normal. Being young and stupid, we decided to try it again. This time we drove through the site. After all, gamma rays have great penetrating power. If they're out there, they'll certainly penetrate the vehicle (not to mention our reproductive organs) and trigger the meter. We drove everywhere, twice. Nothing. Then, the genius of the group decided to look at the meter's instruction manual.[8] Sure enough, the meter was only accurate between 50 and 80 degrees Fahrenheit. The abnormal measurements were an *instrument error.*[9] Needless to say, those measurements never made it into the site investigation report.

---

7 The idea for this section head came from: R. J. Beckman and R. D. Cook (1983). *Outlier. . ..s, Technometrics,* 25, 119–163. If you don't get the joke, don't worry about it.

8 When all else fails—RTFM—read the fricking manual.

9 That is, if you consider the operators part of the instrument. This is called an *ID-10-T* error.

Sometimes, there are deterministic influences that skew some measurements. One time I received a dataset consisting of surface geophysical measurements. I plotted the electromagnetic conductivity measurements and noticed a curious pattern in the data. There was an area of high conductivity that appeared to surround the site. I thought I might have found something important or at least something that merited further study. Never having seen the site, I questioned the field technician about what he might have seen when he conducted the survey. He looked at my maps, smiled, and thanked me for pointing out the location of the site's chain-link fence. He had already found it. He even had pictures. Those data didn't make it into the final report either.

Some outliers aren't errors, but instead are the result of inherent variability or a natural cause. I once developed a model of chloride concentrations in groundwater for a client who operated a landfill. The model fit the data very well and indicated that a small plume of elevated chloride concentrations was associated with the landfill. However, the model also predicted a large chloride plume near the entrance of the facility where there were no operations that might have impacted the groundwater. The model prediction of this plume, though, was based on only a few wells, albeit many samples from the wells, so I thought it might be an outlier. When I finally saw the site, the reason for the outlier was clear. The property across the highway from the landfill was the county's maintenance facility where they stored their road salt for the winter. So, the landfill did have a problem, but it wasn't the main culprit. The client was pleased about that.

So if you run into outliers, try to figure out why they exist. They may mean nothing so that you can delete them from the analysis, or they may be critical to your interpretation of a dataset. You'll probably find that, most of the time, the causes of outliers will be unknown.

Identification

Influential observations and outliers are generally not difficult to detect. Sorting and listing the data will often reveal questionable values, though the best way to identify a potential outlier is by graphing the data. Histograms, box plots, probability plots, time-series plots, or scatter plots of the data will usually reveal any aberrant values. Take Figure 15 for example. Is there any question that there are two outliers?

There are also a great many statistical tests for identifying outliers. Outlier tests follow one of several strategies. Some tests, called deviation/spread tests, are like simple t-tests. They are calculated as the difference between the outlier value and the mean (or other measure of central tendency), divided by the standard deviation (or other measure of data dispersion). Other tests, called excess/spread tests or Dixon-type tests, are calculated as the difference between the outlier and the next closest value (or other observation in the dataset), and the dataset range (or other dispersion statistic). Some statisticians prefer this type of approach because it is not necessary to have good estimates of the mean and variance. Other outlier tests examine sums-of-squares, skewness, and location relative to the center of the dataset.

The truth of the matter is that outlier tests are often superfluous. Identification of

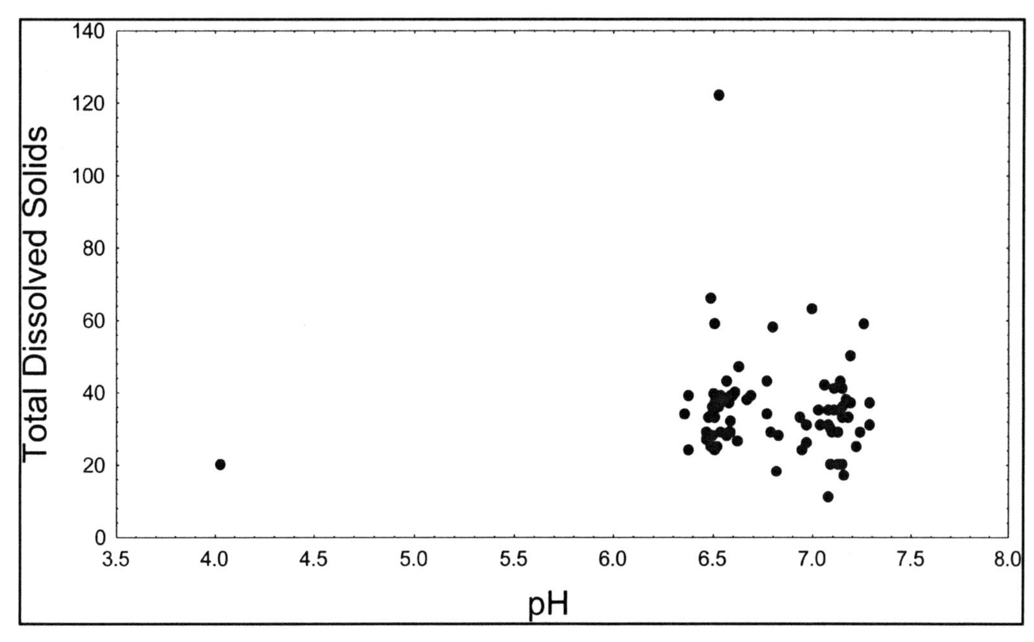

Figure 15. Can you Spot the Two Outliers?

outliers isn't that much of a problem. The real issue is what you do if you find a value you think is an outlier.

Treatment

There are five options for treating outliers:

- **Inclusion**—Inclusion involves keeping the outlier in the dataset. This approach would make sense to use if you're looking to assess the effects of the anomalies. Usually, you're forced to take this approach because an unenlightened reviewer thinks you are trying to slant the analysis. In these cases, it is often beneficial to run your analyses both with and without the outlier so that you can understand its effect.

- **Correction**—Correction involves changing the outlier to the correct value. This doesn't happen often. You might find an outlier to be an error, but you can't correct it because you don't know what the true value should be. In that case, deletion is probably the best option. If you're lucky, though, you might find an outlier to be an error and be able to correct it.

- **Replacement**—Replacement involves changing the outlier to a contingency value. This approach is like the replacement options for missing data. Using the mean or median in place of an outlier will bias the dataset, but not nearly as much as the outlier. This is often the best approach to use for complex statistical calculations.

- **Accommodation—**Accommodation involves keeping the outlier in the dataset but using "robust" statistical procedures that are less sensitive to outliers. Nonparametric statistics are often used for this purpose.

- **Deletion**—Deletion is simply removing the outlier from the dataset. This approach would make sense if you're looking to assess general trends. Once again, it might be beneficial to run your analyses both with and without the outlier.

The option you select should depend on whether you believe the aberrant observation is representative of the population you are investigating. Your objective and the type of analysis you plan to do will also be considerations in this decision.

What Should You Do?

If a statistical graphic or an outlier test suggests that a data value may be an influential observation or an outlier, follow these steps:

1. Examine a variety of graphical depictions of all the data points including box plots, probability plots, bivariate plots, time-series plots, and contour maps to assess possible reasons for the aberrant observation.
2. Review notes and metadata concerning the sample or measurement to determine if any irregularities in the sampling or data collection processes may be responsible for the discordant value.
3. Review documentation related to data quality for the sample or measurement to determine if any irregularities in the sampling, packaging, transport, and analysis or measurement and recording processes may be responsible for the discordant concentration.
4. If any information indicates that the sample is probably not representative of the population being sampled, consider the sample or measurement to be an outlier and replace or delete it from further analysis. If possible, collect a new sample or measurement.
5. If any information indicates that the sample should be representative of the population, review results for other measurements from the same source to determine if other results support the legitimacy of the suspected outlier. Also, review results for the same variable that may have been generated during previous sampling efforts.
6. If prior results for the variable or results for other variables are consistent with results for the suspect sample or measurement, retain the value and evaluate it as an influential observation.
7. If prior results for the parameter or results for other parameters are not consistent with results for the suspect sample or measurement, consider the value to be an outlier and replace or delete it from further analysis. If possible, collect a new sample or measurement.

This procedure works best if both data analysts and reviewers can somehow be involved in the examination process. Be sure to document all findings and decisions during this process.

If you decide to retain the outlier, consider using a nonparametric alternative to the procedure you planned to conduct. If for some reason this is not feasible, consider analyzing the dataset twice, once with the outlier and once without the outlier. Caveat your conclusions on the basis of the outlier and recommend collecting additional samples or measurements to assess its validity. Consultants always recommend additional work anyway, so this should come as no surprise to either clients or reviewers.

If you are assessing data trends, you will probably want to delete or replace any outliers. Even a single outlier can mask significant trends. Be aware however, that this action could bias predicted values and the prediction error if the cause of the outlier is natural.

Given the choice to replace or delete an outlier, consider the number of samples you have and the importance of the variable the outlier is a measure of. Remember, if you delete the outlier, you will end up having to delete either the sample or the variable to conduct your statistical analysis. If the variable is important and you don't have many samples, consider replacing the outlier.

There is also a psychological component to consider when replacing or deleting outliers. Scientists and engineers are taught that it is unethical to delete or change data that might not fit with their expectations. Outliers challenge that notion. Statisticians and reviewers become highly suspicious of each other when the need to judge an outlier arises. Consequently, it is sensible to have a procedure for evaluating outliers in place that everyone agrees to before the need arises. Even so, somebody will criticize you no matter what you do. It's the way things work.

Take heart, you won't have to do all these data scrubbing steps on every dataset you encounter. Experience will tell you when enough is enough. But be prepared, because no matter how hard you try, you won't find every problem. If you're lucky, you won't have to redo much work.

19
Kicking the Data Tires

After all that work preparing the dataset, you're probably anxious to do your statistical analysis. Right? Well, you'll have to be patient just a bit longer. First you have to do a summary analysis (SA) or an initial data analysis (IDA)[1], or an exploratory data analysis (EDA). These are like taking a test drive in a car you plan to buy.

Wake up! I'm hungry. You can get yourself caffeinated after you get me bluefinated.

The amount and type of effort you might put into this step will depend on your plan. If you plan to conduct a specific test or analysis with a narrowly defined hypothesis, this step will be a simple introduction to your dataset. Statisticians call this a summary analysis.

The goals of IDA are a bit more ambitious, and include:

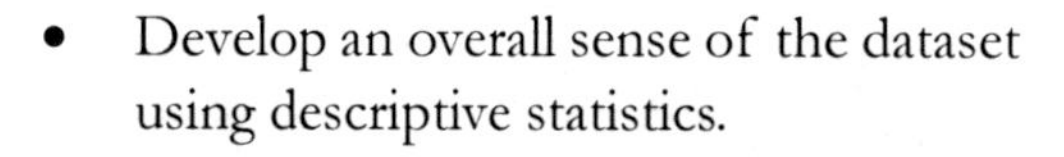

- Develop an overall sense of the dataset using descriptive statistics.
- Identify any interesting patterns or data anomalies that did not surface during data scrubbing.
- Evaluate fundamental statistical requirements of the data.

If, on the other hand, you plan to build a complex model of some object, behavior, or process, or just want to extract whatever knowledge you can from a dataset, this step will be a more thorough examination of your data. Statisticians call this an exploratory data analysis or (EDA). The goals of EDA are the same as for IDA plus:

- Identify important variables and influential observations.
- Explore less obvious relationships and multivariable patterns.

SA, IDA, and EDA are not statistical procedures per se. Rather, they are collections of tools, both computational and graphical, that are used to focus subsequent statistical

1 This is not to be confused with the term *intelligent data analysis*, also termed IDA. Intelligent data analysis involves the use of artificial intelligence techniques, data visualization, data mining, fuzzy logic, statistical pattern recognition, and other sophisticated data analysis procedures.

analyses. Unlike IDA, which assumes some knowledge of the underlying data structure, EDA makes no presumptions and uses the data to explore the data structure. SA and IDA are primarily computational although IDA has some graphical analyses. EDA is primarily graphical with some computational analyses. Conducting either an IDA or an EDA is a prerequisite to conducting almost any type of statistical analysis. EDA is considered a necessary prelude to more sophisticated modeling.

In reality, it doesn't matter whether you call what you do an SA, IDA, or EDA. In fact, these terms mean different things to different statisticians. The important point is simply that you do something like them. This chapter will give you some hints for what to do. There's nothing complicated or arcane about this step. It's just a matter of knowing where to start and what to do. Like driving a car, once you know the basics, it's all just a matter of practice.

A Fistful of Digits

The statistician's job is to draw general conclusions from fragmentary data. Too often the data supplied to him for analysis are not only fragmentary but positively incoherent, so that he can do next to nothing with them. Even the most kindly statistician swears heartily under his breath whenever this happens.

M. J. Moroney, *Facts from Figures: What Happens When We Take a Sample,* 1956.

The first thing you should do to explore your data is to calculate descriptive statistics. They're easy to generate, and they don't consume the reams of paper or hours of screen time that plots will. The most commonly used descriptive statistics are: number of samples, mean, standard deviation, coefficient of variation, skewness, and kurtosis. Other descriptive statistics may also be useful for different datasets. These statistics are demonstrated with two datasets, one Normally distributed and one Lognormally distributed.[2]

I don't care what the descriptive statistics say. It still tastes bad.

If you don't have access to one of the statistical packages described in Chapter 11, now is the time to reconsider buying one. Calculating statistics and preparing plots using spread-

2 For those detail-oriented readers, the lognormally distributed dataset consists of 200 randomly generated values having the characteristics of a lognormal distribution. The normal dataset consists of the logarithms of the lognormal dataset. And no, I didn't use Excel to generate the random numbers.

sheet software will take ten times longer than using statistical software (Figure 16). If you have more than a few variables to explore, you'll make back the cost of the software in the time you save on your first big project.

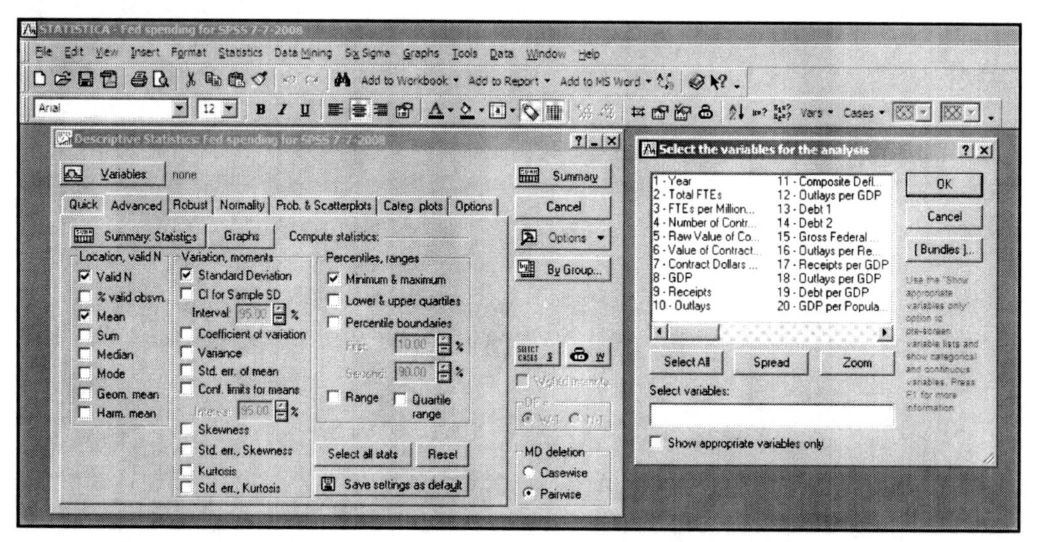

Figure 16. Screen Capture of Statistica's Descriptive Statistics Module Input and Output.

Counts

It may seem like a waste of time looking at such a simple statistic as the number of samples, but if there's a problem with how your software reads your dataset, this is where you'll find it. Check both the total number of samples and the number of samples in the grouping of data you plan to use in the analysis. Look for miscounts. Any discrepancies may be due to:

- Data or format errors in the dataset, like text instead of values
- Problems with how your software reads the data, perhaps from misplaced delimiters
- Missing data that cause the software to ignore some samples

I counted the stairs. There are too many. Bring my toys down here.

If everything looks correct, keep the counts close at hand. You should recheck the counts every time you run additional statistical procedures. Some procedures will delete observations with missing data, and others won't. Only some variables or data groups may be affected so it's not always apparent if there is a problem. Always be sure you are working with the correct number of data points.

Another thing to look for is whether the grouping variables have the same number of samples in each group. Statistical designs in which all groups have the same numbers of samples are said to be *balanced*. Balanced designs are less susceptible to effects of violated assumptions than unbalanced designs. The larger the imbalance, the larger the effect.

Finally, look at the proportions of missing and censored data if you did not address them in the data-scrubbing step. This information will help you decide if it might be necessary to remove variables from subsequent analyses.

Central Tendency

The next descriptive statistics to check are means, both the overall mean for each variable and the means by groups. Look for values that seem too high or too low, especially if they are out of the expected range for a variable (e.g., percentages over 1, negative concentrations). These may indicate the presence of errors or outliers. Compare the means to their corresponding medians. The closer they are, the more likely the sample distribution is symmetrical. The mean will always be on the side of the median where the distribution's tail is longest. If the mean is greater than the median, for instance, the distribution is skewed to the right (i.e., the tail on the right side of the distribution is longer than the tail on the left side).

Dispersion

I'm being a large variance.

After the mean, check the minimum and maximum to be sure none of your data are outside a reasonable range. If a subject's age is 132, for example, you either have a missing decimal point or other data entry error, or a subject who is getting a great return on their social security taxes. Also look at the span of the data, that is, whether your data covers the range of the scale you are interested in. For example, you can't reliably predict a measurement of 1.5 if none of your data are below 5.

The next descriptive statistic to examine is either the standard deviation or the variance.[3] If you're interested in looking at segments of your population, variances have the nice property of being additive. You can add up the variances of individual groups to get the total variance. That's why variances are used in most statistical procedures, like ANOVA. Standard deviations, on the other hand, are in the same units as the original data, so examining them is usually more intuitive than examining variances. Look for differences in the variances or standard deviations for your variables by any grouping factors in the dataset. If heteroscedasticity[4] will be a problem, this is where you'll see the first indications.

Also look at coefficients of variation[5] for the variables. The coefficient of variation can be used in several ways to compare the relative dispersion of:

- Variables that measure the same properties under different conditions or using different methods (e.g., field pH versus lab pH, pH by paper versus pH by meter versus pH by titration; wind speed at 3 meters versus wind speed at 20 meters)
- Variables that measure similar properties in different units (e.g., concentrations expressed as mass/volume, mass ratio, molarity, molality, mole fraction, or normality)
- Related variables (e.g., concentrations of different analytes)
- A variable in groups in which the means are very different (e.g., analyte concentrations or discharges in two streams).

If the coefficient of variation for a variable is greater than 1.0 (100 percent), either the variable has:

- Negative values (which may or may not be errors) and the mean is not significantly different from zero
- No negative values and is right-skewed; and the data might not have come from a Normally distributed population

Compare the coefficients of variation for both different variables and different sample groupings within the same variable. You can also look at nonparametric descriptive statistics for dispersion. The most useful for understanding data spread are quantiles, equally distributed points in a distribution, especially the four quartiles (25 percent of the data) and the ten deciles (10 percent of the data). These statistics as well as the range and the interquartile range[6] are available in most spreadsheet and statistical packages. Table 12 provides descriptive statistics and things you might look for in the two datasets mentioned at the beginning of the chapter.

---

3 The standard deviation is the square root of the variance.

4 Heteroscedasticity means there are significantly different variances in data groupings.

5 The coefficient of variation is the standard deviation divided by the mean. Sometimes the coefficient of variation is expressed as a decimal and sometimes it is expressed as a percentage.

6 The interquartile range is the 3rd quartile (75 percent of the data) value minus the 1st quartile value (25 percent of the data).

Table 12. Descriptive Statistics for Two Datasets

	What to Look For	Normal Data	Lognormal Data
Number of Data Points	This should be the same number you started with.	200	200
Arithmetic Mean	A "reasonable" value based on your expectations. It should be "close" to the median.	0.04	1.59
Lower 95% Confidence Limit for the Mean	The interval shouldn't be too wide. If the interval includes zero, it's likely your mean is not statistically different from zero.	-0.08	1.35
Upper 95% Confidence Limit for the Mean		0.17	1.82
Median	This is the center of your sample distribution. Compare the median to the mean to see if they're close. If not, the mean will be on the side of the distribution with the longer tail.	-0.01	0.99
Sum	Sum usually isn't important unless you are analyzing mixture (constant sum) data.	8.39	317
Minimum	These should be the same numbers you started with.	-2.62	0.07
Maximum		2.47	11.7
Lower Quartile	These will provide indications of the symmetry of the distribution as well as the sizes of the tails.	-0.56	0.57
Upper Quartile		0.61	1.85
10th Percentile		-1.07	0.34
90th Percentile		1.34	3.83
Data Range	These are all measures of the "spread" of the sample distribution. Small numbers indicate less variability.	5.08	11.6
Interquartile Range		1.18	1.28
Variance		0.84	2.92
Standard Deviation		0.92	1.71
Standard Error of the Mean		0.06	0.12

Distribution Indicators

There are a number of ways to take a quick look at a variable's frequency distribution with descriptive statistics, two of which you've already calculated. Compare the mean to the median. If they are about the same, the distribution is probably symmetrical. If the mean is less than the median, the sample is left skewed. It the mean is greater than the median, the sample is right skewed. If the coefficient of variation is greater than 1, you probably have a distribution that is truncated on the left and elongated on the right, such as with a Lognormal distribution. So next, you should look at the skewness and the kurtosis. They are easy to interpret; the farther they are from zero,[7] the more likely you have deviations from the Normal distribution.

Skewness measures distribution symmetry. A value of zero indicates that the distribution is symmetric. Negative values indicate a long tail on the left side of the distribution. Positive values indicate a long tail on the right side of the distribution. Values outside the range of about ± 3 suggest that asymmetry may be a problem.

The kurtosis is sometimes referred to as the peakedness of a distribution, but it's really more of a measure of the proportion of samples in the tails of a distribution versus the proportion in the middle of the distribution. The kurtosis of a standard Normal distribution is 3, so most statistical software subtracts 3 from the kurtosis value so that it is centered on zero. Be sure you know which scale your software uses. Positive kurtosis values indicate a distribution has a high peak and short tails, and is termed *leptokurtic*. Negative kurtosis values indicate a distribution has a flat peak and long or thick tails, and is termed *platykurtic*.

Skewness and kurtosis are power functions, so they can make a small departure from Normality look large. Consequently, they are seldom used alone to assess Normality. They are good places to start if all you have to do is check a box to have your software calculate the values. Distribution statistics can tell you what to look for when you get around to creating distribution plots. Distribution plots, such as histograms, stem-leaf diagrams, and probability plots, are particularly revealing because you can see how your data compares to a Normal distribution. These plots, in turn, support statistical tests that will tell you if the difference you see is big enough to influence your analysis.

Leptokurtic. *Right skewed.*

It's like going to the dentist.[8] The dental receptionist finds out what you want done

---

7 Some statistical software packages calculate kurtosis to be centered at 3 instead of 0. Consult your user's guide to see which value you should be looking for.

8 I'm not implying that initial data analysis is a distasteful experience you might want to avoid but should do anyway. At least I'm not consciously implying that… I think.

(like descriptive statistics), then the dental hygienist examines you in detail (like distribution graphics), and finally the dentist puts on the finishing touches (like statistical tests of Normality). You may not need all three in a particular encounter, but it doesn't hurt to use them if they're there.

Table 13 provides distribution statistics and some things you might look for in the two datasets. The Normally distributed data has a large coefficient of variation because there are negative numbers in the dataset. If a constant is added so that all the data are positive, the coefficient of variation is 0.03, considerably less than 1. Values of skewness and kurtosis are near zero and test probabilities are greater than 0.05 (meaning that the differences could have occurred by chance). The Lognormally distributed data has a coefficient of variation just over 1, somewhat high values of skewness and kurtosis, and test probabilities less than 0.01. All the methods agree that the "Normal" (left) dataset can be modeled with a Normal distribution, but the "Lognormal" (right) dataset cannot.

Table 13. Normality Statistics for Two Datasets

	What to Look For	NORMAL DATA	LOGNORMAL DATA
Coefficient of Variation	If it's over 1 (100%), you might need to use a logarithmic transformation ***unless*** you have negative numbers in your data.	21.8 (0.30 without negative numbers)	1.08
Skewness	A skewness value of 0 indicates that the distribution is symmetrical. The standard error of the skewness can be used to test whether the skewness is significantly different from 0.	0.07	2.70
Standard Error of the Skewness		0.17	0.17
Kurtosis	A kurtosis value of 0 indicates that the distribution's tails are similar to a Normal distribution. The standard error of the kurtosis can be used to test whether the kurtosis is significantly different from 0.	-0.05	9.36
Standard Error of the Kurtosis		0.34	0.34
Kolmogorov-Smirnov Test Statistic*	The K-S test is a nonparametric test (based on ranks) that compares differences between a sample and a Normal distribution. If the probability is smaller than 0.01, the Normal distribution might not be a good fit for your sample.	0.04	0.22
Kolmogorov-Smirnov Probability		> 0.2	< 0.01
Shapiro-Wilk Test Statistic	The S-W test compares quantiles of the sample distribution to quantiles of the Normal distribution. The probability is interpreted in the same way as the K-S test.	0.99	0.71
Shapiro-Wilk Probability		0.87	0.00

* There are quite a few different tests of normality. They'll usually, but not always, give the same answers even though they might approach the comparison to a normal distribution in different ways.

Correlations

If you have more than one variable in your dataset, you'll want to look at correlation coefficients. The strength of the relationship between two variables is usually expressed by the Pearson Product Moment (parametric) correlation coefficient (denoted by r). Correlation coefficients range in value from -1.0 (i.e., a perfect correlation in which all measured points fall on a line having a negative slope) through 0.0 (i.e., absolutely no linear relationship between the variables) to +1.0 (i.e., a perfect correlation of points on a line having a positive slope). The square of the correlation coefficient (called the coefficient of determination) is an estimate of the proportion of variance in the dependent variable that is accounted for by the independent variable(s). The coefficient of determination, therefore, is an excellent statistic for interpreting the strength of the relationship between variables and comparing alternative models. Figure 17 shows four examples of data correlations.

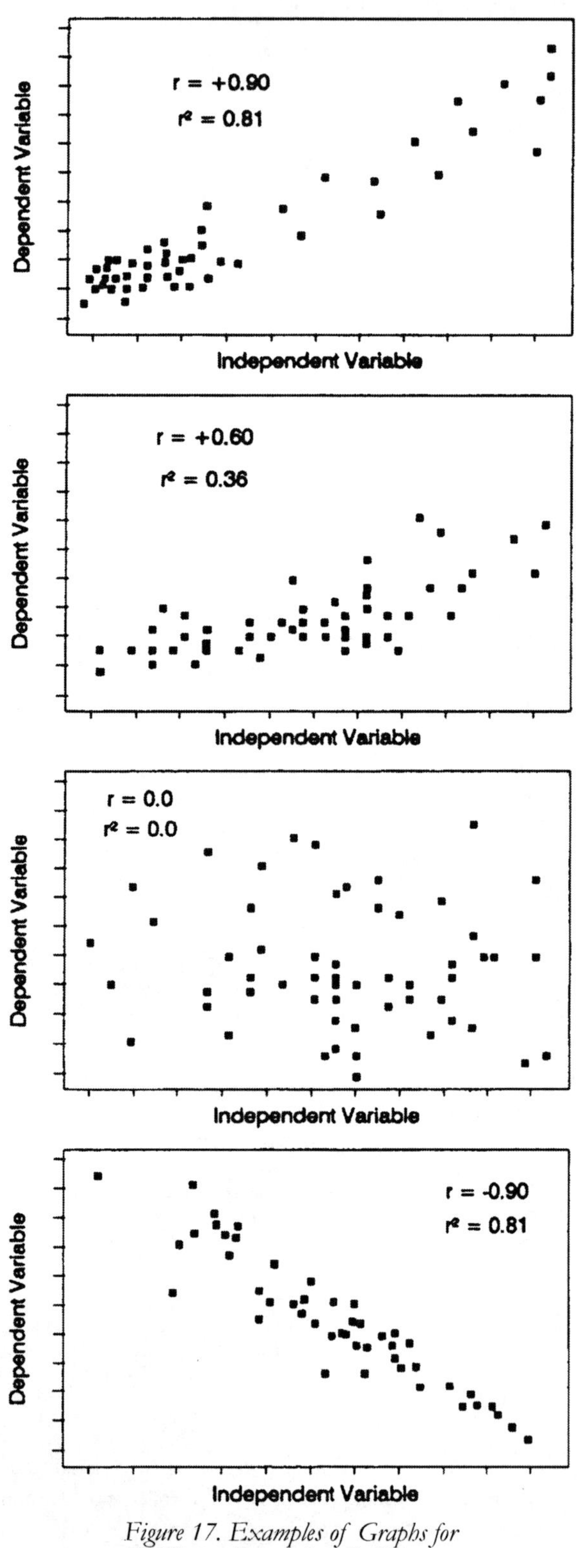

Figure 17. Examples of Graphs for Different Correlation Coefficients.

Correlation coefficients have a few pitfalls to be aware of. A large value of r does not necessarily guarantee that two variables are highly related. This is because the value calculated for r will tend to be inflated if there are only a few data pairs. Furthermore, the magnitude of r is very sensitive to the presence of nonlinear trends and outliers. Nonlinear trends cause the magnitude of the relationship to be underestimated. Outliers (i.e., not representative of the population) that are located perpendicular to the data trend cause the relationship to be underestimated. Outliers parallel to the data trend cause the relationship to be overestimated. A large correlation isn't necessarily a good thing. If you find that your predictor variables are highly correlated with your dependent variable, that's great. But if you find that your predictor variables are highly correlated with each other, that's not good, and you'll have to deal with this multicollinearity in your analysis.

The Pearson Product correlation coefficient is used when both variables are measured on a continuous (i.e., interval or ratio) scale. There are several variations of the Pearson Product correlation coefficients. The multiple correlation coefficient (denoted by R[9]) indicates the strength of the relationship between a dependent variable and two or more independent variables. The partial correlation coefficient indicates the strength of the relationship between a dependent variable and one or more independent variables with the effects of other independent variables held constant. The adjusted (or shrunken) correlation coefficient indicates the strength of a relationship between variables after correcting for the number of variables and the number of data points. There are also correlation coefficients for variables measured on noncontinuous scales. The Spearman R, for instance, is computed from ranks. Other correlation coefficients are mentioned in Chapter 22.

You call that a good correlation?

So, what's a good correlation? It depends on whom you ask.

I once asked a chemist who was calibrating a laboratory instrument versus a standard what value of the correlation coefficient she was looking for. "0.9 is too low. You need at least 0.98 or 0.99." She got the number from a government guidance document.

I once asked an engineer who was conducting a regression analysis of a treatment process what value of the correlation coefficient he was looking for. "Anything between 0.6 and 0.8 was good." His college professor told him this.

I once asked a biologist who was conducting an ANOVA of the size of field mice living in contaminated versus pristine soils what value of the correlation coefficient he was looking for. He didn't know, but his cutoff was 0.2 based on the smallest size difference his model could detect with the number of samples he had.

So, who was right? They all were. The meaningfulness of a correlation coefficient depends, in part, on the expectations of the modeler. But what if you don't have any clear expectations? How do you tell if a correlation is good? The answer comes in three parts:

1. **Value**—Square the correlation coefficient to get the coefficient of determina-

9 R also refers to a programming language, but not here.

tion (R-square). R-square is the proportion of variation in the dependent variable (y) that can be accounted for by the independent variable (x). You might be able to decide how good your correlation is from a gut feel for how much of the variability you wanted your model to account for. For example, correlation coefficient values between approximately -0.3 and +0.3 (less than 9 percent of the variance accounted for) might indicate weak or non-existent relationships; values between -0.3 and -0.6 or +0.3 and +0.6 (between 9 percent and 36 percent of the variance accounted for) might indicate weak to moderately strong relationships; values between -0.6 and -0.8 or +0.6 and +0.8 (between 36 percent and 64 percent of the variance accounted for) might indicate moderately strong to strong relationships; and values between -0.8 and -1.0 or +0.8 and +1.0 (more than 64 percent of the variance accounted for) might indicate very strong relationships.

Table 14. What to Look For in Correlations

What to Look For		Data to Use	Where to Look			
			Sign of r	Value of r	Plot	Test
Nature of Relationship	Is the trend linear?	All data from pairs of variables			X	
Direction of Relationship	Is the trend slope positive (rising) or negative (decreasing)?		X		X	
Strength of Relationship	How closely related are data points for two variables?			X	X	X
Outliers	Are there data points that are not representative of the population trend?				X	
Hidden Trends	Are there correlations in some data groupings but not others?	Data from pairs of variables from a single relevant group	X	X	X	X
Multi-collinearity	Are independent variables correlated with each other?	All data from pairs of independent variables	X	X	X	X
Auto-correlation	Are values of a dependent variable correlated with the order in which the variable was measured	A single dependent variable and the order the data were generated	X	X	X	X

2. **Significance**—Every calculated correlation coefficient is an estimate. The "real" value may be somewhat more or somewhat less. You can conduct a statistical test to determine if the correlation you calculated is different from zero (or some other number if it's relevant). The larger the calculated correlation and the greater the number of samples, the more likely the correlation will be significantly different from zero. For example, a correlation of 0.59 (R-square

of 0.35) would be significantly greater than zero based on about 25 samples, but a correlation of 0.01 wouldn't be significantly different from zero with 250 samples.

3. **Plots**—You should always plot the data used to calculate a correlation to ensure that the coefficient adequately represents the relationship. The data should be linearly related and free of outliers.

What makes a good correlation, then, depends on what your expectations are. If you are conducting exploratory research, maybe 0.2 isn't bad. If you are testing a known process having some natural variability, 0.6 might be acceptable. But if you are calibrating instruments, maybe 0.9 isn't good enough. So you have to consider correlations on a case-by-case basis. Remember too that "no relationship" may also be an important finding. But you can't just look at the numbers. You have to consider the plots and the tests too. Table 14 provides a summary of what you can look for when assessing correlations.

It was Professor Plot in the Diagram with a Graph

Correlation is not causation but it sure is a hint.

Edward Tufte in *Beautiful Evidence,* p. 159

You probably were taught how to graph data in high school. Depending on your work, you may frequently plot data yourself or look at graphs prepared by others. Even if you don't use graphs on your job, you may run into them during your leisure time, reading the newspaper, managing your finances, or playing D&D.[10] But there's a big difference between looking at someone else's graph and preparing one yourself. When you were learning how to graph in school, the teachers told you what kind of graph to use. They gave you carefully selected data that was matched to the graph you were supposed to create. There was help available if you had any questions. Now, it's just you and your computer. If you have no clue as to where to begin, here are a few tips that may help.

First let's get past the jargon of plots, charts, graphs, and diagrams. All of these terms are defined as visual representations of data. All are used synonymously. All are used as both nouns and verbs. All have other meanings. To split hairs:

- Plots tend to place more emphasis on individual data points.
- Charts tend to involve lines and areas more than individual points.
- Graphs tend to be more mathematically complex than charts and plots.
- Diagrams tend to be more artistic and fill the entire data space.

Not everyone would agree with this, of course. That being said, you can usually refer

---

10 Dungeons and Dragons, the first and most successful fantasy role-playing game distinguished for its integral use of statistics (for examples, see www.monkeysushi.net/gaming/DnD/math.html).

to visual representations of data by any of the four terms without being called out by a smart-aleck critic. If you're referring to a *specific* kind of visual representations of data, one of the four terms usually is preferred, for example, bar charts, scatter plots, and block diagrams. Most specific kinds of visual representations of data are called plots or charts, and to a much lesser extent, diagrams. The term graph is used mostly in a general sense, which is how it is used in this chapter.

A Graph a Minute

The first thing you'll need to do is figure out what kinds of graphs you could draw. Start by answering these questions:

- Is your focus on variables or samples? Do you want to show how a number of samples are related to each other on the basis of one or more variables or do you want to show how a number of variables are related to each other for a very small number of samples?
- Will you plot individual points or group means? How many data points do you have to plot? Do you want to show the points individually or do you want to show the averages of groups of data points (this is useful when you have a large number of data points)?
- What is the aim of the graph? There are many reasons to plot data and most graphs have multiple goals. For simplicity, decide whether the primary aim is to show:
    - » Data frequency and distribution
    - » Relative proportions of the components of a mixture
    - » Properties or values of a variable
    - » Trends, patterns, or other relationships among variables.
- How many axes will you need? How many variables do you have? Are they measured on the same or different scales? Are the scales discrete or continuous?

Once you can answer those questions, you can use Table 15 to help you choose some of the more common kinds of graphs to try with your data. There are, of course, a virtually uncountable number of kinds of graphs, subspecies of graphs, variations and extensions of graphs, and combinations of graphs. For now, focus on simple graphs you can get from the software you have available. Later, you can prepare the Piper plots you used to justify your purchase of that specialized piece of software you wanted. Table 16 provides descriptions of the graphs listed in Table 15 and some tips for what to look for in interpreting the graphs.

Table 15. Types of Graphs for Analyzing Data

Chart	Used to Show	Chart Axes	Data Scales		
			Horizontal Axis	**Vertical Axis**	**Additional Axes**
Box Plot	Distribution	Rectangular	Categorical, continuous (sample size)	Continuous	
Dot Plot	Distribution	Rectangular	Ordinal, continuous	Ordinal	
Histogram	Distribution	Rectangular	Ordinal, continuous	Ordinal	
Probability Plot	Distribution	Rectangular	Ordinal, continuous	Continuous	
Q-Q Plot	Distribution	Rectangular	Ordinal	Ordinal	
Stem-Leaf Diagram	Distribution	Rectangular	Ordinal	Ordinal, continuous	
Ternary Plot	Mixtures	Triangular	Continuous (percentages)	Continuous (percentages)	Continuous (Percentages)
Pie Chart	Mixtures	Circular	Categorical	Continuous (percentages)	
Area Chart	Properties	Rectangular	Ordinal, continuous	Continuous	
Bar Chart	Properties	Rectangular	Categorical	Continuous	
Candlestick Chart	Properties	Rectangular	Continuous	Continuous	
Control Chart	Properties	Rectangular	Continuous	Continuous	
Deviation Plot	Properties	Rectangular	Continuous	Continuous	
Line Chart	Properties	Rectangular	Categorical, ordinal	Continuous	
Map	Properties	Rectangular	Continuous	Continuous	Any
Matrix Plot	Properties	Rectangular	Nominal	Nominal	Text
Means Plot	Properties	Rectangular	Continuous	Continuous	
Spread Plot	Properties	Rectangular	Continuous	Continuous	
Block Diagram	Properties	Cubic	Nominal	Nominal	Nominal

Chart	Used to Show	Chart Axes	Data Scales		
			Horizontal Axis	Vertical Axis	Additional Axes
Rose Diagram	Properties	Circular	Ordinal, continuous	Continuous	
Multivariable Plot	Relationships	Rectangular, circular, other	Any	Continuous	Continuous
Bubble Plot	Relationships	Rectangular	Continuous	Continuous	Continuous
Contour Plot	Relationships	Rectangular	Continuous	Continuous	Continuous
Icon Plot	Relationships	Rectangular	Continuous	Continuous	Multivariable plot'
Scatter Plot: 2D	Relationships	Rectangular	Continuous	Continuous	
Scatter Plot: 3D	Relationships	Cubic	Continuous	Continuous	Continuous
Surface Plot	Relationships	Cubic	Continuous	Continuous	Continuous

Table 16. What to Look for in Charts

Chart	Description and Interpretation
Area Chart	Stacked bars and area charts can represent raw quantities, raw percentages, or percentages normalized to 100%. Be sure you know which is which. The chart will look different depending on the positions of the data series. Examine all possibilities. Avoid unnecessary additional dimensions.
Bar Chart	Bar charts generally represent quantities. Side-by-side bar charts are used to compare quantities in different groups. Limit bar graphs to two dimensions and limit the number of bars per cluster of groups as it is difficult to detect patterns with more than a few bars.
Block Diagram	Block diagrams usually represent qualitative information, like a three-dimensional map. Look at areas between points of confirmed information to assess the validity of the diagram.
Box Plot	Box plots represent a variable's central tendency and spread using a rectangular box and protruding lines, called whiskers. They are very good for showing data symmetry and outliers and have many options for incorporating other descriptive statistics. A long box indicates a platykurtic distribution. Short whiskers indicate truncated tails. Look for outliers plotted as asterisks.
Bubble Plot	These graphs attempt to represent a third dimension on a 2D scatter plot. Bubble plots use the size or area of circles to represent the third dimension. Patterns may be easier to identify than in 3D scatter plots.
Candlestick Chart	These are charts in which the x-axis represents time and the y-axis represents the some variable with vertical lines at each data point to represent data spread. In stock charts, the data point represents the average stock price and the vertical lines represent high and low prices or open and close prices.

Contour Plot	Contour plots use lines of constant value, called isopleths to represent a third dimension on a 2D scatter plot. Patterns are usually easier to identify in contour plots once you're accustomed to viewing isopleths.
Control Chart	Control charts are used to monitor the consistency of some process, usually an industrial process like manufacturing. There are many varieties of control charts, most of which use time on the x-axis and some measurement of quality on the y-axis. Horizontal lines represent the limits the process is allowed to vary within. There are many patterns to look for in control charts to evaluate process drift, shocks, and defects.
Deviation Plot	This is a plot of the differences between data and some reference point (e.g., zero or the data mean). Look for large positive and negative values.
Dot Plot	Dot plots use dots instead of histogram bars or stem-leaf data values. Each data value can be seen as in stem-leaf diagrams and sometimes the dots are plotted on both sides of the x-axis to provide a different view of data frequency.
Histogram	Histograms depict the frequency distribution of a variable using bars to show the number of data points in an interval. Most people are familiar with histograms. Look for the bars to roughly follow the form of some theoretical distribution.
Icon Plot	Icon plots are multivariable plots placed on a scatter plot to provide additional information. Look for patterns.
Line Chart	Line graphs have a continuous-scale variable on the y-axis and a discrete-scale variable on the x-axis. Be cognizant of the measurement scale of the x-axis variable (i.e., nominal vs. ordinal) Look for trends, both linear and nonlinear, and possible outliers. If appropriate, color-code data groupings to identify hidden trends and data distribution patterns. Label samples to check for patterns attributable to the collection order.
Map	A map is nothing more than a scatter plot of two geographic location coordinates on which is represented location-dependent data, like topographic or cultural features. Everybody has seen a map and some people even know how to interpret them.
Matrix Plot	A matrix plot is like a spreadsheet having only a few rows and columns, into which is placed information. Usually the information is qualitative, summary-level, and textual. Matrix plots are sometimes called windowpanes.
Means Plot	This is a plot of group means for two variables. Compare locations of the group means. This is essential for any ANOVA model.
Multivariable Plot	These plots try to represent more than three dimensions with axes that are less than 90 degrees apart. They can be full-sized plots or icons placed on a 2D graph. Generally, only patterns can be discerned rather than data values.
Pie Chart	Must represent 100% of a quantity. Be wary of tilted, 3D, or exploded pies as they can exaggerate areas. Too many slices can be confusing especially if not labeled adequately
Probability Plot	Probability plots use data values to compare a data distribution to a model distribution. If the data fits the model, the plot should lie in a straight line. Look for where there are deviations from the line.
Q-Q Plot	Q-Q plots use small groupings, called quantiles, of both the data and the model to compare a data distribution to a model distribution. They are a robust alternative to probability plots.
Rose Diagram	Rose diagrams are like bar charts for orientation data. Like other polar graphs, they are designed to show patterns more than data values. They are most commonly used to depict wind directions and other types of data measured on a compass, but have been applied to other types of repeating scales (even music, http://c0573862.cdn.cloudfiles.rackspacecloud.com/1/0/4829/72772/Keys2.5-webfull_o.gif).

Chart	Description and Interpretation
Scatter Plot: 2D	Scatter plots have continuous-scale variables on both the y-axis and the x-axis. Look for trends, both linear and nonlinear, and possible outliers. If appropriate, color-code data groupings to identify hidden trends and data distribution patterns. Label samples to check for patterns attributable to the collection order. Compare data plot to correlation coefficient.
Scatter Plot: 3D	3D scatter plots represent data points on three (i.e., x, y, z) axes. Look for the same patterns as for 2D scatter plots from a variety of perspectives. Many people have difficulties understanding these plots, so consider your audience if you plan to use one in a report.
Spread Plot	This is a plot of the standard deviation, range, or some other measure of data spread versus a variable. Usually the purpose is to ensure that the variability is relatively constant over the range of the variable (homoskedasticity).
Stem-Leaf Diagram	Stem-leaf diagrams are similar to histograms and interpreted the same way but they use the actually data values to provide more information on data modes.
Surface Plot	Surface plots use three dimensional grids to facilitate recognition of x,y,z data point locations. The "topography" of the grid represents the values of the z variable being portrayed. If x and y are location coordinates, the plot is a type of relief map in which the z variable can be land elevation, a contaminant concentration, or any other regionalized variable. Like any three-dimensional figure presented in two dimensions, surface plots can be difficult to interpret at first.
Ternary Plot	Ternary plots are triangles in which each leg of the triangle represents a variable axis. The catch is that the three variable values for the samples must add up to a constant. These plots are usually used to represent mixtures of three components. Ternary solid plots add a fourth axis to represent an additional linear dimension. Don't expect anyone to understand them.

* (e.g., Radar Plot, Sun Chart, Star Plot, Side-by-side bar charts, Polygon Plot, Sparklines, Chernoff faces)

Figures 18, 19, and 20 show some of the graphs available in commercially available software. Figure 18 shows the types of graphs in Excel 2007. As with most software that provide graphing capabilities, Excel provides pie charts, bar charts, line charts, and scatter plots. Excel also provides surface plots, bubble plots, radar charts, and stock (a.k.a. candlestick) charts, which most people don't learn about in school because they are harder to draw by hand. Figure 19 shows the types of graphs in SPSS 14. SPSS provides most of the graphs Excel does plus box and whisker plots, probability plots, 3-D scatter plots, time-series charts, and control charts. Most major statistical software provide these graphs. Figure 20 shows the types of graphs in Statistica 9. Statistica provides most of the plots SPSS does plus matrix plots, icon plots, and categorized plots. Specialized software for statistical graphics, such as Origin and Sigmaplot, provide a few more specialized types of charts and many more capabilities for customizing the plots.

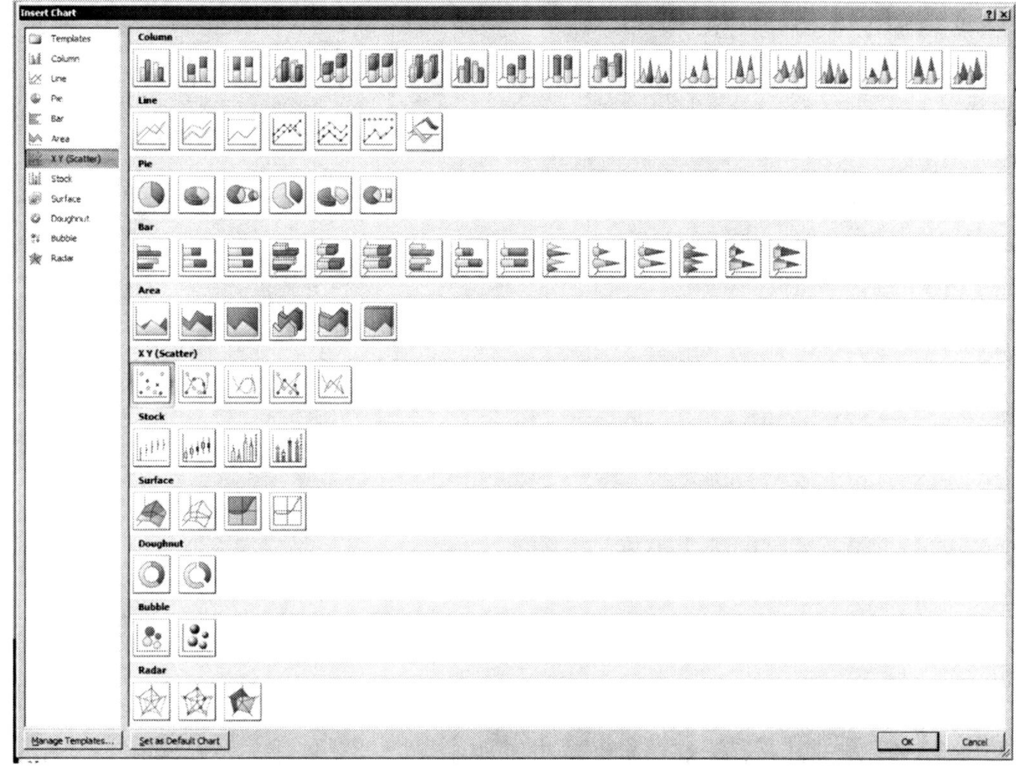

Figure 18. Types of Graphs Available in Excel 2007.

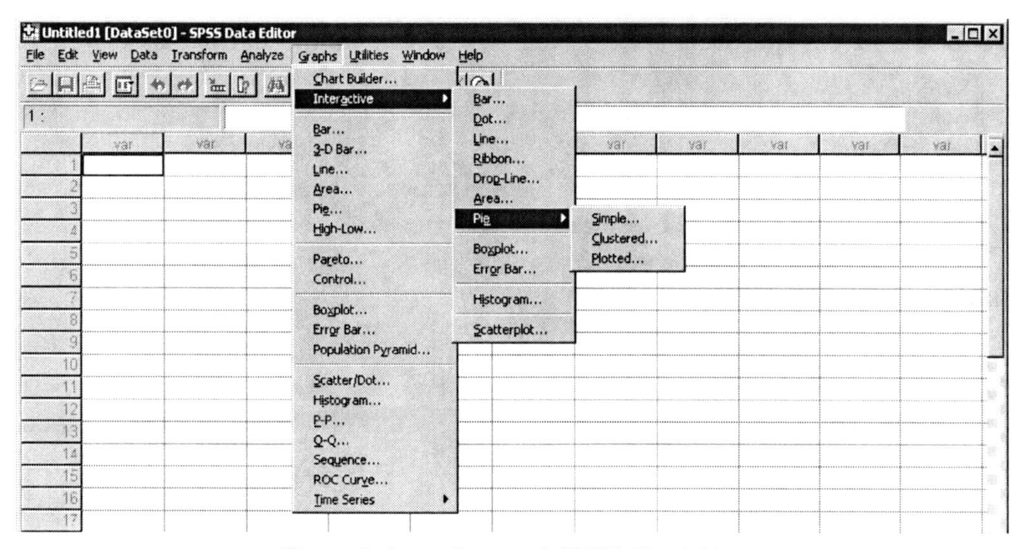

Figure 19. Screen Capture of SPSS's Graph Menu.

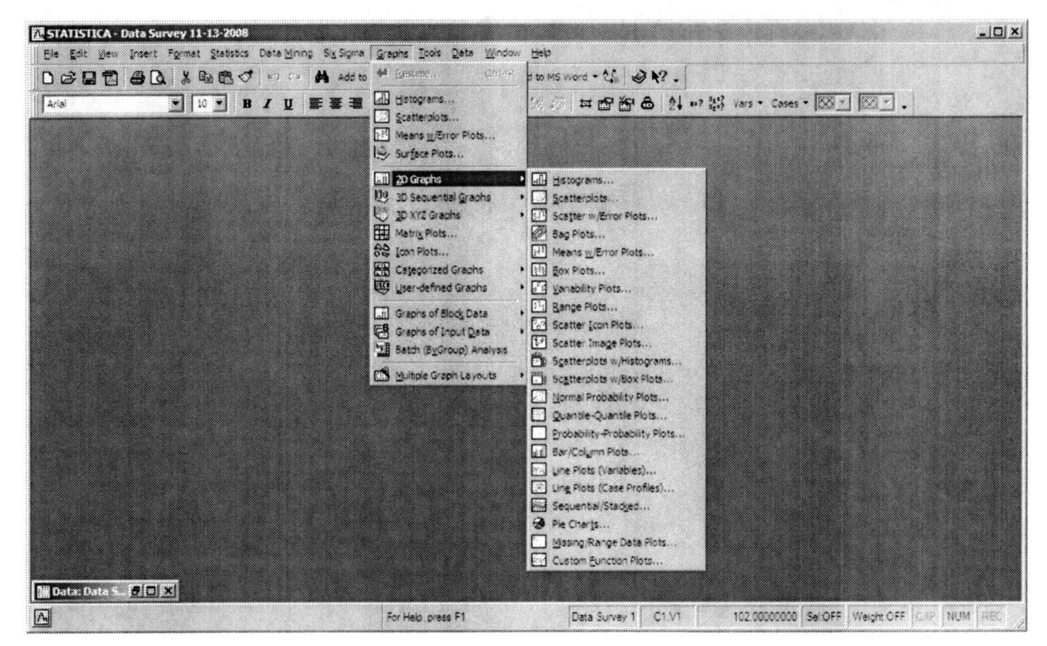

Figure 20. Screen Capture of Statistica's Graph Menu.

You Can't Spell Chart without Art

There are competing philosophies of graphing, divided to some extent by perceptions about the audience for a graph. The philosophy of many art directors of newspapers and magazines is to keep the graph simple, interesting, and attractive in order to engage the reader. Look no further than *USA Today*, *Newsweek*, or *Time* to see three-dimensional exploded pie charts and bar charts made of little soldier icons or dollar bills or some other cutesy graphic.

In contrast, Edward Tufte, perhaps the preeminent expert in informational graphics, espouses a philosophy that assumes the audience is knowledgeable and interested. Graphs should provide as much information as needed as efficiently as possible. Tufte makes many good points in his books[11] such as:

- The dimension of a chart must not be greater than the dimension of the data. For example, if you're plotting two variables on a Cartesian (rectangular) graph, don't add an extra axis (dimension) for depth. It may be visually appealing but it's scientifically misleading.

- Data must be presented in context. You shouldn't show just part of a dataset. Label everything you need to make sure the data are presented accurately and meaningfully.

- Maximize the data density and the data-ink ratio. Put enough data in your graph

11 Books on information graphics written by Edward R. Tufte include: *The Visual Display of Quantitative Information (1983, 2001), Envisioning Information (1990), Visual Explanations (1997), and Beautiful Evidence (2006).*

to make it worthwhile. Eliminate everything on the chart that isn't data or contributes to the interpretation of the data.

- Eliminate chart junk, the unnecessary pictures, dimensionality, grid lines, fill patterns, and other objects that clutter a graph while adding no scientific value.

Tufte believes he has the audience's attention while the art directors believe they have to compete for it. Then there are authors like David McCandless (www.informationisbeautiful.net) who look at presenting data from an artistic perspective. Their graphics are truly works of art though the graphs are based on data and aimed at certain audiences. All of these graph developers make valid points. They simply have different perspectives, different audiences, different aims, and different data.

Hit Me with Your Best Plot

Sometimes you'll use graphs to illustrate attributes of a single sample or variable. More often, a graph will involve many samples and several variables. Occasionally a graph will contain all the information needed to tell the story of an entire analysis. Graphs are tools. If you only have a hammer and a screwdriver in your toolbox, you won't be able to accomplish very much. You'll want to be able to pull out a 14-inch Stillson wrench and 45-degree external snap ring pliers when you need them.

At this point in your analysis, don't worry about the attractiveness of your graphs. Produce them honestly so that you can understand your data. When you decide which ones you need to tell your data's story, you can focus on making them appropriate for your audience. That's where the art of the chart comes in.

In building your graphs into a story, there are a few challenges you might face, including:

- Too many samples
- Too many variables
- Too many variable scales.

Too Many Samples

There aren't many instances in which statisticians get to complain about having too many samples, but trying to graph thousands of data points is one of those occasions. You have two choices. You can take a sample of your sample or graph the means of data groups.

Some software, like SPSS and SAS, have utilities for taking a sampling of a dataset, but you can do it easily in a spreadsheet. Sort your data by the variables you plan to plot and then select data points at an interval that will provide a workable number of points. Prepare several of these graphs using the same interval but different starting points. Any patterns you see should be replicated in all the graphs. If you plan on labeling each data point, factor the space requirements into your selection of an interval. Two hundred points may fit on your graph, but you may not be able to fit even fifty with labels.

If your objective is to focus on data groups rather than individual samples, the solution to having too many samples is easy. Calculate the means for your group classes and plot them. You could even put means for the classes of several different groupings on the same graph. For example, the means for sex, race, education level, and income bracket still wouldn't be very many points to plot, hence the statistician's old adage *He who graphs class, graphs less.* If you go this route, don't limit yourself to class means for two variables. Try plotting other descriptive statistics, or for a single variable, the mean versus the standard deviation or the coefficient of variation versus the sample size.

Too Many Variables

Having many variables to explore simultaneously is a common challenge. If you don't have too many data points, you can try matrix plots. Matrix points are small scatter plots of a set of variables, taken two at a time, arrayed as a matrix. Because of their small size, the plots turn to mud if you have a lot of data points. They can be good, though, for variable triage, quickly wading through many variable relationships so that you can focus on just a few with the strongest trends.

Table 17. Sample Data for Icon Plots

	Variables									
Samples	**1**	**2**	**3**	**4**	**5**	**6**	**7**	**8**	**9**	**10**
1	3.22	4.37	5.05	8.23	9.53	13.16	8.40	15.92	13.17	15.28
2	3.02	4.15	7.41	11.28	13.85	10.33	13.31	13.48	16.44	23.59
3	4.09	8.88	7.75	12.89	10.94	10.63	13.40	12.24	22.88	16.07
4	9.39	8.28	10.60	10.40	16.04	11.86	19.52	19.62	21.78	24.23
5	10.26	12.08	14.40	13.24	19.11	13.77	23.50	17.50	14.52	27.31
6	13.62	10.82	10.68	19.52	12.74	22.70	14.31	17.48	20.86	30.29
7	9.46	10.11	11.31	11.17	18.90	16.36	17.21	20.18	17.65	19.48

If you have a few variables that you want to show on the same graph, there are several methods you can use. The easiest and most readily available approach is to vary the types, sizes, and colors of symbols used to represent points. When doing this, be sure to decide what the focus of the graph is, in other words, what is the order of importance of the variables. Your most important variable should be represented by large, solid, more pointed, red symbols, while your least important variables should be represented by small, open, more rounded, blue or light-colored symbols.

Table 18. Types of Icons Used to Display Sample Characteristics.

Description	Icons
Pie-chart icons show proportions as angles of the circle. It's difficult to identify patterns if there are more than a few variables. If you can't use color, you're at an added disadvantage.	
Sun ray icons can show actual values as well as percentages. Each variable is assigned a ray of the sun as an axis. The rays are set apart at equal angles so information is conveyed using shape rather than color or pattern.	
Star plot icons are similar to sun ray icons except that the axes (sun rays) outside the data values are omitted so every ray may be a different length. This difference allows patterns to be identified more easily at the expense of some perception of the values of the variables.	
Polygon icons take the objective of star plots one step further by eliminating all axes and filling in the resulting shape. Polygons show patterns very well but provide little sense of the values of the variables.	
Bar chart icons have the same problem as pie chart icons; as the number of bars increase, it becomes harder to distinguish patterns in the data especially if you can't use color.	
Profile icons make pattern detection easier by eliminating the bars and replacing them with the shape the bars formed. This difference is analogous to the difference between star icons and polygon icons.	
Sparklines are profile icons with the frame and the fill eliminated. They emphasize data patterns over data values.	
Chernoff face icons replace conventional graph axes by facial characteristics. It's easy to detect differences between faces but difficult to interpret patterns in the differences.	

As the number of variables on a graph increases or the data points begin to overwrite each other, using different symbols becomes ineffective. One alternative is to place one or more plots within the main plot, the best example of which are icon plots. An icon plot consists of a foundation graph, usually a Cartesian graph like a scatter plot or a map, on which the data points are represented by multivariable icons instead of simple symbols. Data icons can be tiny bar graphs, polygons, profile lines, pie charts, star plots, sun plots, or other data representation.

Consider the dataset shown in Table 17. This dataset could be represented by tiny pie charts or bar charts for each sample, but with 10 variables it would be difficult to see patterns. Pie chart icons can be simplified as sun, star, and polygon icons. Bar chart icons can be simplified into profile icons and sparklines. Examples of these types of icons are shown in Table 18 (previous page).

Data patterns can also be represented by faces. Because humans are accustomed to distinguishing subtle differences in human faces, zany statistician Herman Chernoff reasoned that representing data values using facial characteristics would facilitate identifying patterns. He developed a procedure to replace conventional graph axes by such characteristics as head size and shape, and the positions, lengths, and angles of the eyebrows, eyes, nose, ears, and mouth. Hence, Chernoff faces... proof that the pre-disco 1970s were a lot more fun than anyone can remember.

So if you need to express the relationships between many variables for a relatively small number of samples, consider an icon plot. Pick the two most important variables and make them the axes for the foundation graph. Then pick an appropriate icon type for the remaining variables and plot them on the foundation graph in place of data points. Figure 21 illustrates an icon plot for the data in Table 17.

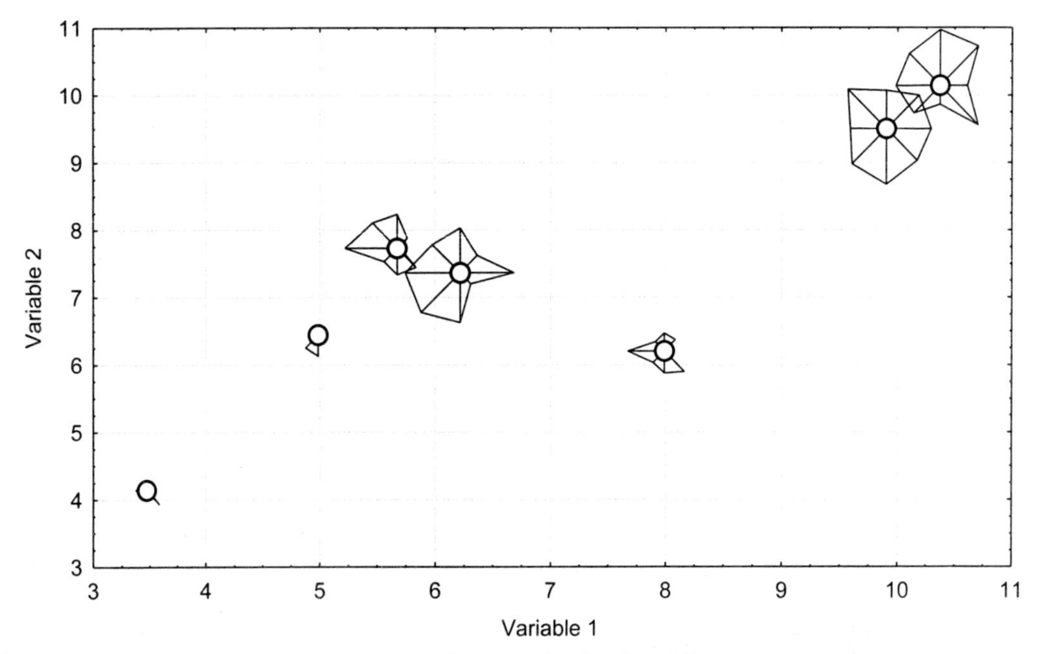

Figure 21. Example of a Star-Icon Plot.

The number of samples on the foundation graph must be small or the graph will be too confusing. If you do have many samples, plot the means of some relevant data groupings, or limit the span of your graph so that the information is readable. Just be sure you don't mislead your readers or yourself by limiting the data you show.

Too Many Scales

Maybe you don't just have a lot of variables, but each one has a different scale of measurement. There are a variety of things you can try. The simplest thing you can do is add a second axis to the graph space, like a second y-axis. Most spreadsheet software supports this capability. If appropriate, you could also add a third dimension with a z-axis. Some people have a hard time understanding data presented in three dimensions, so this shouldn't be your first course of action. Another approach to consider is converting the variables to the same units. The conversion might involve familiar units, such as converting cubic feet of a liquid to gallons, or specialized relationships, such as converting inches of snowfall to inches of rainfall. Calculating percentages is another way to convert values to a common scale.

Another problem that occurs is that two variables use the same unit of measurement but have values on very different ranges of the scale. For example, if you were to plot your personal monthly expenditures over time, you might find that the mortgage or rent payment is so large compared to everything else you can't see the increasing trend in your automobile expenses. Again, there are several things to try. You could add a second axis. You could try using a logarithmic scale, although sometimes this just shifts problems from one end of the scale to the other.

A different approach to consider is to multiply or divide variables by constants so that all the variables fall in the same range. Factors of ten are used commonly for this purpose, although the constants could just as well be 5 for one variable and 20 for another. Dividing each data value by the variable's minimum data value will produce a scale that ranges from one (assuming there are no negative values) to the maximum divided by the minimum. Dividing each data value by the maximum data value will produce a scale that ranges from near zero (assuming there are no negative values) to one. You can also convert your values to z-scores by subtracting the average of the data from the value and dividing by the standard deviation. Z-scores don't have a limited range, but if your data are Normally distributed, 99 percent of the values will fall into the six-sigma range -3 to +3. These scale changes will not affect patterns of relationships between variable even though the data values change.

It may not be easy at first to sort through all these options, but with practice, you'll be able to pick the approach that will make the data most understandable for the audience.

Graphs You Gotta Look At

In practically every case, you'll need to look at two and maybe three types of graphs: distribution graphs, variable scatter plots, and sample location or time-series plots. The next sections explain what these graphs are and why you have to look at them.

Distribution Graphs

Even if all you plan to do is calculate a few descriptive statistics, you ought to check the frequency distribution[12] of the data. You want to look for:

- **Shape**—Are there an equal number of values in each interval (a uniform distribution) or is there a cluster of data around a particular value (a unimodal distribution), or is there more than one cluster of data modes?
- **Symmetry**—Do the data fall symmetrically around the distribution's center? Symmetry is the property measured by skewness.
- **Central Tendency**—Where is the center of the distribution? This is the property measured by the median. If your frequency distribution isn't symmetrical, your calculated mean may be misleading.
- **Dispersion**—How much are the data spread out. Dispersion is the property measured by the variance or standard deviation. If your frequency distribution isn't unimodal and symmetrical, your calculated variance may be misleading.
- **Outliers**—Are there any data points located far away from the rest? Outliers will affect virtually any statistic you calculate.
- **Model**—If you know what frequency distributions of mathematical models look like, you can get a sense of which model might be a good approximation for your data. If you can identify a theoretical distribution, like the Normal distribution, that is a good fit for your data you can use that distribution's equation to make predictions about the population your sample is from. This makes statistics a lot of fun.

Frequency Diagrams. There are quite a few different types of graphs that will provide information about your data's frequency distribution. Many are only available in statistical software. The most readily available type of graph for frequency distributions is the histogram. Histograms are available in even low-end statistical software as well as some spreadsheet programs like Excel.

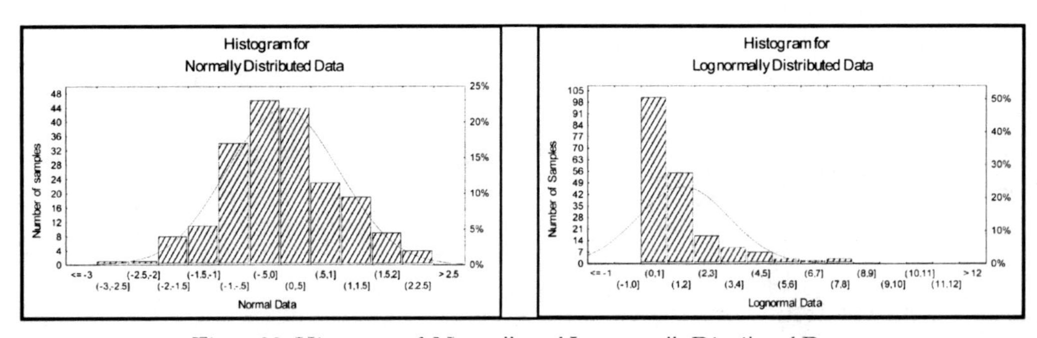

Figure 22. Histograms of Normally and Lognormally Distributed Data.

12 Sort your data, divide the data into equal-sized intervals, and then plot the number of data in each interval versus the average value of the interval. That's a frequency distribution.

Figure 22 presents two histograms for the data characterized in Tables 12 and 13. In the figure, the horizontal axis represents intervals of the variable scale. The vertical axis represents the number of samples. The heights of the bars represent the number of samples in each interval. The line represents the number of samples expected in a Normal distribution. The closer the bars follow the shape of the line, the more likely it is that the sample was drawn from a Normally distributed population.

In the histogram on the left, representing the raw data, the histogram bars are unimodal, mostly symmetrical, and there are no apparent outliers. The distribution has a crude bell shape. Some people would interpret this as "close enough" to a Normal distribution. Others would disagree citing the lack of symmetry.[13] In contrast, the histogram on the right, representing the logarithms of the raw data, is unimodal but the majority of the data are bunched on the left side of the histogram. This is termed *skewed left* or *skewed low* (in reference to the lower data values). These data are clearly not Normally distributed.

A dataset won't necessarily have a unique histogram. If the dataset is large enough, there may be considerable flexibility in the selection of the data intervals for the bars. Depending on whether there are two practical sets of data intervals or two dozen sets of intervals, the histogram can look deceptively different. That's why their interpretation can be very subjective.

Another type of distribution plot is a stem-leaf diagram. It's a frequency diagram like a histogram in which the bars have been replaced with the individual data values. The process of creating a stem-leaf diagram is somewhat complicated and is usually done by software. What happens is that each data value is split into leading (leftmost) digits, called the stem, and one or two trailing (rightmost) digits, called the leaf. In the stem-leaf diagram shown in Figure 23, for example, the third row of the top box represents the data values 2.0, 2.1, 2.3, 2.3, 2.3, 2.4, 2.4, and 2.4. Thus, stem-leaf diagrams provide more information than histograms even if they are incredibly ugly.

There's nothing that says you have to use bars or columns to express the distribution of values. Dot plots replace the histogram bars with a single dot for each scale interval. Because dot plots are less cluttered, you can add a lot of additional information to the graphs.

Box Plots. Histograms are good for visualizing the form of data distributions and for comparing them to a theoretical model like the Normal distribution. They're not quite so good for looking at central tendency, dispersion, and outliers. That's what box plots do particularly well.

Box plots consist of two main parts, a rectangular box with lines extending from the ends. Hence, they are also known as box and whiskers plots. The length of the box represents some measure of data dispersion, usually the standard deviation or the interquartile range, and is drawn to the scale of the variable. Within the box, there may be a line or a point representing the median or the mean, or both. Sometimes the width of the box is drawn to a scale representing the sample size. The lines expending from the

---

13 Be careful in selecting data intervals for your histograms. Different interval sizes and starting points can alter the appearance of the graph.

	Number in Class	Percentiles
-3° 3 · · · ·	1	min = -2.62
-3° 9 · · · ·	1	
-2° 01333444 · · · ·	8	
-2° 55556788889 · · ·	11	
-1° 0000001111111122222222333334444444 ·	34	25%
-1° 5555555555666667777777888888888888899999999999	46	median
0° 00000000000011222223333333333333444444444444	44	
0° 55555666666777777888999 · ·	23	75%
1° 0000011122233333444· · ·	19	
1° 555667889· · · ·	9	
2° 0024 · · · ·	4	max = 2.46
one leaf = 1 case	**200**	**Total**

	Number in Class	Percentiles
0° 0111222223333334444444 · ·	43	min = .07280
0° 55555566666677777888888999999 ·	58	25%
1° 00000012233344444 · · ·	33	median
1° 55556667999 · · ·	22	75%
2 ° 001234 · · · ·	10	
2° 56899 · · · ·	7	
3° 1234 · · · ·	5	
3° 5799 · · · ·	5	
4° 012 · · · ·	4	
4° 55 · · · ·	3	
5° 13 · · · ·	2	
5° 8 · · · ·	1	
6° 1 · · · ·	1	
6° 6 · · · ·	1	
7° 0 · · · ·	1	
7 89 · · · ·	2	
8° · · · ·	0	
8° · · · ·	0	
9° 1 · · · ·	1	
9° · · · ·	0	
10° · · · ·	0	
10° · · · ·	0	
11° · · · ·	0	
11° 7 · · · ·	1	max = 11.7265
one leaf = 2 cases	**200**	**Total**

Figure 23. Stem-and-Leaf Plots for Normal (top) and Lognormal (bottom) Data.

box usually indicate the data range, although sometimes they represent some statistical interval and points or stars are used to indicate outlier data values.

Box plots are especially useful for comparing a variable's central tendency and dispersion between different groupings of samples. Depending on how the box plot is drawn, you might be able to compare means to medians, the median to the 75 percent quartile, or the 95 percent confidence interval for the mean to the maximum.

I've got the whiskers if you have the box.

Box plots for the example data are shown in Figure 24. These boxes represent the interquartile range (i.e., 75 percent quartile minus the 25 percent quartile) and the whiskers represent the data range. The vertical axis represents the variable's scale. The box plot for the Normal distribution, on the left side of Figure 24 is symmetrical from top to bottom while the box plot for the Lognormal distribution, on the right side of the figure, is noticeably asymmetrical.

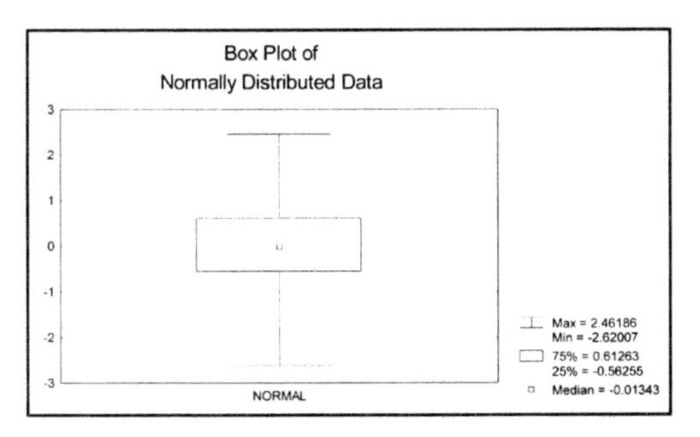

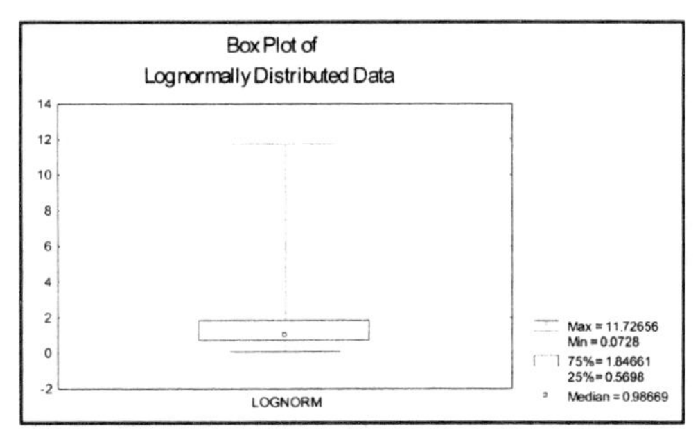

Figure 24. Box Plots for Normal and Lognormal Data.

Probably the major obstacle to using box plots is that you need specialized software to create them. All major statistical software packages will create box plots, but they don't all provide comprehensive options, like scaling for sample size. You can create box plots, or at least something approaching a box plot, with some spreadsheet packages, but it's not as easy as just checking a box like with statistical software.

Probability Plots. If you want a visual comparison of your data points to some theoretical distribution, look at a probability plot. In a probability plot, your data are ranked and plotted as a cumulative frequency versus what would be expected from the theoretical distribution you are comparing your data to. *Huh*, you say? Don't worry about it. There are two reasons why. First, like box plots, you're not likely to be able to create a probability plot without statistical software. Second, if you have statistical software that can generate probability plots, the interpretation is relatively easy. Your plotted data points should fall on the straight line representing the theoretical distribution you are comparing the data to. The farther away the data points are, the poorer the fit.

Figure 25 provides probability plots for the example data. The plotted points for the Normally distributed data fall very close to the line representing the theoretical Normal distribution. In contrast, the plotted points for the Lognormally distributed data do not consistently fall close to the line representing the theoretical Normal distribution. If you are mostly interested in central tendency, make sure the data don't deviate too much from the theoretical distribution line (as the Lognormal data do) or your calculated means may be inaccurate. If you are looking at statistical tests, look at the ends (i.e., the tails) of the plot. If the data and the distribution don't match in the tails, the calculated probabilities will be inaccurate.

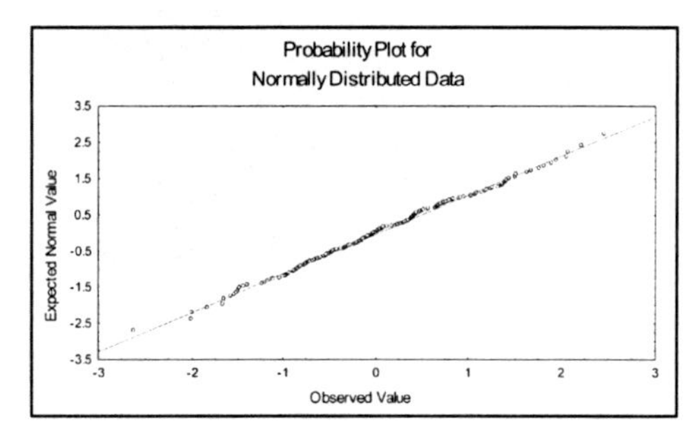

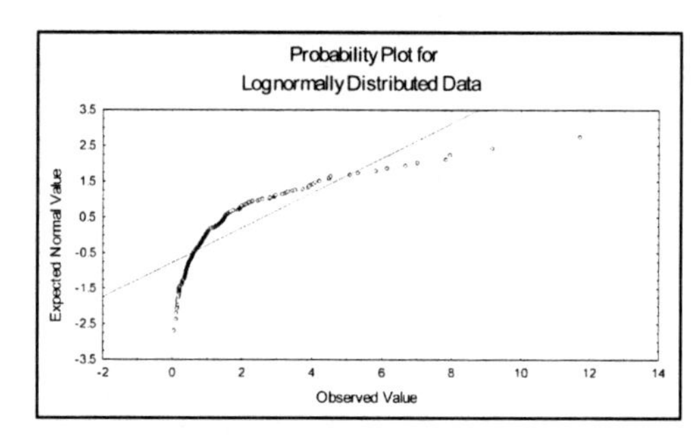

Figure 25. Probability Plots for Normal and Lognormal Data.

Now for the more scary part. Like box plots, there are many variations of probability plots. The term *probability plot* usually refers to what is called a Q-Q plot or quantile-quantile plot because data are grouped into regular intervals along the variable scale called quantiles. Also, probability plots usually compare a data distribution to the Normal distribution. That's not always the case, however. Probability plots can compare any two distributions, either empirical data or theoretical models. There are also important variations on probability plots. Detrended probability plots show the differences between two distributions on the y-axis. The deviations should plot along a horizontal line at 0.0 on the y-axis. If there are any patterns in the plotted points, excessive scatter, or shifts from the zero line, there are likely to be important differences between the distributions.

Bivariate Plots

Bivariate plots are plots of one variable versus another, each with its own scale on a separate axis. You might know them as two-way plots, x-y plots, 2D plots, scatter plots, line plots, or another of their many monikers. In an exploratory data analysis, you should first plot your dependent variable on the y-axes against your independent variables on x-axes. Then plot your independent variables against each other. Here's what to look for in the plots:

- **Types of Relationships**—Start simple. Do the data form any pattern? Straight lines are easiest to discern followed by simple curves. The more complex the curve, the harder it is to identify in data. This step will tell you if you might need to include some transformed variables in your analysis.

- **Strength of Relationships**—Compare correlation coefficients to their respective plots. Make sure the coefficients are not adversely influenced by nonlinear trends or outliers.

- **Complex Relationships—**So much for the simple stuff. Now you need to look for less typical, more complex patterns, like cycles. Trends may also be noncontinuous. There may be shifts suddenly up or down. Trends may appear or disappear or change within portions of the data range. There may also be trends in some groupings of the data that don't appear in the overall set.
- **Weird Stuff—**You may also see things that you won't see in a Statistics 101 textbook. Some things are important and others not so much so, and sometimes it's what you don't see that is important.
  - » ***Censored Data***—You'll know if you have censored data, but you might not know what it does to data plots. If you get a pattern in which the data look like they hit a wall and bounced off, that wall is the effect of the censored data limit. Don't be misled by such distracting patterns.
  - » ***Missing Data*—**Sometimes it's not what's there that counts. Remember the discussion in Chapter 18 on missing data? If you have data that are MNAR (missing not at random) or even MAR (missing at random), you might see their effects it in these data plots.
  - » ***Hidden Data*—**If you plot data measured on ordinal scales, such as from opinion surveys, you're likely to have hidden data. When you plot the data, you won't be able to tell if a point on the ordinal scale axis represents one data point or a hundred. It's the pattern you *don't* see that's important. Bubble plots can circumvent this problem.
  - » ***Missed Cutoffs*—**Ever pay a bill late? Your records will show no payment in one month and two payments the next. With financial records, these patterns occur most often near holidays and fiscal quarters. Any measurements collected in batches or intervals can show such a dip-peak or high-low pattern. Finding the pattern may be an important result in your analysis or just a distraction you'll want to filter out.
  - » ***Clusters*—**Not every pattern you see will be in the form of a line or a curve. Sometimes data will plot together in clusters. This pattern is different from random scatter, in which all the data is scattered on the graph. There will usually be several groupings, circular or elongated, separated or partially overlapping. If the clusters are important to your analysis, you might consider conducting a discriminant analysis (Chapter 22).
  - » ***Variance*—**Look for a relatively consistent amount of scatter in the points. Sometimes the scatter might increase or decrease over time or at the limits of a measuring device's calibrated range. This is something you'll have to fix or at least be cognizant of when you evaluate your statistical tests and intervals.

Space-Time Plots

If you collected your data at specific locations or times, you should examine some specialized plots to evaluate the possibility of autocorrelation.

Plots for Location-Dependent Data. The first and simplest things to check are the sample locations. Plot the location coordinates of each point and verify that all the samples are in the correct places. If you're not using special mapping software, be sure that the scales of the axes are the same. If you can, plot the data on a map, or better, an aerial photo. Next check the data density and distribution. Take note of any differences in habitats, ecology, hydrology, geology, and soils. Look for any boundaries to natural conditions, such as roads, pipelines, and so on. You can also produce icon plots using the base maps and the sample locations.

Try plotting each location coordinate (i.e., northing, easting, depth) against the variables you are analyzing. Make sure there are no patterns or else your data are sure to be autocorrelated. If you do more spatial analysis, you later want to construct sophisticated plots, called variograms, which assess spatial autocorrelation based on distances between samples rather than location coordinates.

Plots for Time-dependent Data. Even before you plot anything, make sure you understand how your software handles date and time data. Most software uses days after a set start date for date data. Time data is also usually cumulative although the format may not look like it. Start plotting by producing time-series graphs of all your variables. Look for data gaps; trends, cycles, changes in trends and cycles; shifts, pulses, and any other identifiable pattern. As with location-dependent data, if you do more sophisticated temporal modeling, you'll prepare plots based on the time intervals between samples called correlograms. You'll need specialized software to construct variograms and correlograms, which is why you generally don't produce the plots at this point in the analysis.

Once you have assessed the dataset and know what you have to work with, the next step is to see if you can make it better.

20 Teaching Old Data New Tricks

In exploring your dataset, you may have found and addressed problems with samples like censored data and replicates. You should also have found and addressed problems with individual data points like errors and outliers. Now you also have to do something about problems with variables. Rehabilitating variables involves transformations, methods of changing the scales of your variables that might further your analyses.

Here's a great trick. I pull on this tablecloth and a whole bunch of things fall down here where I can play with them.

As part of this process, you should consider what other information you can add that might be relevant to your analysis. This is especially important if you are planning to develop an exploratory statistical model. Experience will tell you when expanding your dataset might make a difference and when it won't. If you don't have that experience yet, start by learning about why you might transform variables and how it can be done. Then practice; try a variety of different techniques and learn along the way. But first you need to understand some of the pros and cons of what you might do to your dataset.

Transformations—Yes But No But Yes

> *Every time we've looked beyond [what] we were familiar with, we've found things that we wouldn't have thought were there.... It's hubris to think that the way we see things is everything there is.*
>
> Lisa Randall, *Discover*, vol. 27, no. 7, 2006.

There is some controversy over the use of transformations (and other methods of creating new variables) that has caused a few statisticians to argue forcefully for or against their use. Here's the gist of what these statisticians dream about after eating jalapeño chili before going to bed.

Why Transformations *Should Not* Be Used

The three most common arguments that have been made against the use of transformations involve:

- Measurement philosophy—*stick to what the instrument tells you.*
- Presentation of results—*report in the same units you used to measure.*
- Capitalizing on chance—*your results should work on other samples.*

The measurement philosophy argument states that the scale of measurement should be a direct result of the measurement process and not a derivation of it, otherwise the measurement is artificial. In other words, use what your instruments tell you, don't reinvent the scale. The counter argument to this position is that since we don't know what scale God (or nature) uses for a measurement, any scale that works for an analysis is acceptable. Transforming a variable may be making it artificial or it may be bringing it closer to reality. For example, the concentration of hydrogen ions in a water sample is never expressed that way. Instead, it is reported as the negative logarithm of the hydrogen ion concentration, which is called pH. Why should it be acceptable to apply a logarithmic transformation to hydrogen ion concentrations but not other concentrations or other types of measurements for that matter? This argument is like religion. People are either solidly on one side or the other. There are no fence sitters.

The presentation of results argument states that you should present results in the same units as the original measurement so that "consumers" (e.g., clients, reviewers) can understand the results more easily. This argument does make a lot of sense. The problem, however, is that some transformations do not have simple back-transformations. You're stuck with the transformed scale. As a consequence, some statisticians maintain that any statistic that uses the same scale as the original data should not be calculated by back-transforming the statistic calculated from the transformed data. For example, means and standard deviations, which use the same scale as the original data, should not be calculated from transformed data and then back transformed to the original scale. Dimensionless statistical results, such as descriptive statistics (e.g., coefficient of variation, skewness, kurtosis) and the results of statistical tests (e.g., expressed as t-values, F-values, or probabilities) do not use the scale of the original variable and so would not have to be back transformed in any case. So the presentation of results argument accepts some transformations but not others. This argument is like politics. Some people are solidly on one side and some people are on the other, but most people just ignore it.

Talk to the paw, the rest of me is snoozing.

The capitalizing on chance argument states that using a transformation because it works well for a particular dataset may make the results inapplicable to the population being studied if the dataset is not truly representative of the population. In other words, the transformation may work for the data but not the population. Overfitting is a problem that is often cited in relation to this argument. Overfitting involves building a statistical model solely by optimizing statistical parameters, and usually involves using a large number of variables and transformations of the variables. The resulting model may fit the data almost perfectly but will produce erroneous results when applied to another sample from the population. This argument certainly has merit, but it is a big slide down a slippery slope from using a simple transformation to overfitting a model. Overfitting is like becoming too muscular from weight training. It doesn't happen suddenly or simply, so it shouldn't come as a surprise. Furthermore, there are several techniques that can be used to assess the effects of overfitting.

Why Transformations *Should* Be Used

There is a single simple argument for using transformations—they work better than the original variable scales. If they don't work better, you don't use them. William of Ockham would have liked that argument. So what constitutes working better? Consider these examples of the three ways that transformations are used.

One, perhaps the most important use of transformations is to reduce the effects of violations of statistical assumptions. If you plan to do any statistical analysis that involves using a Normal (or other) distribution as a model of your dependent variable, it's important to use a scale that makes the data fit the distribution as closely as possible. If the model isn't a good fit for the data, probabilities calculated for some tests and statistics will be in error. Because costly or risky decisions may be made from these probabilities, inaccuracies can be a big deal.[1] So using a transformation to correct violations of statistical assumptions is perhaps their most valuable use.

Two, perhaps the most common use of transformations is to find scales that optimize the linear correlation between data for a dependent variable and data for independent variables. Statistical model building almost always benefits from this use of transformations. Everybody does it.

Three, perhaps the most overlooked use of transformations is, in a word, convenience. Sometimes transformations are used to convert measured data to more familiar units, improve computational efficiency, eliminate replicates, reduce the number of variables, and other actions that facilitate, but not necessarily improve, the analysis.

Now that you've been warned, here are four things you can do that might further your analyses:

- **Sample Adjustments**—methods for fixing missing, erroneous, or unrepresentative data points.
- **Dependent Variable Transformations**—methods for changing the scale of the dependent variable to minimize the effects of violations of statistical assumptions.
- **Independent Variable Transformations**—methods for creating new variables

1 Or as Joe Biden would say, "A big f***ing deal."

from the original independent variables, which have better correlations with the dependent variable.

- **Supplemental Variables**—methods for creating new variables from untapped data sources.

There is nothing sacred about this list. Some of the categories might overlap or omit other ideas, so use these examples to stimulate your own thinking. In time, you'll develop a sense of what you need for a particular analysis.

Sample Adjustments

The whole is more than the sum of the parts.

Aristotle

Sample adjustments involve changing individual data points for a variable, unlike most transformations which result in the creation of a new variable. They are a good place to start the discussion of enhancing your dataset for two reasons. First, if you read the last few chapters, you're already familiar with the concepts, if not the methods, for making adjustments to individual samples. You've read about the nature of outliers and errors, missing data, and replicates and about how to address some of the problems. And it's always a good idea to start with something you know. Second, sample adjustments are the first thing you'll want to do to improve your dataset. These are like the things you do to dig out weeds and fill in holes before you plant a new lawn. Here are five kinds of sample adjustments you can make.

Deletion

Simple enough. If the data point offends thee, pluck it out. Just be sure you have a good theoretical reason for the deletion and aren't just doing something that will give you an answer you want to see. Remember too if you delete a data point, you will also have to delete either the sample or the variable from the analysis. You can't analyze a matrix with empty cells.

Replacement by Fiat

This has nothing to do with Italian cars. Replacement by fiat is when you manually change a data point to compensate for a misleading value in an original set of measurements. For example, financial data (e.g., payments, expenditures, sales) sometimes miss a cutoff date for inclusion in a periodic report. As a consequence, two of the same monthly financial action might appear within one month with no action the previous or subsequent month. You as the data analyst recognize that some adjustment should be made to reflect typical patterns of actions. Using your knowledge of the data, you rearrange the financial data to preserve the pattern of actions. Your replacement of the data is based solely on your judgment that the data are not representative of the population or process. Furthermore, you must use your judgment to decide what values should be substituted. As data analyst, your replacement may be arbitrary, but it is also certainly authoritative, hence a replacement by fiat.

Think of your mortgage or rent payment. Usually you pay it on the first day of each month. One month, you know you will be out of town on the first day of the next month, so you pay your next month's mortgage or rent a week early. The result is two payments one month and none the next. Since you are more interested in patterns in the data rather than variability, you rearrange the monthly payments in the dataset to reflect one payment per month. By fiat, the date of the individual payments would change, but the sum of the payments would be the same.

Replacement with Constant

Replacement with a constant involves using the same value in place of suspect data points. For example, you can replace a censored data point with the value of the censoring limit. Or you can replace missing values with the mean of the variable. Replacement with a constant isn't as drastic as deletion because you aren't left with the problem of removing a sample or a variable from the analysis. Also, using a constant appears less suspicious to reviewers than just making up a replacement value by fiat. Remember, though, almost every replacement technique tries to get a missing data point to be simpatico with the other data points. That means that central tendency statistics will probably be OK, but dispersion will be underestimated; some statistical tests may yield false positive results.

Replacement from Same Variable

If you want to use (or at least delude yourself into thinking you're using) replacement values that are more likely to represent the missing value, you can try interpolating values from the same variable. In a time series, you can average several values before and after the missing value, weighting them more heavily the closer they are to the missing value. If time isn't relevant, you could use a grouping factor to define where to get the values to average.

Replacement from Different Variable

If you have a variable that has the same information as a variable with missing data, you can substitute the data from the second variable for the missing data. For example, you might have two variables that measured the same characteristic with different measuring devices, or are from different sources of information. Further, sometimes a variable with missing values will be highly correlated with another variable in the dataset. When this happens, you can use the second variable to create a regression equation for predicting the variable with the missing data. This approach is more work, but it'll probably provide a better estimate of the missing data than simpler replacement schemes. This is like hot deck imputation.

Dependent Variable Transformations

The problem is all about your scales, she said to me
The R-squares will be better if you've matched 'em mathematically
It's just a way to make your model fit nicely
There must be fifty ways to fix your data

She said it's really not my preference to transform
'Cause sometimes, the new scales confuse, overfit, or misinform
But I'll Box-Cox 'em all if it means they'll fit the norm
There must be fifty ways to fix your data
Fifty ways to fix your data

Take the tails for a trim, Kim
Try a replace, Grace
You can use the rank, Hank
Just try 'em and see
Make it more smooth, Suz
Lots of functions you can choose
A higher degree, Dee
Will get you more fee.

Sing to the tune of "Fifty Ways to Leave Your Lover"
by Paul Simon

After you've filled all the holes in your data matrix with sample adjustments, the next thing you should do is to make sure the dependent variable approximates a Normal distribution. If you haven't looked at histograms and the other indicators of Normality, always do that first. Then if your data distribution differs enough from the Normal distribution to make you nervous about your analysis, try a transformation of the dependent variable. Try several, in fact. Transformations of dependent variables create new variables but only one of the candidate dependent variables is retained for an analysis. You want to pick the candidate dependent variable that fits a theoretical distribution best so that calculations of test probabilities are most accurate.[2]

---

2 Dependent variable transformations can be problematical if your aim is prediction. Usually, you'll want to predict something that is measurable, not a transformation of what you measure. If you select a transformation that is algebraically simple, you can convert the model back to the measurable criterion. That's why more sophisticated transformations, like Box-Cox transformations, are sometimes shunned in predictive modeling. Prediction is difficult enough without adding back-transformations to the mayhem.

Common Normality Transformations

Figure 26 illustrates transformations that can be used to correct problems with data distributions. If your data distribution is positively skewed (i.e., long tail is on the right), try using a square root or logarithmic transformation.[3] If your data distribution is negatively skewed (i.e., long tail is on the left), try using a power transformation. Create the new dependent variable by applying the transformation you select to every value of the original dependent variable. Evaluate those transformed values for Normality. If they are a good fit for the Normal distribution, use the variable with the transformed values as your new dependent variable. If the fit of the transformed data to the Normal distribution is better than your original data but still not as good as you would like, your next option is to try a Box-Cox transformation.

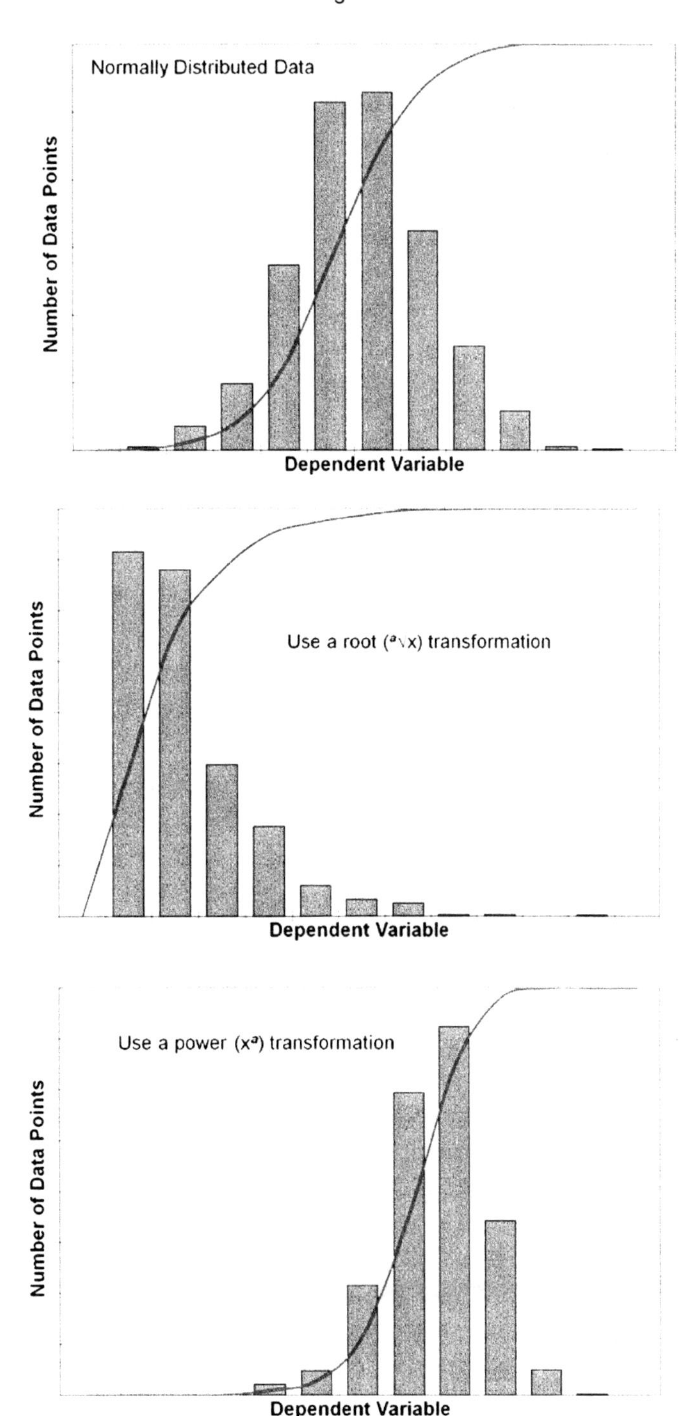

Figure 26. Transformations for Correcting Skewed Data Distributions.

---

3 You might have to add a constant to all of the values of the dependent variable to make them positive before applying log or root transformations. Similarly, you might have to divide all of the values of the dependent variable by a constant to reduce their value before applying power or exponential transformations.

Box-Cox Transformations

The Box-Cox family of transformations may look mathematically intimidating, but they are really not so bad. Let me explain why.

Box-Cox transformations have the general formula:

$$C_T = [(C_O^{\lambda} - 1)/\lambda]$$

where:

C_T = the transformed data (e.g., the values of your new dependent variable)
C_O = the original data (e.g., the values of your original dependent variable)
λ = a constant used in the transformation, usually between -5 and +5

The only part of the equation that should be at all mysterious is the λ (lambda). Aside from that, the transformation boils down to your original dependent variable to a power (that lambda). Lambda is just a number, but you need to figure out what that number should be to have the best transformation toward Normality.

You might think that looking for a good value of lambda would be a lot of work or would require special software. If so, you would be right, but only for one of the two.

It's possible to find a Box-Cox transformation in Excel or another spreadsheet application. First, make sure all your dependent variable values are positive and not too big. Then, generate transformed variables using the spreadsheet formula:

= ((original_value ^ lambda_value) – 1) / lambda_value

where you change the lambda_value in small increments between, say -2 and 2. Assess each transformation for Normality, pick new values of lambda, and repeat the process until you find an acceptable transformation or you begin gnawing on your keyboard. Yes, it's a lot of work.

If you didn't think it was worthwhile buying statistical software back in Chapter 11, you might want to reconsider that decision now. Some, not all, statistical packages have utilities that will automatically do the calculations and magically tell you what the optimum value of lambda would be. Once you know the best you can do, all you have to decide is if the transformed data fit the Normal distribution model well enough to be used in your analysis.

Figure 27 is a screen capture of one such utility for Box-Cox transformations. If you use the default settings, which will be fine most of the time, the utility would require three clicks of the mouse—two clicks to select the dependent variable you want to transform and one more click to start the utility. Figure 28 shows what you get with a few more clicks: a graph of the lambda values that were tested, and before and after histograms and probability plots for the best Box-Cox transformation. That's quite a difference from the laborious spreadsheet approach. Statisticians need statistical software just as much as cabinetmakers need precision woodworking tools. If you plan to do a lot of data analysis, you will too. Don't expect to be able to make a Norm Abram roll-top desk if you buy your tools at the Dollar Store.

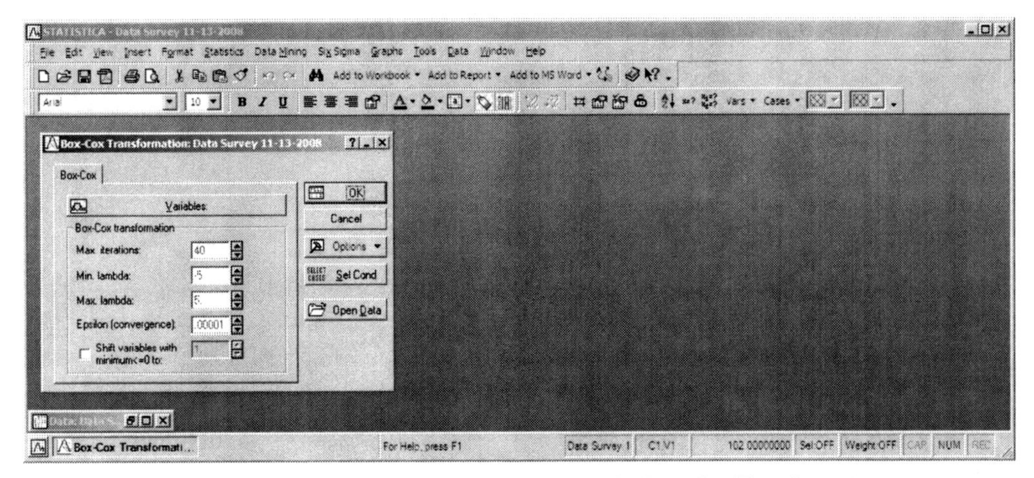

Figure 27. Screen Capture of Statistica's Module for Box-Cox Transformations.

If you know you have to transform your dependent variable, you might as well start by looking for a Box-Cox transformation. That's because Box-Cox transformations include the most commonly used transformations as well as the infinite number of minor variations in between. It all depends on the value of lambda:

- Values for λ of 2 or more correspond to power transformations.
- A value for λ of 0.5 corresponds to a square root transformation.
- A value for λ of 0.0 corresponds to a logarithmic transformation.
- A value for λ of -0.5 corresponds to the reciprocal of a square root transformation.
- A value for λ of -1.0 corresponds to a reciprocal transformation.
- Values for λ of -2 or less correspond to reciprocals of power transformation.

So, if you can't find a Box-Cox transformation that will Normalize your dependent variable, start thinking about nonparametric statistics.

If you do find a transformation that achieves Normality, you can conduct your statistical analysis as you had planned only using the transformed data instead of the original data for your dependent variable. But before you transform the results back to the original scale of measurement, consider this. In transforming the data, you are asserting that the scale of measurement you originally used was not what nature intended. As part of the analysis, you found what the naturally correct scale should be, that is, the transformed scale. So why would you want to transform back to a scale that you decided wasn't the best way to measure the phenomenon? It doesn't make sense unless the old scale is so well-known to your audience that you would confuse them by not back-transforming.

Then there's the special case of the Lognormal distribution. If you used a logarith-

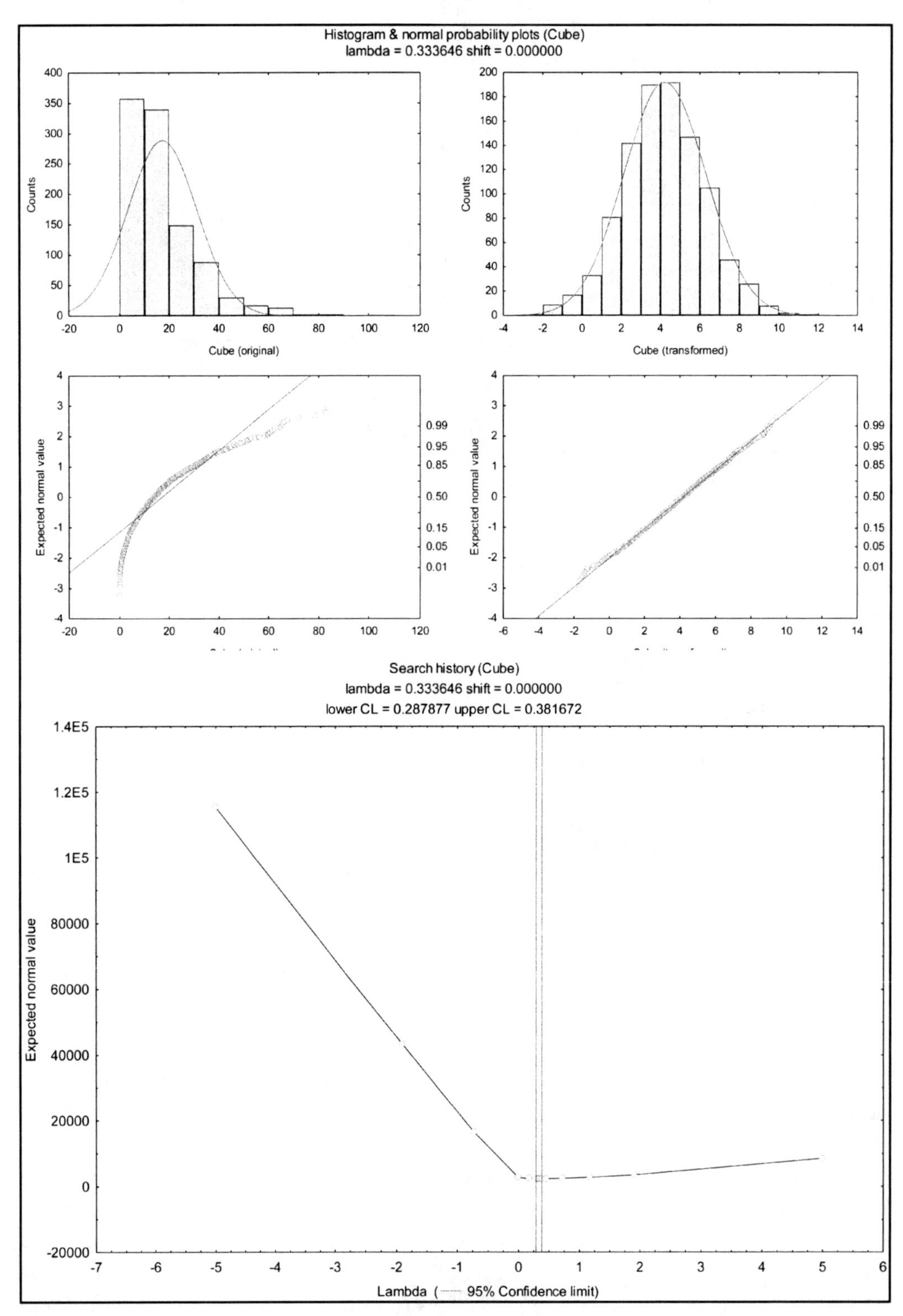

Figure 28. Results of Statistica's Box-Cox Transformation Utility.

mic transformation (lambda = 0.0), you are asserting that your data follow a Lognormal distribution model rather than a Normal distribution model. That's fine; it's just that the back transformation is a bit more complex because the Lognormal distribution has different properties than the Normal distribution. If you want, you can transform back to the original scale using the relationships:

$$\text{Mean} = e^{(\text{geometric mean} + (\text{geometric variance}/2))}$$
$$\text{Variance} = (\text{Mean})^2 \, (e^{\text{geometric variance}} - 1)$$

Where the geometric mean is the arithmetic average of the natural log-transformed data and the geometric variance is the variance of the natural log-transformed data.

Independent Variable Transformations

We changed the crude nominals, turned groupings into scores
I was hoping for no nitpicks, but reviewers wanted more
The scale imparted order, as the data 'came arrayed
When we pointed out the benefits, the reviewers then OKed
And so it was that later, as the data told their tale
That their spread at first quite grainy, turned a righter kind of scale.
Sing to the tune of "Whiter Shade of Pale" by Procol Harum

Once you have the dependent variable you want to work with, you can go on to examine all the relationships between that dependent variable and the independent variables. While the target for transforming a dependent variable is the Normal frequency distribution, the target for transforming independent variables is a straight-line correlation between the dependent variable and each independent variable. This can be a lot of work. Remember, you have to look at correlations and plots, perhaps even for special groupings of the data. That's the reason you always start by finding a scale for the dependent variable so that it fits a Normal distribution. You wouldn't want to repeat this process for more than one dependent variable if you didn't have to. For the same reason, transformations that are more refined than those discussed here are usually pursued later in your analysis.

Variable Adjustments

Variable adjustments are changes, some quite minor, made to all the values for a variable (as opposed to just modifying specific samples as in sample adjustments). All variable adjustments create new independent variables for analysis. Here are four examples of variable adjustments—differencing, data shifting, smoothing, and standardizing.

Differencing

Important sources of variability can often be explored by subtracting the value of a variable from a subsequent value of the same variable. Calculating temporal duration, for example, is perhaps the most common application of differencing and is usually a simple subtraction depending on the time units of the variable.

Data Shifting

If your samples are autocorrelated and collected on constant intervals, you can assess the dependency by shifting data up or down one or more rows in your data matrix. Although this may sound bizarre, data shifts produce new independent variables called lags (when previous times are shifted to the current time) or leads (when subsequent times are shifted to the current time). Shifting all the data by one row is called a first-order lag or lead. Shifting data by two rows is called a second-order lag or lead. In general, shifting data for a variable by *k* rows is called a *k*-order lag or lead. Determining which lags to use in an analysis can be accomplished by trial-and-error, but there are more efficient albeit sophisticated methods. If you're interested, search the internet for autocorrelogram and partial autocorrelogram. The downside of data shifting is that you'll lose as many observations as the order of your lag or lead from the top or bottom of your dataset.

Smoothing

Smoothing is the opposite of differencing. Whereas differencing highlights differences, smoothing suppresses differences. Mathematical smoothing to reduce misleading variance is most often applied to data involving locations and times. The most common example of a smoothing procedure for a time-dependent variable is the rolling average, in which data values are replaced by the average of a set number of previous values. The most common example of a smoothing procedure for a location-dependent variable is the inverse-distance average, in which data values are replaced by the average of a set number of nearby values weighted by the reciprocal of the distances between the points.

Standardizing

You might have been taught in school that you can't compare apples with oranges. Well, when you're analyzing data you have to compare variables with different scales and units all the time, apples and oranges, fruits and grains, food production and nutrition, health care and population, staff and work load, productivity and compensation, 1980 dollars and 2010 dollars, and on and on and on. Different units don't matter much if you're putting all your variables in a regression equation, but if you want to get everything on one graph, that's a different story.

Equating some measures is relatively easy, like adjusting currency for inflation (e.g., a 2010 dollar could only buy 38 cents of goods in 1980; see http://data.bls.gov/cgi-bin/

cpicalc.pl). Go to the U.S. Bureau of Labor Statistics website and find the appropriate consumer price index (or other relevant index) for the years you want to convert and multiply your dollars by the CPIs. Percentages are another way to standardize variables if you can find a common basis for the variables.

You might see a pattern here; standardization involves ratios. If you want to standardize a value, divide by a reference point. With percentages, for example, the ratio is usually an individual value divided by the sum of the values.

To statisticians, the best standardization procedure is to convert the data for a variable to z-scores. Z-scores standardize how much a data value deviates from the average of a variable by using the standard deviation as a point of reference. Perhaps the best thing about a variable consisting of z-scores is that the variable will have a mean of zero and a standard deviation of one. Z-values will fall in the range -3 to +3 about 99 percent of the time, so it's easy to tell if a data value is trying to tell you something. They are very useful for converting variables having very large or very small values, or large ranges that would otherwise cause computational inefficiencies. Most importantly, perhaps, they are also valuable if you want to compare variables having different units of measurement. For example, you can use z-scores to compare changes in sales (dollars) to the number of customers (people) and the amount of precipitation (inches).

Z-scores are calculated in two steps:

1. Calculate the mean and standard deviation of the variable you want to rescale.
2. Subtract the mean from each data value and divide the remainder by the standard deviation.

The result of the division is a z-score. To be sure you did the calculations correctly, calculate the mean and the standard deviation of the z-scores. They should be 0 and 1, respectively.

Z-scores add information about the mean and standard deviation. However, because the mean and standard deviation are constants and z-scores are the result of simple arithmetic, results for statistical tests will be the same for z-scores as the original variable. Still, many statistical software packages offer options for calculating z-scores because they offer computational advantages, especially in complex statistical procedures such as cluster analysis (Chapter 22).

Variable Rescaling

Sometimes it can be useful to change the scale or units of a variable to simplify or facilitate an analysis. Changing the scale of a variable is different from changing the units of a variable. Both are important. Changing units usually involves only simple mathematical calculations with or without the addition or deletion of information. Rescaling variables involves adding or removing information or changing a point of reference. Rescaling usually involves making changes based on logic but may include mathematical calculations as well. Recoding is perhaps the most common way of rescaling a variable. Some statistical software have utilities to facilitate recoding (Figures 29 and 30).

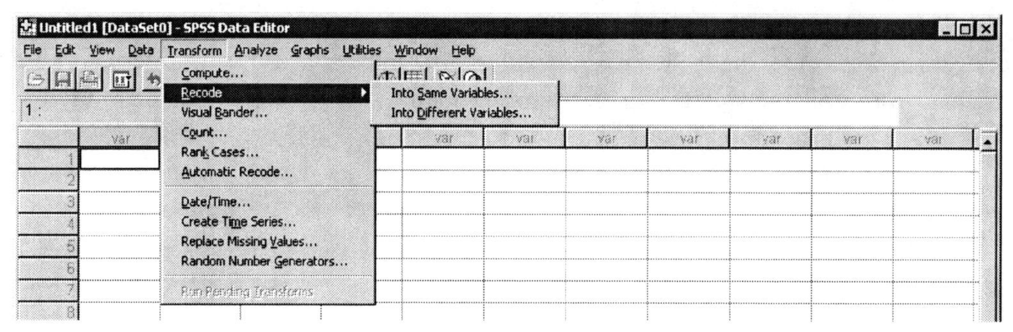

Figure 29. Screen Capture of SPSS Module for Recoding Variables.

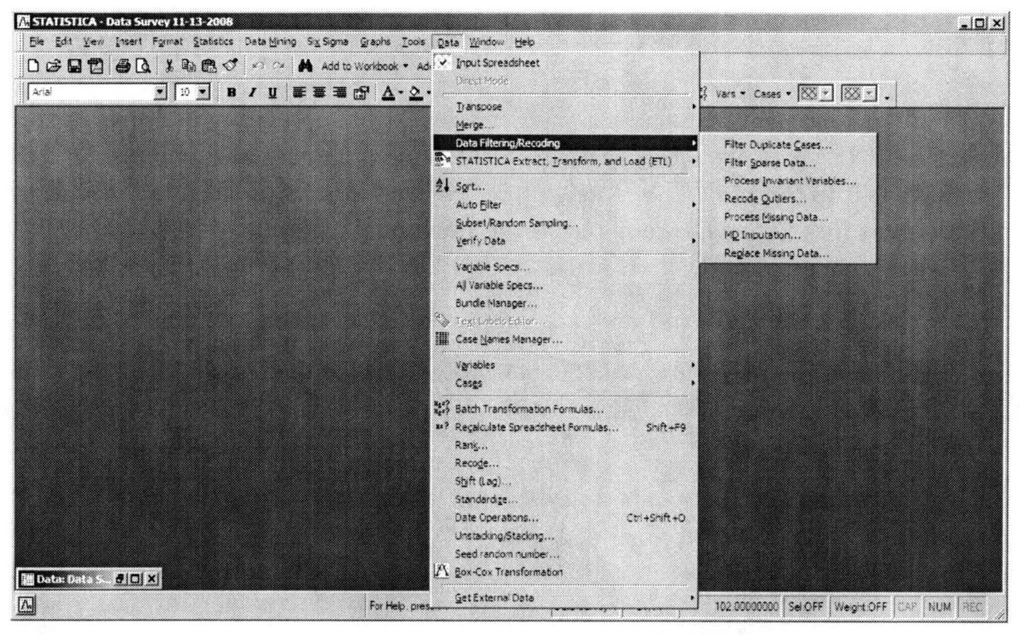

Figure 30. Screen Capture of Statistica's Module for Recoding Variables.

Convert Units: Simple

A simple conversion of units involves arithmetical operations—adding, subtracting, multiplying, and dividing—of constants. The information content of the scale doesn't change. For example, changing temperature measurements from degrees C to degrees F is a change in units, but both provide the same information. Converting concentrations from parts per billion (ppb) to parts per million (ppm) changes the units but not the information content. This is important because so long as two variables have the same information content, they will produce the same results for statistical tests. If a variable you have rescaled has a correlation of 1.0 with the original variable, the two variables will produce the same re-

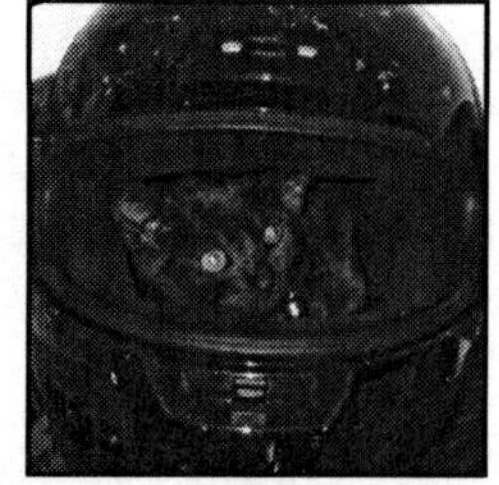

Like a cat out of hell, I'll still yawn when the morning comes.

sults. This is not to say that different units are not important. NASA lost the $125 million Mars Climate Orbiter in 1999 because one contractor used English units instead of metric units.[4]

So why would you want to apply a simple conversion of units? Probably the most basic reason is convenience. For example, if you have velocity data measured in furlongs per fortnight, you might convert the units simply because you and your audience are more familiar with miles per hour. If you need to make a simple conversion, just do it and don't worry about it.

Convert Units: Add Information

You can also convert to different units and add information in the process. This transformation will provide different results for statistical tests in terms of calculated test probabilities although not necessarily different decisions based on the tests.

A good example of changing units by adding information involves chemical concentrations. Concentrations can be expressed in different units as:

- **Mass/Volume**—the mass of a solute (e.g., contaminant) dissolved in one liter of a solution (e.g., water); for example, milligrams per liter (mg/l).
- **Mass/Mass**—the mass of a solute dissolved in one kilogram of a solution, for example, parts per million (ppm).
- **Moles/Volume (Molarity)**—the number of moles[5] of a solute dissolved in one liter of a solution.
- **Moles/Mass (Molality)**—the number of moles of a solute dissolved in one kilogram of a solution.
- **Moles/Moles (Mole fraction)**—the ratio of the number of moles of a solute to the total number of moles present in a solution.

These scales are not identical because they provide different information, in terms of mass, volume, or moles. However, they are likely to be correlated with each other, and if these correlations are high, it won't help to retain more than one of the variables. In this case, the scale that is most familiar to the audience and is most highly correlated with the dependent variable is the logical variable to select.

Rescaling for Computational Efficiency

Rescaling simply to improve computational efficiency was essential in the days when mainframes had the computational efficiency of today's blackberry. Now it's rarely used

---

4 See www.cnn.com/TECH/space/9909/30/mars.metric.02/ or mars.jpl.nasa.gov/msp98/news/mco990930.html.

5 A mole is the mass of 6.022×10^{23} (Avagadro's number) atoms. The atomic weight in grams is equal to one mole of a substance. So even though moles represent a mass, it incorporates new information on atomic weight.

except in cases of large numbers of samples and variables, and variables with large ranges. Many statistical packages have options to rescale variables for this purpose.

There are instances where you will want to rescale variables for computational efficiency. If you plan to use logarithms or roots, for instance, you'll want to add a constant to your variable to eliminate any negative values. If you plan to use exponents or powers, you'll want to divide by a constant so that the variable variables are not too large. Other than that, don't worry about computational efficiency.

Converting Quantitative Scales to Qualitative Scales

Variables on an ordinal or nominal scale can be created from one or more continuous-scale variables by recoding a range of values to a single value. For example, say you ask a question on a survey about how much time a respondent spends watching television. Since you can be pretty sure the responses will be estimates if not wild guesses, it may be more revealing to reduce the times to just a few categories. In this example, you probably wouldn't lose much information, but you would reduce the apparent variability. Similarly, a sampling-round variable may combine samples that were collected over several dates.

Rescaling to Fewer Divisions

Sometimes it's useful to simplify even an ordinal scale. Say, for example, that you conducted an opinion survey on satisfaction with a consumer product in which one of the questions used the five-level scale: excellent, good, average, fair, and poor. You could recode excellent and good responses as positive and fair and poor responses as negative. You would lose the information differentiating the levels, but you would have three groups to analyze instead of five.

Rescaling to fewer divisions happens all the time in sports when a game score is converted into a simple won or loss. In this conversion, the ordinal-scale measure of the number of points a team scored in a game is lost, but the relation of the team's points to its opponent's points is preserved on a nominal scale.

Rescaling to More Divisions

You can also add information to split out new data groups or even convert an ordinal scale into an interval or ratio scale. In the example of the opinion survey, you could create more levels by adding information on frequency of use. Say you set up an ordinal-scale variable representing frequency of use with the four levels frequently, sometimes, seldom, and never. An excellent response would carry more weight if the respondent used the product frequently rather than seldom. Combining the five-level satisfaction scale with the four-level frequency scale would create a new rescaled variable that would have twenty levels.

Variable Linearization

Improving the correlation between a dependent variable and an independent variable is a big part of statistical modeling. The objectives of this type of transformation are (1) to persuade the data to follow a straight line, and (2) minimize the scatter of the data around the line. There are four types of nonlinearity that you can correct at this point; the first three are illustrated in Figure 31:

- **Convex (r-Shaped) Curves**—Data increase in the value of the dependent variable (on the y-axis) quickly and then rise slowly with increasing values of the independent variable (along the x-axis). These curves can be corrected using power and exponential transformations. Some curves can also be corrected using negative values of root and logarithmic transformations.
- **Concave (j-Shaped) Curves**—Data rise slowly at first then rise at an increasing rate with increasing values along the x-axis. These curves can be corrected using root and logarithmic transformations. Some curves can also be corrected using negative values of power and exponential transformations.
- **L-Shaped (Axis-Hugging) Curves**—Data decrease rapidly for small values of x then flatten out. These curves can be corrected using a reciprocal.
- **Negative Slopes**—These can be corrected by using a negative sign in front of the appropriate concave or convex curve transformation,

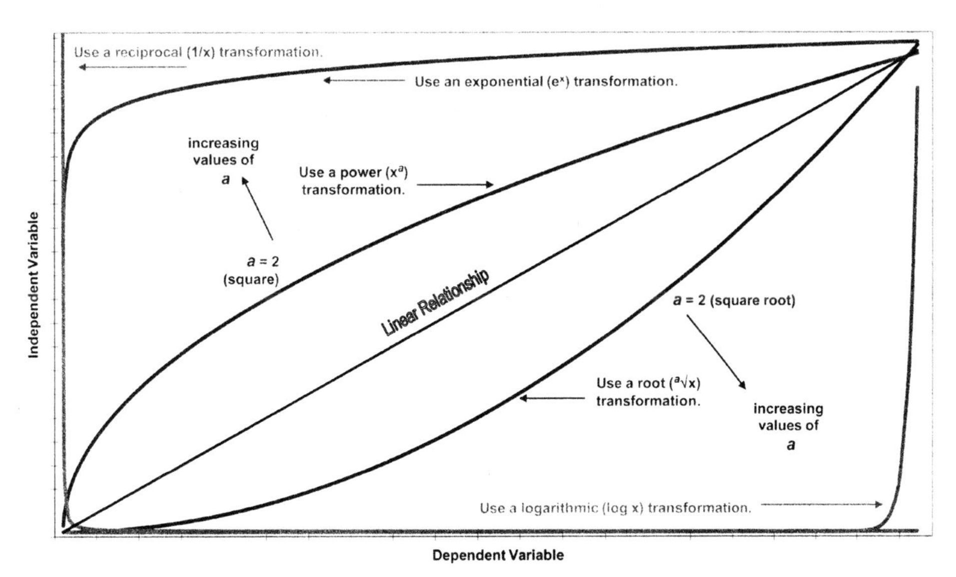

Figure 31. Transformations for Correcting Nonlinear Relationships.

These transformations can be fine tuned by adding or multiplying by constants.

Variable Combinations

Variable combinations are new variables created from two or more existing variables using arithmetic operations or more complex mathematical functions. Variable combinations should be based on theory rather than created for convenience. Usually the variables combined should have the same units (e.g., dollars), although units can be standardized using z-scores.

Simple Arithmetic

You might be surprised how informative adding, subtracting, multiplying, or dividing two or more variables to create a new variable can be to your analysis. The only problem is that there will be far too many possible combinations to evaluate. That's why you have to depend on the theoretical underpinnings of your analysis. Without that background, it would be impossible to even know where to start. But with knowledge of your phenomenon and your variables, it's not so intimidating. Here are a few hints to get you started.

- **Sums**—Sums are not only a great way to get an overall impression of several variables together, but they also may reduce how many variables you have to work with. Examples of summed variables include totals like total expenditures (the sum of several types of expenditures), total concentration (the sum of the concentrations of a group of chemical analytes), total votes (the sum of votes from individual precincts), and so on.
- **Differences**—This is not the same as differencing, which involves taking differences between values of the same variable. Calculating differences between variables can be used to explore changes and to make comparisons. For example, differences between wind speed at 10 meters and 30 meters, radiation counts from samples minus background counts, and so on.
- **Products**—Taking the product of two or more variables assumes that all the variables are needed for a response. If one variable is zero the effect is zero. For example, the model for heat index has several terms involving products of temperature and relative humidity. If the temperature or relative humidity is low, there is little synergistic effect on the heat index. Multiplicative models, models in which the variables are multiplied by each other instead of being added, assume that the dependent variable is conditional on the presence/magnitude of all the variables in the model.
- Division—Ratios can be used in several ways. They can be used as measures of proportion, change, or difference when the variables measure the same quantity. They can also be used as indices when the variables represent different quantities, although usually, in the same units. Ratios are used commonly in business and economics.

More Complex Mathematics

There are many ways you can use mixtures of arithmetic operations and mathematical functions to create new variables. Because there are so many possible combinations, most complex combinations usually follow a recipe. For example, you can average several variables that measure the same information either to obtain more accurate values or reduce the number of variables to consider. This is done often in activities that are graded by several judges, such as school projects and sports like ice skating. In chemistry, you can calculate an ionic balance as the sum of the concentrations (in milliequivalents per liter) of cations minus the sum of the concentrations of anions. If the balance isn't close to zero, one or more of the analyses of a sample is probably in error.

Spatial distances can be calculated from location coordinates (e.g., northings and eastings) using the formula:

$$\text{Distance} = \sqrt{[(\text{Difference in Eastings})^2 + (\text{Difference in Northings})^2]}$$

That will give you the two-dimensional, *as-the-crow-flies* distance.[6] You can calculate the straight-line three-dimensional distance by adding the depth (or height), using the formula:[7]

$$\text{3D Distance} = \sqrt{[(\text{2D Distance})^2 + (\text{Depth/Height})^2]}$$

Location coordinates can be rescaled by rotating and translating (i.e., moving the origin) the axes. The formulas to use are:

$$\text{Realigned Easting} = [(\text{Original Easting} - \mathbf{T_E})\cos(\alpha)] + [\text{Original Northing} - \mathbf{T_N})\sin(\alpha)]$$

$$\text{Realigned Northing} = [(\text{Original Northing} - \mathbf{T_N})\cos(\alpha)] + [\text{Original Easting} - \mathbf{T_E})\sin(\alpha)]$$

Where:

α is the angle of rotation applied to the coordinate axis.
$\mathbf{T_E}$ is the amount of translation along the original Easting axis.
$\mathbf{T_N}$ is the amount of translation along the original Northing axis.

If coordinates are translated but not rotated, there will be no difference in statistical tests on the variables. On the other hand, if coordinates are rotated, the results of

6 If you want driving distances, you'll need specialized software, or *a plan wheel* and some free time, or a string and ruler and a *lot* of free time.

7 Yes, the two formulas are based on the formula for the lengths of the legs of a triangle, $a^2+b^2=c^2$, which you may remember from high school trigonometry.

statistical tests will change because the rotation involves nonlinear functions, namely the trigonometric functions sine and cosine.

Supplemental Variables

I can see now that a concept or even a feeling makes no sense unless out of our substance we spin around it a web of references, of relationships, of values.
Ella Mailla, Swiss adventurer and writer

You won't necessarily just add variables at the beginning of your analysis. You may add them continuously throughout your analysis as you learn more about your data. Some variables may turn out to be critical to the analysis and others will just facilitate reporting or some other ancillary function.

Variables Created by Stratification

Grouping variables are straightforward categorizations of your samples that may help reduce variability. You may already have created grouping variables if you used stratified sampling or experimental control in your statistical design. Grouping variables are measured on either a nominal or an ordinal scale. Common examples of grouping variables include:

- Sex of survey respondent
- Organizational unit of a business entity
- Sampling round (instead of using multiple dates)
- Upstream versus downstream
- Land use, county, state
- Species, soil/rock type, or other classification.

Grouping variables are usually easy to create and are a fundamental component in analysis of variance designs. It is usually best to keep the number of categories for a grouping variable to a manageable few.

Variables Created by Concatenation

Concatenated variables are new variables made from one or more variables in the dataset, usually with a special purpose in mind. The best examples of concatenated variables are specialized IDs especially for identification on outputs having limiting formats. A one-digit identifier, for instances, might be created for use on plots. Eight-character identifiers might be used to label group IDs on computer outputs.

Variables Created from Metadata

Every project seems to have reams of data that fail to get involved in the analysis. Most commonly, they are hard to find, extract, decipher, or interpret, are qualitative, and are sporadically recorded for all the samples. They may be buried in hand-written notes and logbooks, paper files, reference books, or even the failing memories of those who used to be in the know. Even so, you may be able to make some use of the information so long as you can create some entry for all or most of the samples. The information may even be quite important if it represents sources of variability that were not adequately considered in planning. Remember, the data don't have to be on an interval scale, or even quantitative. They can represent groupings or even presence/absence. You can expect to have to decipher some cryptic handwriting to get to the data.

Examples of variables created from metadata include:

- From Samplers' Logbooks
  - Nominal Scale Groupings—sampler, sampling or measuring device, weather.
  - Continuous Scales—date and time of sample collection.
- From Environmental Logbooks
  - Nominal Scale Groupings—rock or soil type, aquifer or flow zone, habitats, species identified, land use.
  - Continuous Scales—percentages of gravel, sand, silt, clay, and organic material; distance from surface water.
- From Topographic Maps
  - Nominal Scale Groupings—land use.
  - Continuous Scales—land slope; distances from roads, water bodies, cultural features.
- From Aerial Images and Photos
  - Nominal Scale Groupings—historical land use, stressed vegetation, moisture content of soil, unexplainable pixel coloration.
- From Industrial Facility Records
  - Nominal Scale Groupings—types of facility activities.
  - Continuous scales—distances from pollution sources, pipelines, fence lines, and so on.

Variables Created from References

Reference variables come verbatim from well-established sources, usually government agencies. They are most often used to characterize some background conditions relevant to the analysis, especially in time series. Examples of reference variables include:

Census Bureau for demographic statistics, National Center for Education Statistics for school data, Bureau of Labor Statistics for employment data, Internal Revenue Service for tax data, Bureau of Economic Analysis for economic data, Bureau of Justice Statistics for crime data, Energy Information Administration for energy consumption data, Environmental Protection Agency for environmental data, Geologic Survey for hydrologic data, National Oceanic and Atmospheric Administration for climate date; and many more at www.fedstats.gov.

The Matrix Resolutions

Everything that has a beginning has an end. I see the end coming…
The Oracle in *The Matrix Revolutions*

In Statistics 101, when your instructor gave you a dataset, that was it. You did what the assignment called for, got the desired answer, and you were finished. But it doesn't work that way in the real world overflowing with data but lacking in wisdom. Sometimes you have to put more effort into making sense of things. Statistics is the duct tape that brings data together to make information, knowledge, and wisdom.

There are many tools you can use to assemble building blocks of data to a temple of wisdom. Consider Table 19. Transformations are like sculptors' tools. They can smooth, reshape, adjust, add texture, augment, condense, and on and on. Suffice it to say that with transformations, there must be at least *fifty ways to fix your data.*

Table 19. Examples of Transformations Used in Data Analysis

<table>
<tr><th colspan="2">Type of Transformation</th><th>Example of Use</th><th>Number of Variables</th><th>Information Added</th><th>Result</th><th>How It Helps</th></tr>
<tr><td rowspan="5">Sample Adjustments</td><td>Deletion</td><td>Trimming tails, deleting outliers</td><td>One</td><td>No</td><td rowspan="5">Changed variable</td><td rowspan="2">Reduce variation</td></tr>
<tr><td>Replacement by Fiat</td><td>Changing misleading values</td><td>One</td><td>No</td></tr>
<tr><td>Replacement with Constant</td><td>Filling in for censored or missing data</td><td>One</td><td>No</td><td rowspan="3">Keeps sample & variable in analysis</td></tr>
<tr><td>Replacement from Same Variable</td><td>Averaging replicates</td><td>One</td><td>No</td></tr>
<tr><td>Replacement from Different Variable</td><td>Regression estimates</td><td>Any</td><td>Yes</td></tr>
</table>

Type of Transformation		Example of Use	Number of Variables	Information Added	Result	How It Helps
Dependent Variable Transformations		Simple	One	No	New variable	Achieve Normality
		Box-Cox				
Independent Variable Transformations	Variable Adjustments	Differencing	One	No	New variable	Explore data patterns
		Data Shifts				
		Smoothing				Reduce variation
		Standardization		Yes		
	Variable Rescaling	Convert Units: Simple	One	No	Changed variable	Convenience
		Convert Units: Add Information	One	Yes	New variable	Explore data patterns
		Computational Efficiency	One	No	Changed variable	Convenience
		Rescale to fewer divisions	One	No	New variable	Simplify Analysis
		Rescale to more divisions	One	Yes	New variable	Explore data patterns
	Variable Linearization	Reciprocals, roots, powers, logs	One	No	New variable	Explore data patterns
	Variable Combinations	Simple Arithmetic	Any	Yes	New variable	Explore data patterns
		More Complex Mathematics				
Supplemental Variables	Created by Stratification	Groupings from demographic data	Any	Yes	New variable	Explore data patterns
	Created by Concatenation	Sample ID, plot IDs	Any	Yes	New variable	Convenience
	Created from Metadata	Sampler, sampling device	Any	Yes	New variable	Explore data patterns
	Created from References	Government data	Any	Yes	New variable	Explore data patterns

Statistical analyses are like skyscrapers. The main statistical analysis (e.g., ANOVA, regression analysis) is like the observation deck at the top of the skyscraper. It's what attracts all the attention. The main statistical analysis, however, rests on a number of preparatory steps (e.g., distribution assessments, outlier evaluations) that are like the lower floors in the building. All these analyses use the same dataset. Consequently, the dataset is like the skyscraper's foundation. If the foundation is flawed, the skyscraper may not be safe. If a dataset has errors, the analyses based on it may lead to incorrect conclusions. While it's possible to repair a skyscraper's foundation after it is completed, it's a lot more expensive and time-consuming than if it had been done correctly in the beginning. The same is true of errors in a dataset. This skyscraper analogy is a gentle way of saying make sure the dataset is error-free before you begin your analysis. If the budget is tight or the schedule is short (and they always are), you better make sure you won't have to do your analysis over again because of data errors.

So now you know why it takes so long and costs so much to prepare a dataset for a statistical analysis. It will probably consume the majority of your project budget and schedule. Different professionals will probably do it differently from each other, so remember to document everything you do. You may have to justify your technical reasoning to a client or a reviewer. You may also need the documentation for contract modifications, presentations, and any similar projects you plan to pursue. Be sure to save your files often, giving them descriptive names in the process. Back the whole mess up, then come back tomorrow and do it all over again.

PART 5

A Model for Modeling

The dataset is ready to go. Now it's time to get to the really fun stuff.

Back in Chapter 6, you read about what models are and how they can be both inputs to and outputs from statistical analyses. In Chapter 7 you read that models make assumptions, and in particular, that statistical models make at least five assumptions. Models aren't mysterious and don't even have to be complicated. But, to make best use of a model, you have to understand where they come from and how they work. Once you understand that, you'll be able to build your own statistical models and critique models prepared by others.

Part V aims to give you some ideas that will help you work with statistical models:

Chapter 21—Modelus Operandi describes five key elements that go into creating a statistical model. There are also some hints on how to select a statistical analysis method.

Chapter 22—The Land beyond Statistics 101 provides abstracts of statistical methods that can be used to analyze data and build models, from simple regression to sophisticated techniques like canonical analysis.

Chapter 23—Models and Sausages describes some of the wild and wacky things that can happen in trying to build statistical models and what you might do about them.

Now understand, you probably won't be able to use all these concepts and techniques right away, but this part will point you in the right direction. With practice, you'll become a model modeler.

21 Modelus Operandi

If you've never created a statistical model before, fear not. This section will describe the things you have to consider. You might be surprised to find that building a statistical model involves a lot more than statistics. It's like traveling. You don't start by thinking about your transport, the plane, train, or bus you might take. You start by knowing where you are and where you want to go. Only then do you create your itinerary, select a carrier, buy a ticket, pack your belongings, and make the trip. Likewise, modeling starts with the phenomenon you're trying to model and ends with the model. Between those two points, though, there are many possible routes.

I know where I am and where I want *to go.*

Many Paths Lead to Models

Nothing unconnected ever occurs.

Emanuel Swedenborg, *Heaven and Hell,* pph 37, 1758.

Creating a model always starts with the phenomenon you plan to model. You have to know a lot about the phenomenon because every decision you make during the modeling process will rely on it. But that's actually a good thing. Whether you are modeling something for school or work, or even some personal interest like your home energy usage, you probably know at least something about the phenomenon. If you need to know more, you can look into the information sources described in Chapter11 and get smart.

Once you're comfortable with your knowledge of the phenomenon, you get to the core of model development. This is where it gets much more complicated. Consider the journey you might take to create a model, shown in Figure 32. There are many possible scenarios. For example, after studying a phenomenon, you might decide how you would use a model, and from that, decide what the model should focus on, what data you'll need, what statistical method you'll use, and how you'll calibrate the model. Or, you may be given a dataset by a client, and from the samples and variables, you determine what models could be created, and what statistical methods would be required. Sometimes, you decide what you want the model to do, but the variables would be too difficult or costly to collect, so you have to revise the model specifications and reconsider the samples and variables. Similarly, you might find that the statistical method you want to use will require different data or model boundaries, so you have to reconsider your plans. Sometimes you iterate through these considerations several times before you're ready to advance to the actual modeling process. As you model more and more phenomena, you're likely to take these paths and many more. Each excursion through the maze of modeling elements will be a new and different adventure for you to learn from. If you get lost, find someone to give you directions. Here are some hints for how to address the five elements of model development.

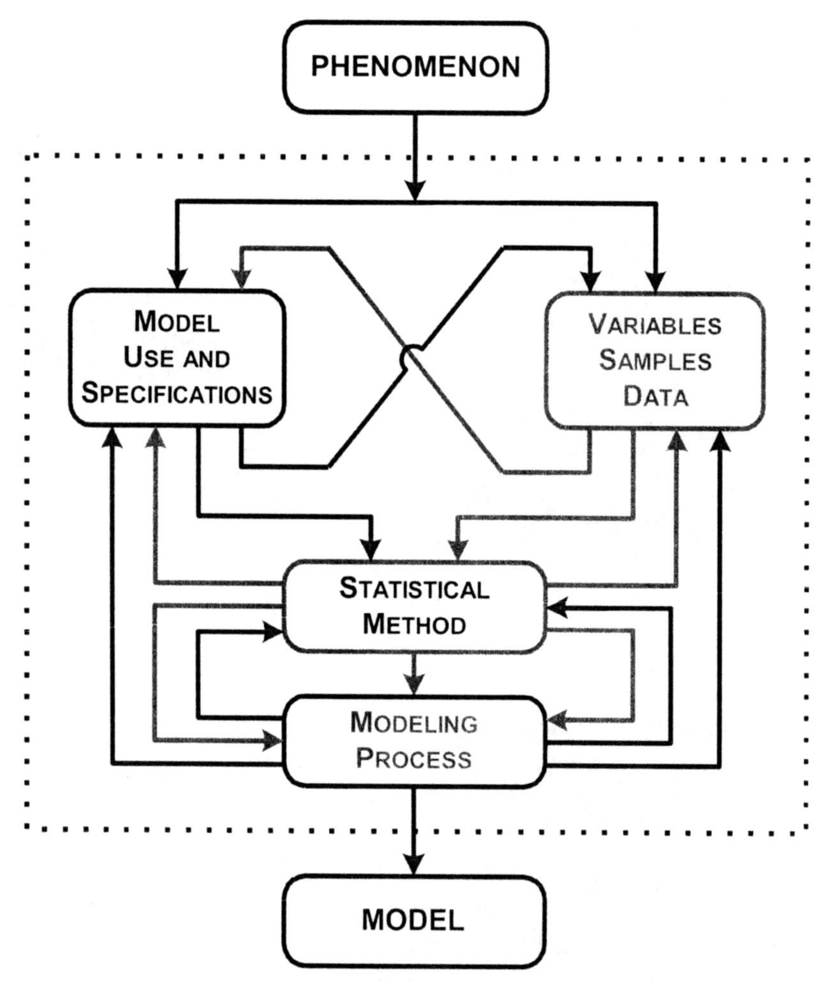

Figure 32. Key Elements in the Creation of a Statistical Model.

The Phenomenon

The usual approach of science of constructing a mathematical model cannot answer the questions of why there should be a universe for the model to describe. Why does the universe go to all the bother of existing?

Stephen Hawking, A Brief History of Time, 1988

The first thing you'll have to do is to think about the phenomenon you want to model. This may sound trivial, but it's not. Even if you were assigned the work by your boss or academic advisor, you're going to have to make a lot of decisions on your own. If they were going to make all the decisions, they wouldn't have given the project to you.

The nature of the phenomenon has to do first with how tangible the phenomenon is. Is the phenomenon an object that can be seen and touched? Is it a process that can be watched and interacted with or a behavior that can be observed but not necessarily manipulated? Is it a condition that can be monitored, or if not visible, at least measurable (like radioactivity)? Or is it an opinion that can't be seen or touched, and may not even be measurable directly?

The nature of the phenomenon also has to do with how changeable the phenomenon is. Is it something that is fixed and unchangeable? If it changes, what is the rate of change? Is it too slow or too fast to be observable? Does the phenomenon exist in states of equilibrium and disequilibrium? Can changes be manipulated by an experimenter?

Thinking about the nature of the phenomenon will help you narrow your options for what form might be appropriate for the model based on its intended use. For example, would it be possible to build a physical model or will the model have to be a less tangible written model, blueprint, computer application or mathematical equation? It's not uncommon for several types of models to be developed to display, manipulate, or substitute for the phenomena. Automakers, for example, make many types of models of the automobiles they sell, from the styling, to the performance, to the marketing.

Model Use and Specifications

The division between those who try to learn about the world by manipulating it and those who can only observe it had led, in natural science, to a struggle for legitimacy. The experimentalists look down on the observers as merely telling uncheckable just-so stories, while the observers scorn the experimentalists for their cheap victories over excessively simple phenomena.

Richard C. Lewontin, "Sex, Lies and Social Science: An Exchange" in *New York Review of Books,* May 25, 1995

After the phenomenon, you'll need to think about what you want to do with the model and how it will be designed. You can do this before, after, or at the same time that you consider your samples and variables, and statistical methods.

The first thing you should think about is how the model will be used. You can use a model to:

- **Display**—use of the model to describe or characterize the sample or the population.
- **Substitute**—use of the model in place of the phenomenon, such as for prediction.
- **Manipulate**—use of the model to explain aspects of the phenomenon.

As a point of reference, most analyses involve the simple display of descriptive information. That's the simplest use of statistical models. If you plan to use them for substitution or manipulation, you'll have to know more about the phenomenon, more about modeling, and more about statistics.

Whatever your planned use, you'll have to think about how you want to approach the modeling. Here are three considerations to get you started:

- The viewpoint you'll take to develop the model
- The boundaries of the model relative to the phenomenon
- The level of detail of the model

Modeler's Viewpoint

Top down! *Bottom up!*

Your viewpoint in modeling is how you plan to approach the effort, that is, either from the top down or the bottom up. A top-down viewpoint will require you to understand the big picture, things like what the phenomena is associated with. This viewpoint is more correlative and is commonly used in statistical models, especially predictive models. A bottom-up viewpoint will require you to understand the details, the conditions that cause or affect the phenomena. This viewpoint is more deterministic and is commonly used in statistical models for explanation and in theoretical models.

Top-down models usually don't require as many variables as bottom-up models so long as they are the right variables. The problem with top-down models is that sometimes relationships appear to be oversimplified or obscure. Why should skirt length predict stock prices, for example? It makes no sense, but a high correlation has been found between the two measures.

Bottom-up models tend to require more variables to characterize all the facets of a phenomenon. Larger numbers of variables, in turn, require greater levels of effort than for top-down models. Furthermore, many of the details included in a bottom-up model are often found not to have a significant impact on the overall model. Hence, bottom-up modeling tends to be labor intensive and inefficient, but in the end, at least you know how everything fits together.

Some modelers take their viewpoint as an extension of their own personalities. Big-picture people think of a phenomenon in terms of general concepts, mechanisms, trends, and patterns and tend to model from the top down. They don't care if their favorite team has weaknesses as long as the team's winning percentage is high. Details people think of discrete parts or elements that make up a phenomenon, and tend to model from the bottom up. They believe the whole is equal to the sum of the parts. Their team could be in first place, but they're concerned about one player who is in a slump.

Often both viewpoints work equally well for modeling a phenomenon. Sometimes, though, one or the other viewpoint will work better, be easier, or even be the only feasible approach. For example, say you want to model the performance of an automobile. Using a top-down viewpoint, you might focus on acceleration, gas mileage, top speed, and so on. You might be able to model how the automobile will perform under certain driving conditions, but you won't learn anything about how the components of the automobile work together. Using a bottom-up viewpoint, you might focus on number of cylinders, gear ratios, timing, and so on. You might be able to model how changing a component could boost or diminish its function, but you won't know if the change would provide the same effect to the automobile's overall performance. You have to be sure that your viewpoint is appropriate for how you plan to use the model or else the model won't be useful. At every step in your modeling effort, ask yourself, "Will I be able to do what I need to do with the results of the model?"

Model Details

Every phenomenon complex enough to have to be modeled assuredly has many levels of detail. You have to decide how much detail to put in your model, especially if your viewpoint is bottom-up. Still, there are practical limits imposed by restrictive budgets and schedules or by what is known about the phenomenon. For example, if you want to model the performance of an automobile, do you concentrate on the engine or also consider aerodynamics, steering, braking, and other components? If you concentrate on the engine, do you focus on the internal combustion components or also consider the pollution control devices, the electrical system, and other components? If you concentrate on the internal combustion components, do you focus on the pistons or also consider the spark plugs and the fuel?

Model Boundaries

Where will your model end? This is easy to visualize with location and time; you can draw a line on a map or block out dates on a calendar. Many phenomena aren't so easy to isolate, though. Processes, in particular, often use inputs from other processes or contain subprocesses that can't be isolated. In modeling the performance of an automobile, for example, do you include different makes (e.g., Ford, Honda), different models (e.g., sedans, SUVs), different options (e.g., engines, transmissions), different drivers, different types of road conditions, and so on.

These determinations will affect everything else you do.

Other Model Specifications

There are many other things about your model that might have a bearing on the variables and samples you select, the statistical methods you use, and how you go about optimizing the model. Here are a few specifications that may be relevant to your model:

- **Users**—Who will be using the model? If it's just you, the model may not need to have a polished appearance and extensive documentation. If others will be using the model, though, consider that audience. You may not have to build a comprehensive user interface, but you'll at least need to try to make it understandable and sufficiently documented. Don't try to make it idiot-proof, though, it's not worth the effort. God is just too good at making idiots.
- **Frequency of Use**—If the model will be used on a recurring basis, make sure there will be some provision for you or some other qualified individual to review the model periodically to ensure it is being used correctly and is still appropriate for representing the phenomenon.
- **Accuracy and Precision**—As a general rule, statistical models tend to be fairly accurate but never as precise as you need them to be. Have some notion of the accuracy and precision you want. That way you'll know when to quit. A good way to specify the precision you want is to start from a gut feel and specify the precision as a percentage, for example, ±5 percent or ±10 percent. Then you'll have to control variance and manipulate the number of samples and the confidence level so that $\mathbf{t_{(\alpha,n-1)} s/\sqrt{n}}$ is close to your target precision.[1] If the n is too big, try again with a larger allowable Type I error (α).
- **Limit of Complexity**—Some models were not meant to be. If you can't fit the model to the data, you have to be prepared to call it quits. In a way, this is equivalent to a Do Not Resuscitate order in medicine, and likewise, it can be a sensitive subject. It's usually easier to create new variables or try some other statistical manipulation than it is to give the bad news, and the bill, to the client.

---

1 Just in case you were wondering:

$t(\alpha,n-1)$ is the value of the t-distribution with an acceptable Type I error rate equal to α (i.e., a confidence of 1-α) and degrees-of-freedom equal to 1minus the number of samples

s is the standard deviation of the sample

$\sqrt{n}$ is the square root of the number of samples.

Variables, Samples, and Data

Does not any analysis of measurement require concepts more fundamental than measurement? And should not the fundamental theory be about these more fundamental concepts?

John Stewart Bell, "Quantum Mechanics for Cosmologists"
published in Quantum Gravity, 1981, p. 611–637

Perhaps the most complicated and time-consuming aspect of model building is selecting the components of your model, the variables, the samples, and the data. This effort is discussed ad nauseam in Part III. Here are a few tips on how to apply the ideas in Part III to developing a statistical model.

Variables

Here's a quick review. The key variable or variables that characterize the phenomenon to be modeled is called the criterion variable, or more commonly, the dependent variable. Variables (usually, but not necessarily, more than one) that will be used to test, predict, or explain the dependent variable in the model are called grouping variables, predictor variables, explanatory variables, or more commonly, independent variables. At this point, you could represent your prototype model as:

Criterion variable(s) that characterizes the phenomenon	=	Independent variable 1 + Independent variable 2 + Independent variable 3 + . . . and so on

By convention, the criterion or dependent variable is always placed to the left of the equals sign, and the independent variables are placed to the right. This model can also be written, in terms only a mathematician could love, as:

$$y = a_0 + a_1x_1 + a_2x_2 + a_3x_3 \ldots a_nx_n + e$$

where:

y is the dependent variable that characterizes the phenomenon

$\mathbf{x}_1$ through $\mathbf{x}_n$ are the independent variables that test, predict, or explain the dependent variable

a_0 through a_n are constants[2] (called coefficients or parameters of the model) that are estimated using the statistical procedure

2 If a_0 through a_n aren't constants, you have a nonlinear model.

e is the uncertainty or error term of the model, without which, this just wouldn't be a statistical model.

In this representation of your model, the ***y*** and the ***x***s are the variables you create and measure on your samples. The ***a***s and the ***e*** are the constants the statistical procedure estimates.

Usually, statistical models have only one dependent variable. These are called univariate statistical models. If more than one dependent variable is needed to describe a phenomenon, the model is called a multivariate statistical model. Some statistical textbooks, particularly in the social sciences, refer to statistical procedures that analyze more than one variable, either dependent or independent, as multivariate.

Select as many dependent variables as you feel you'll need to characterize the phenomenon. If you think you'll need only one dependent variable, that's great. It will make for a fairly straightforward analysis. If you need more than one dependent variable, try to limit the number. Multivariate analyses quickly become very complex. If you have more than a few dependent variables, here are a few things you can do to reduce the number of candidate dependent variables.

- **Focus on Aspects of the Phenomenon**—Some phenomena are very complex or at least multifaceted. You may be able to reduce the number of dependent variables you are considering by focusing on just one aspect of the phenomenon. For example, in the workload example discussed in Chapter 13, you might limit your attention to either the staff or work available.
- **Narrow the Objective**—If you are trying to do too much in one study, you might try to reduce your aims, or break up the project into parts and conduct the subprojects sequentially.
- **Focus on Hard Information**—Hard information involves measurements of tangible, observable demonstrations as opposed to measurements of intangible beliefs or opinions. Focus on dependent variables that involve hard information.
- **Focus on Direct Information**—Direct information involves measurements specifically of the phenomenon being investigated, as opposed to measurements of factors associated with the phenomenon. Focus on dependent variables that directly measure the phenomenon.
- **Eliminate Correlated Variables**—If several candidate dependent variables are highly intercorrelated, pick the best and eliminate the rest.
- **Create Multiple Models—**If you have to have more than one dependent variable, create a different model for each one. This is like subdividing the objectives—not optimal but sometimes a necessary evil.
- **Conduct a Factor Analysis**—You might be able to reduce the number of dependent variables using factor analysis to combine the multiple variables into one (Chapter 22).

If you can't do any of these things, you're probably headed for a multivariate analysis. Consider looking for help.

Your selection of independent variables will hinge on what you plan to use the model for. Here are a few tips for identifying candidate measures and scales:

- **Variables for Characterizing, Classifying, Identifying, and Explaining**—Select enough variables to address all the theoretical aspects of the phenomenon, even to the point of having some redundancy. Sometimes two differently measured or differently scaled variables that address the same theoretical concept will make dissimilar contributions to the model. When you calibrate the model, the extra variables will drop out.
- **Variables for Comparing**—Test what you want to know, not everything under the sun. Keep the number of variables to an absolute minimum or your analysis will become intractable. Try to use conventionally recognized variables and scales rather than creating new ones if you can. This will facilitate replication studies.
- **Variables for Predicting**—Be sure that the variables and scales you select are relatively inexpensive and easy to create or obtain. A prediction model won't be very useful if the prediction variables cost more to generate than the prediction is worth. For example, if you plan to use the model repeatedly, say to make monthly forecasts, you'll want the model inputs to be simple enough that you could generate all the data you would need in a couple of weeks. If the inputs were so complex that they take months to generate, you wouldn't be able to use the model as you wanted. Stress precision in selecting variables. Accuracy tends to come easy while precision is elusive. Prediction models usually keep only the variables that work best in making a prediction, so the number of variables you select initially isn't that important. Recognize, though, that the more variables you have in your conceptual model, the more work it will be to winnow out the ones you don't need.

Some of the variables may have several possible scales. If these extra scales are related to each other by a linear algebraic relationship, keep only one. This is because the variables will be perfectly correlated, and thus, will add no new information to the model. For example, if you measure temperature in degrees Fahrenheit, you don't need to also include temperature in degrees Celsius.[3] Pick the scale that will give you the best resolution. In the example of temperature, Fahrenheit-scaled thermometers can be read with greater precision than Celsius-scaled thermometers because they have smaller divisions.

If two measures have unrelated scales or can be measured differently, keep them all at this point. You will sort out the best measures when you calibrate the model. For example, if a concept that you want to evaluate with your model were a person's size, you could use a height scale and a weight scale. However, you wouldn't need to include weight in both pounds and kilograms because the two scales are linearly related (1 kg = 2.2 lbs). You

---

3 The temperature in degrees Fahrenheit is equal to 1.8 times the temperature in degrees Celsius plus 32.

could include weight measured by a balance beam, a strain gauge, a spring scale, or even a circus weight-guesser because they use different techniques to measure weight (although they would probably be highly correlated).

Samples and Data

You should remember from Chapter 14 how the samples you select must represent the population you want to analyze. A lot of thought must go into defining the population and finding samples that will fairly represent that population. Then all those mental maneuvers go into fitting considerations, like the sample hierarchy, resolution, and the sampling scheme, into a comprehensive sampling plan. So the last thing you want to have happen is to have the sampler, the person who will generate the data, stray from the carefully thought-out plan. You don't want field technicians moving sampling locations so that they don't have to walk so far from their truck. You don't want doctors reassigning their friends to experimental groups that will get preferential treatment. You don't want your survey takers concentrating on attractive members of the opposite sex. You get the idea. When it comes to samples, samplers should have little or no discretion to stray from the plan. Follow the map that will find the population you're looking for.

Then there's the process of generating the data. As much as you plan to minimize variance with reference, replication, and randomization, there will always be opportunities at the point of data collection to improve the process. A dropped meter may require recalibration that's not called for in the sampling plan. A survey taker might ask a clarifying question, check spelling, or point out a math error before a respondent forever disappears. A surveyor can correct a map with an incorrectly located sampling point. An accountant can adjust misclassified debits in financial records. As the data analyst, you are mostly powerless to make such corrections and clarifications until it's too late, and you have to puzzle over the cause of an outlier. You need to rely on the knowledge and experience of the people collecting the data. So when it comes to data, samplers should have considerable discretion to use their initiative to ensure the quality of the data, minimize variance, and achieve the intent, if not the letter, of the sampling plan.

Now that you know what you want your model to do and you know what you need to measure, consider conducting a review with your project team to go over what you have in mind for the model. If you're working alone, you might consider meeting with the client or at least writing them a letter. If you have missed something or are going in an errant direction, it's better to find out sooner than later.

Statistical Methods

The right method in any particular case must be largely determined by the nature of the problem.

Arnold Toynbee, *Lectures on the Industrial Revolution in England*, 1884

This is the step where you really have to know statistics. If you have a degree in statistics or a lot of experience, this step will take minutes at most, and you'll be fairly confident that you know what you have to do. If you've only taken Statistics 101, this step will still probably only take a few minutes, but you'll be unsure about whether you picked the right procedure. That's to be expected. If you're unfamiliar with the procedure you've selected, now is the time to look into getting reference material because you'll have to learn a lot before you go any further.

Conducting a statistical analysis is like preparing food. When you prepare a dish, you start with a recipe. You get together everything you'll need, the major ingredients and the spices, as well as your kitchen tools. Then you mix the ingredients and cook them. In a statistical analysis, the objective is like a picture of the finished dish. The recipe is the plan for sampling and data analysis. The data are like the primary ingredients. They represent the core of the analysis. If you don't have the right primary ingredients (data), you won't be able to prepare the recipe. And if the primary ingredients (data) are spoiled, your dish will be bad. The sampling strategy is like the herbs and spices. If you forget about them, the dish may look good, but it will taste bad. The statistical procedures are like the kitchen tools and culinary techniques. If you don't mix the ingredients and cook the mixture correctly, your dish will be bad. Like cooking, there's a lot that goes into a statistical analysis. But with the appropriate knowledge and some experience, it's not that hard to be a statistical gourmet.

If you haven't been trained in statistics, selecting which technique to use may seem bewildering. You can usually get in the right ballpark, though, if you understand your variables and your objectives. Consider the hierarchy for selecting a statistical analysis method summarized in Table 20.[4] The table has five major decision points:

1. How many variables do you have?
2. What is your statistical objective?
3. What scales are the variables measured with?
4. Is there a distinction between dependent and independent variables?
5. Are the samples autocorrelated?

The first decision is a no-brainer. How many variables do you have—one or more than one? It doesn't get any simpler than that. In fact, by the time you get to this point in building your model, you should already have determined the answers to the first four questions. Don't worry about any of the statistical methods mentioned in Table 20 that you haven't heard of before.[5] Chapter 22 provides summaries of most of the techniques. If you decide you might want to use one of these techniques, read more about it on the Internet and go from there.

---

4 If you search the Internet, you'll find quite a few sites on this topic. Many of them are very specific to certain types of analysis compared to Table 20, which is very general. Look at a few of these other methods to see if any fit your needs better or augments the categories suggested by Table 20.

5 I don't want to scare you, but these methods are just some of the more commonly used. There are scores more.

Who needs tools when you have these?

As with most generalized procedures, there are exceptions. Perhaps the most notable are variables based on cyclic scales. Orientations and months of the year are two common examples. There are two options for treating these types of scales. Either you can transform the variable into a nonrepeating linear scale or use specialized techniques. The first option is usually easier, but the second option usually provides better results. Also, if you have more than one dependent variable and you want to analyze all the variables simultaneously, you have to use multivariate statistics. Multivariate statistics are a quantum leap more complex than univariate (i.e., one dependent variable) statistics, and are probably best left to experienced statisticians.

So if you've made it this far, you have some notion of what statistical techniques you might apply. Just remember, describing all the statistical techniques you might use in an analysis would be like trying to describe all the tools used in carpentry. There are some very common tools, such as saws and hammers, as well as very specialized tools, the ones that aren't likely to be on the shelves at your local Home Depot. Don't worry about the very specialized tools. You can accomplish quite a lot with these off-the-shelf statistical techniques. The other thing to bear in mind is that method selection guides such as those presented here can help you decide what you *could* use but not what you *should* use. You can use a sledge hammer to drive a nail, but you'd probably be better off using a smaller hammer. That's a matter of experience, or at least, trial and error.

Table 20. Hierarchy for Selecting Statistical Methods

Number & Type of Variables	Dependent & Independent Variables	Goal of Your Analysis	Methods for Dependent Variables Having	
			Discrete and Grouping Scales	**Continuous Scales**
Just One	—	Description and Characterization	Counts and proportions	Descriptive statistics, Statistical graphics, Distribution fitting
		Comparison, Detection, and Testing	Nonparametric tests	t-tests
		Identification and Classification		Outlier tests

Number & Type of Variables	Dependent & Independent Variables	Goal of Your Analysis	Methods for Dependent Variables Having: Discrete and Grouping Scales	Methods for Dependent Variables Having: Continuous Scales
More Than One	No	Description and Characterization	Counts and proportions, Tables, Correlations	Tables, Descriptive statistics, Statistical graphics, Distribution fitting, Correlations
		Comparison, Detection, and Testing	Tables, Nonparametric tests	Tables, Statistical tests
		Identification and Classification	Association rules	Outlier tests, Cluster analysis
		Explanation	Correspondence analysis	Principal components, Factor analysis, Multidimensional scaling
	Yes	Description and Characterization	Counts and proportions, Tables, Correlations	Tables, Descriptive statistics, Statistical graphics, Distribution fitting, Correlations
		Comparison, Detection, and Testing	Nonparametric tests, Analysis of variance/covariance	Statistical tests
		Identification and Classification	Discriminant analysis, Association rules, Classification trees, Logistic regression	Regression methods, Canonical analysis
		Prediction and Explanation	Discriminant analysis, Association rules, Correspondence analysis, Classification trees, Logistic regression	Regression methods, Canonical analysis, Neural networks
Way Too Many		Description and Characterization	Counts and proportions, Tables, Correlations	Tables, Descriptive statistics, Statistical graphics, Distribution fitting, Correlations
		Comparison, Classification, Prediction, and, Explanation	Discriminant analysis, Association rules, Correspondence analysis, Classification trees, Logistic regression, Data mining techniques	Regression methods, Canonical analysis, Neural networks, Data mining techniques
Time-Dependent		All Objectives	Smoothing interpolation methods, Time-series regression, Autoregressive integrated moving average modeling, Spectral analysis, Neural networks	
Location-Dependent			Smoothing interpolation methods, Trend surface analysis, Geostatistics	

The Modeling Process

All models are wrong; some models are useful—so choose the least hopelessly wrong model.

G. E. P. Box, "Robustness in the strategy of scientific model building." In R. L. Launer, and G. N. Wilkinson, (eds.) *Robustness in Statistics*, 1979

The process of developing a statistical model involves finding the mathematical equation of a line, curve, or other pattern that faithfully represents the data with the least amount of error (i.e., variability). Variability and pattern are the yin and yang of models. They are opposites yet they are intertwined. Improve the fit of the model's pattern to the pattern of the data, and you'll reduce the variability in the model and vice versa.

Say you have a conceptual model with a dependent variable (**y**) and one or more independent variables ($\mathbf{x}_1$ through $\mathbf{x}_n$) in the fear-provoking-yet-oh-so-convenient mathematical shorthand:

$$y = a_0 + a_1x_1 + a_2x_2 + a_3x_3 \ldots a_nx_n + e$$

Estimating values for the model's parameters (i.e., a_0 through a_n) and the model's uncertainty (i.e., the e) so that the model is the best fit for the data with the least imprecision is a process called calibrating or fitting a model. Every statistical method has criteria that the procedure uses to calculate the parameters of the best model given the variables, data, and statistical options[6] you specify. Your job is to specify those variables, data, and statistical options.

This is how it works:

1. You collect data that represent the y and the xs for each of the samples.
2. You make sure the data are correct and appropriate for the phenomenon and put the values in a dataset.
3. Using the software for the statistical procedure you selected, you specify the dependent variable, the independent variables, and any statistical option you want to use.
4. Magic happens.[7]
5. You evaluate the output from the software and, if all is well, you record the parameters and the error, and you have a calibrated statistical model. If the model

---

6 Every statistical procedure has a variety of options that can be specified. If you're doing a factor analysis, for instance, you can try different extraction techniques, different communalities, different numbers of factors, and so on. If you're a statistician, you know what I mean. If you're not a statistician, don't worry about this.

7 This is what you learn about if you major in statistics.

fit isn't what you would like, which is what usually happens, you make changes and try again.

What changes could you make? Here are a few hints. If you are well acquainted with statistics, you can try making adjustments to the variables and the statistical options, and perhaps even the data, to see how the different combinations affect the model. For example, you can try including or excluding influential observations, filling in missing data, changing the variables in the model, or breaking down the analysis by some grouping factor (if you have enough data). If you are well acquainted with the data but not statistics, you might rely more on your intuition than your computations. Look for differences between the different models as well as between the results and your own expectations based on established theory.

If you specify only one way that you want to combine the variables, data, and statistical options, the statistical method will give you the best model. However, if you specify more than one combination, you have to have your own criteria for selecting which of the best models to use as your final model. The two most commonly used criteria are the coefficient of determination and the standard error of estimate.

- **Coefficient of Determination**—also called R^2 or R-square, is the square of the correlation of the independent variables with the dependent variable. R-squared ranges from 0 to 1. It is thought of as the proportion of the variation in the dependent variable that is accounted for by the independent variables, or similarly, the proportion of the total variation in the relationship that the model accounts for. It is a measure of how well the pattern of the model fits the pattern of the data, and hence, is a measure of accuracy.
- **Standard Error of Estimate**—also called s_{xy} or SEE, is the standard deviation of the residuals. The residuals are the differences (i.e., errors) between the observed values of the dependent variable and the values calculated by the model. The SEE takes into account the number of samples (more is better) and the number of variables (fewer is better) in the model, and is in the same units as the dependent variable. It is a measure of how much scatter there is between the model and the data, and hence, is a measure of precision.

For a set of models you are considering, the largest coefficient of determination usually will correspond to the smallest standard error of estimate. Consequently, many people look only at the coefficient of determination because it is easier to understand that statistic given its bounded scale. It's essential to look at the standard error of estimate as well, though, because it will allow you to evaluate the uncertainty in the model's predictions. In other words, R-square will tell you which of several models may be best while SEE will tell you if that best is good enough for what you need to do with the model.

"If all is well" in step 5 of the modeling process means that all the statistical tests and graphics that the software provides indicate that the model will be satisfactory for your needs. This, of course, is the crux of statistical modeling that statisticians write all those books about. You'll want to get at least one reference for the type of analysis you want to do and maybe another one for the software you plan to use. Then you actually have to read them.

The best results you can hope for, in a way, are the mundane conclusions that confirm what you expect, especially if they add a bit of illumination to the dark places on the horizon of your current knowledge. Expect that there will be some minor differences between simulations. They'll probably be inconsequential. But be cautious if the results are a *big surprise*. Be skeptical of anything that might make you want to call a press conference. It's OK to get surprising results, just be sure you aren't the one surprised later to find an error or misinterpretation.

After you're done with model calibration, you're ready to implement the model in a process called deployment or rollout. You'll find a lot of information about deployment on the Internet, particularly in regards to software. Most data analyses give birth to reports, mostly shelf debris. Statistical models that perform a function, though, usually involve software. These models can be programmed into a standalone application or integrated into available software like Access or Excel. Consider your audience. Perhaps the best advice is to keep a deployed model as simple as possible. Most users won't have to know the details of the model, only how to use it. Be sure you provide enough documentation, though, so that any number crunchers in the group can marvel at your accomplishment.

The Statistical Modeler's Ten Commandments[8]

I. *State your objectives and targets for accuracy and precision in clear quantitative terms.*
II. *Use a sampling plan that is consistent with the population and your objective.*
III. *Scrub and explore the dataset before you begin the analysis; know your data.*
IV. *Select a statistical method that is suitable for your variables, your samples, and your objectives.*
V. *Evaluate the validity of the assumptions on which the model is based.*
VI. *Calibrate the model to obtain the best reliable fit possible.*
VII. *Interpret the structure of the model in terms of possible underlying conditions.*
VIII. *Understand the uncertainty in the model and the precision of the model's predictions.*
IX. *Be cautious of surprising results; use the model to make decisions that are sensible.*
X. *Reevaluate the model with new data when conditions change.*

8 From: Kufs, Charles. 1989. Making Sense of Statistical Models, unpublished seminar notes.

22 The Land Beyond Statistics 101

You've probably figured out by now that statisticians don't make the big bucks just for calculating averages and drawing pie charts. Maybe the tip off was all the strange terms in Table 20. This chapter tries to make those terms a little less strange. If you see something that interests you, you can find more information on the Internet. The chapter won't tell you about all the places in the land beyond Statistics 101, just the largest busiest places. But once you get familiar with the culture and trappings of mainstream statistics, you'll be able to, and want to, explore some of the more far-flung and extraordinary realms.

Methods for One Variable

Whoever wants to reach a distant goal must take small steps.
Saul Bellow, Canadian-born American writer

Let's start simple. Say you have only one variable, no groupings, no demographic data, no metadata. There are only a few things you can do. You can characterize the sample or the population the sample came from using descriptive statistics and distribution-fitting techniques. You can also test whether the average (or some other descriptive statistic) of the samples is different from a constant, such as zero. But that's about it. You can't do any nontrivial prediction or explanation with only one variable. Figure 33 shows the wealth of descriptive statistics you can obtain with just a few mouse clicks if you have statistical software like Statistica. Figure 34 shows some results from SPSS. Figure 35 shows how easy it is to test your variable in SPSS.

Most statisticians wouldn't consider methods involving only one variable to be modeling because there's no equation. It's more of a display function, like what a fashion model does. Don't worry about it. It's not a big deal for you at this point.

The statistics you use for one variable will depend on the scale of measurement of the variable. This is where all that information in Chapter 4 comes into play. First, decide what category your variable falls into, from:

- **Continuous**—variables measured on interval or ratio scales
- **Maybe continuous**—variables measured on ordinal scales

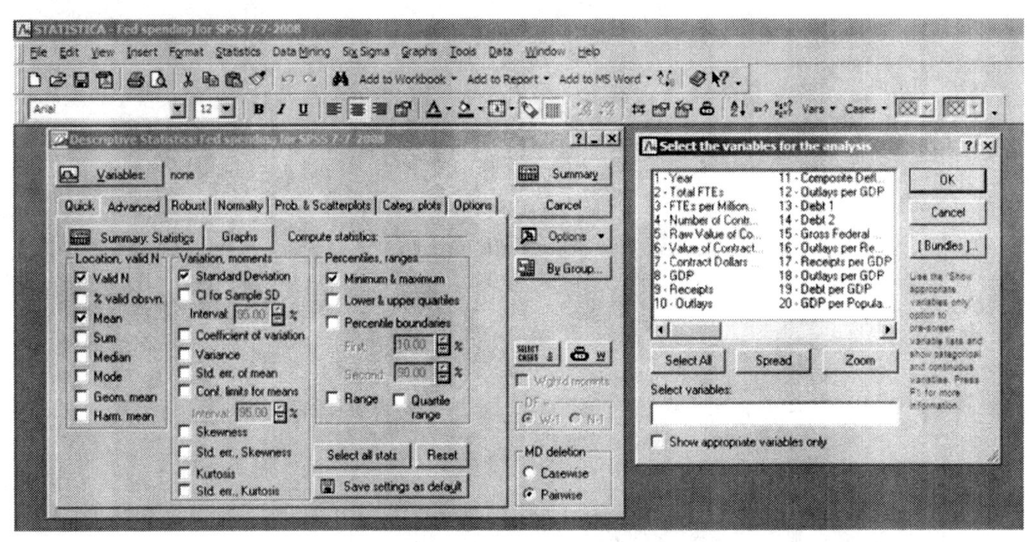

Figure 33. Screen Capture of Statistica's Descriptive Statistics Module.

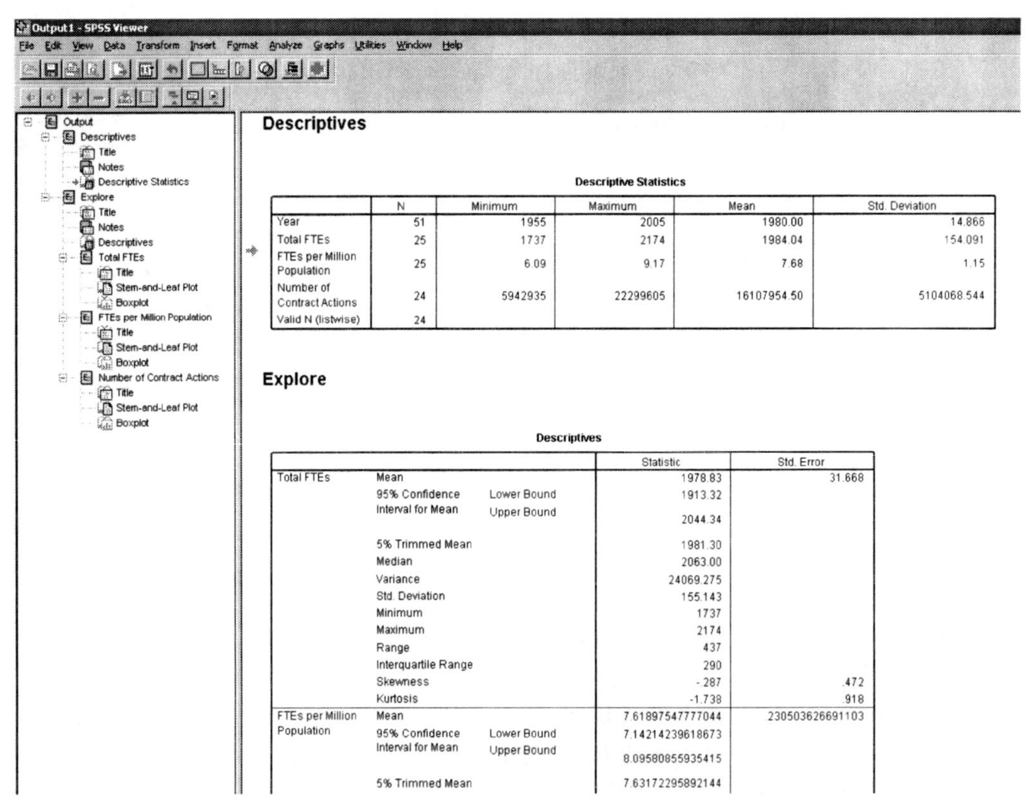

Descriptives

Descriptive Statistics

	N	Minimum	Maximum	Mean	Std. Deviation
Year	51	1955	2005	1980.00	14.866
Total FTEs	25	1737	2174	1984.04	154.091
FTEs per Million Population	25	6.09	9.17	7.68	1.15
Number of Contract Actions	24	5942935	22299605	16107954.50	5104068.544
Valid N (listwise)	24				

Explore

Descriptives

			Statistic	Std. Error
Total FTEs	Mean		1978.83	31.668
	95% Confidence Interval for Mean	Lower Bound	1913.32	
		Upper Bound	2044.34	
	5% Trimmed Mean		1981.30	
	Median		2063.00	
	Variance		24069.275	
	Std. Deviation		155.143	
	Minimum		1737	
	Maximum		2174	
	Range		437	
	Interquartile Range		290	
	Skewness		-.287	.472
	Kurtosis		-1.738	.918
FTEs per Million Population	Mean		7.61897547777044	.230503626691103
	95% Confidence Interval for Mean	Lower Bound	7.14214239618673	
		Upper Bound	8.09580855935415	
	5% Trimmed Mean		7.63172295892144	

Figure 34. Screen Capture of Output for SPSS' Descriptive and Explore Modules.

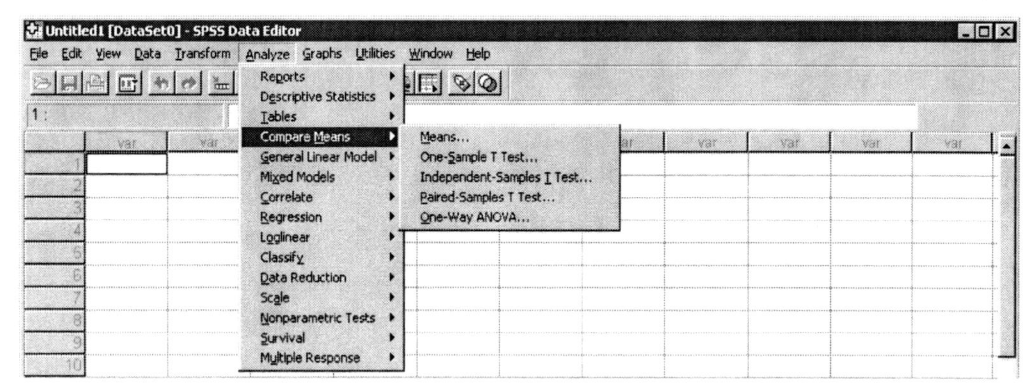

Figure 35. Screen Capture of SPSS Compare Module.

- **Discrete**—grouping variables measured on nominal scales
- **Special**—cyclic scales like orientations and restricted range scales like some proportions[1]

If your variable falls into the "maybe continuous" category, examine all the levels of the scale and decide if each is more like a discrete group or a continuous, albeit irregular, progression. Knowledge of the theory behind the measurement should be your first consideration. If that is inconclusive, here's a rule of thumb. If the scale has only two or three levels, consider it a grouping variable. If the scale has six levels or more, consider it a continuous variable. Scales with four or five levels can be problematical, but you'll have to make a choice and probably defend it later. The trade-offs are these. It's a safer bet to downgrade the scale to nominal if you're just not sure about the theoretical basis of the scale and are concerned about hostile reviewers. On the other hand,[2] you'll have more analysis choices if you treat the scale as continuous.

If your variable is measured on a discrete, nominal, or grouping scale, you'll have to use nonparametric statistics. Nonparametric descriptive statistics include the median for central tendency, and quartiles, deciles, the interquartile range, and the full data range for dispersion.

If your variable is measured on a continuous scale, you can use parametric and nonparametric statistics. Parametric descriptive statistics include the mean for central

---

1 If the scale for a proportion has endpoints, for example 0 percent or 100 percent, the scale has a restricted range. Restricted range scales may have to be treated differently than continuous-scales that don't have a restricted range. However, if the values for a variable are measured on a scale that is theoretically restricted but not limited in practice, the variable can usually be treated as you would any continuous variable. For example, the percent oxygen in ambient air is theoretically limited to the range of 0 percent to 100 percent. In practice, though, measured values would usually be about 21 percent plus or minus a few percentage points. Because the actual measurements do not approach the 0 percent or 100 percent endpoints, the variable can be treated as though it were measured on an unrestricted scale.

2 It's reasonable to assume that any assertion that is "on the other hand" is correct 50 percent of the time. Likewise, "rules of thumb" are presumably correct only 20 percent of the time. So why is it taken as a virtual certainty when somebody is fingered as a criminal in court?

tendency, and the variance (or standard deviation) for dispersion, as well as the skewness and kurtosis for distribution fitting. You can conduct a t-test to compare the mean to constant, or an F-test to compare the variance to a constant, or a χ^2 (chi-square) test to compare the sample distribution to a specific distribution model.

Methods for More Than One Variable

The most merciful thing in the world ... is the inability of the human mind to correlate all its contents.

H. P. Lovecraft, "The Call of Cthulhu," in *Weird Tales,* 1928

If you have more than one variable, the next decision point is whether you distinguish between dependent and independent variables or not. If there is no distinction between dependent and independent variables, you can use the same types of analysis as for single variables plus two other kinds of data evaluations. For describing variables by groups, you can use many types of tables or comparison tests. Figure 36 shows options for creating tables in SPSS.

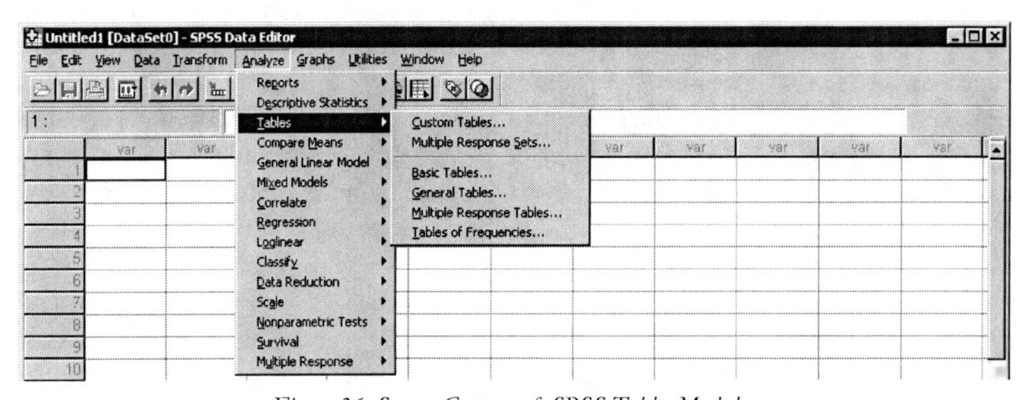

Figure 36. Screen Capture of SPSS Tables Module.

For characterizing relationships between variables, you can examine several types of correlation coefficients. The selection of a type of correlation coefficient depends mostly on the scale of the variables being investigated. Table 21 summarizes some of the most commonly used (and misused) correlation statistics. Figure 37 shows the input screen and output from Statistica's correlation module. Figure 38 shows output from SPSS's correlation module.

Table 21. Statistics for Characterizing Relationships between Variables

<table>
<tr><th colspan="2">Type of Correlation</th><th>First Variable</th><th>Second Variable</th><th>Comments</th></tr>
<tr><td colspan="2">Pearson</td><td>One variable; continuous scale</td><td>One variable; continuous scale</td><td>This is the correlation coefficient you learned about in your introductory statistics course. It is used to quantify the relationship between two variables.</td></tr>
<tr><td rowspan="3"></td><td>Multiple</td><td>One variable; continuous scale</td><td>More than one variable; continuous scales</td><td>Used to quantify the relationship between a single variable and a set of several variables.</td></tr>
<tr><td>Partial</td><td>One variable; continuous scale</td><td>More than one variable; continuous scales</td><td>Used to quantify the relationship between two variables while holding the effects of a set of other variables constant.</td></tr>
<tr><td>Canonical</td><td>More than one variable; continuous scales</td><td>More than one variable; continuous scales</td><td>Used to quantify the relationship between two sets of several variables.</td></tr>
<tr><td colspan="2">Spearman Rho, Kendall Tau</td><td>Interval or ordinal scale</td><td>Ordinal scale</td><td rowspan="7">Used to quantify the relationship between two variables that are not measured on continuous scales.</td></tr>
<tr><td colspan="2">Polyserial</td><td>Interval scale</td><td>Ordinal scale</td></tr>
<tr><td colspan="2">Polychoric</td><td>Ordinal scale</td><td>Ordinal scale</td></tr>
<tr><td colspan="2">Point-biserial</td><td>Continuous scale</td><td>Binary scale</td></tr>
<tr><td colspan="2">Biserial</td><td>Interval scale</td><td>Binary scale</td></tr>
<tr><td colspan="2">Rank biserial</td><td>Ordinal scale</td><td>Binary scale</td></tr>
<tr><td colspan="2">Phi, tetrachoric</td><td>Binary scale</td><td>Binary scale</td></tr>
</table>

For explaining patterns in a set of more than a few variables, you could apply one of several statistical classification techniques to the variables (or the samples). If you want to maintain the original identities of the variables, you might try cluster analysis.

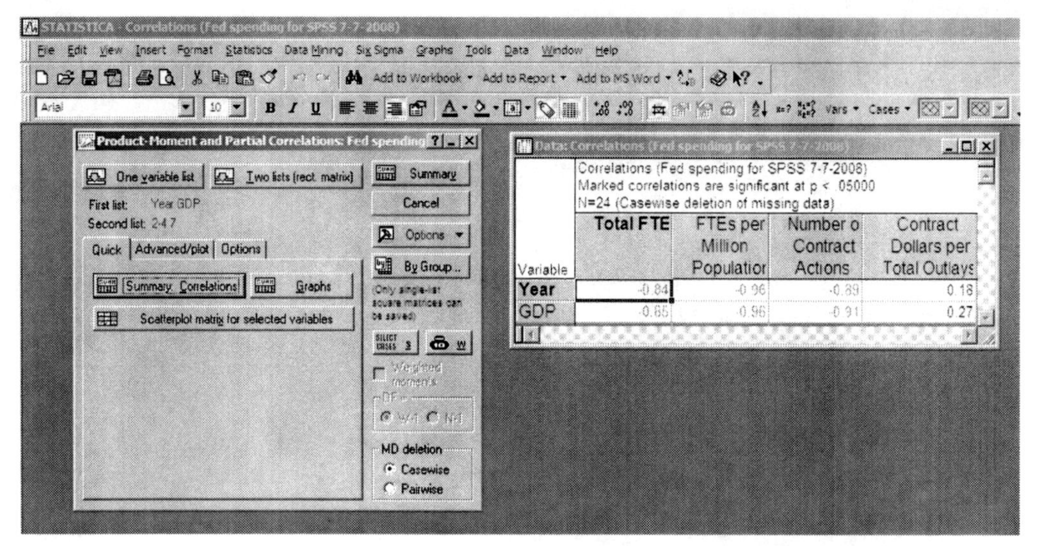

Figure 37. Screen Capture of Input and Output to Statistica's Correlation Module.

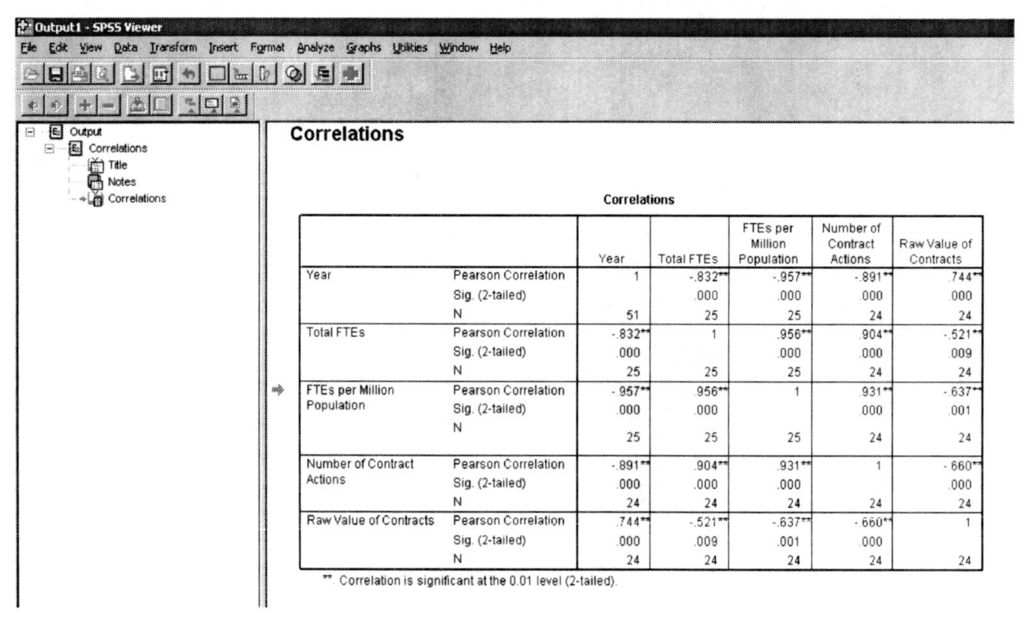

Correlations

Correlations

		Year	Total FTEs	FTEs per Million Population	Number of Contract Actions	Raw Value of Contracts
Year	Pearson Correlation	1	-.832**	-.957**	-.891**	.744**
	Sig. (2-tailed)		.000	.000	.000	.000
	N	51	25	25	24	24
Total FTEs	Pearson Correlation	-.832**	1	.956**	.904**	-.521**
	Sig. (2-tailed)	.000		.000	.000	.009
	N	25	25	25	24	24
FTEs per Million Population	Pearson Correlation	-.957**	.956**	1	.931**	-.637**
	Sig. (2-tailed)	.000	.000		.000	.001
	N	25	25	25	24	24
Number of Contract Actions	Pearson Correlation	-.891**	.904**	.931**	1	-.660**
	Sig. (2-tailed)	.000	.000	.000		.000
	N	24	24	24	24	24
Raw Value of Contracts	Pearson Correlation	.744**	-.521**	-.637**	-.660**	1
	Sig. (2-tailed)	.000	.009	.001	.000	
	N	24	24	24	24	24

**. Correlation is significant at the 0.01 level (2-tailed).

Figure 38. Screen Capture of Output from SPSS's Correlation Module.

Cluster Analysis

Cluster analysis refers to a variety of procedures for arranging ungrouped items into statistically similar collections on the basis of variables you specify. Either samples or variables can be clustered. The clusters can be used to describe the data or they can be coded as grouping variables for other types of analysis. Perhaps the most commonly used type of cluster analysis is hierarchical cluster analysis. Results from a hierarchical clustering are usually expressed as a tree diagram, which looks a bit like a company's organization chart. Figure 39 is a screen capture from SPSS's cluster analysis module.

The challenge in cluster analysis is to decide how many distinctly different clusters there are. Think about how a company's organization chart might look since it's similar to a tree diagram from a cluster analysis. Each level of the organization would have a different, ever increasing from top to bottom, number of groups. At the CEO level, there would be one group consisting of everyone in the company. At the individual level, there would be as many groups as there are employees, each consisting of one individual. In the same way, hierarchical cluster analysis provides clusters from the individual sample (or variable) to the all encompassing cluster at the top of the tree. Somewhere between those two levels is a clustering that makes sense for your analysis. You just have to decide what it is. This decision can be as much an intuitive art as it is a mathematical science. Most of the more sophisticated statistical methods are like this. If you like to use both sides of your brain, you'll enjoy exploring this land beyond Statistics 101.

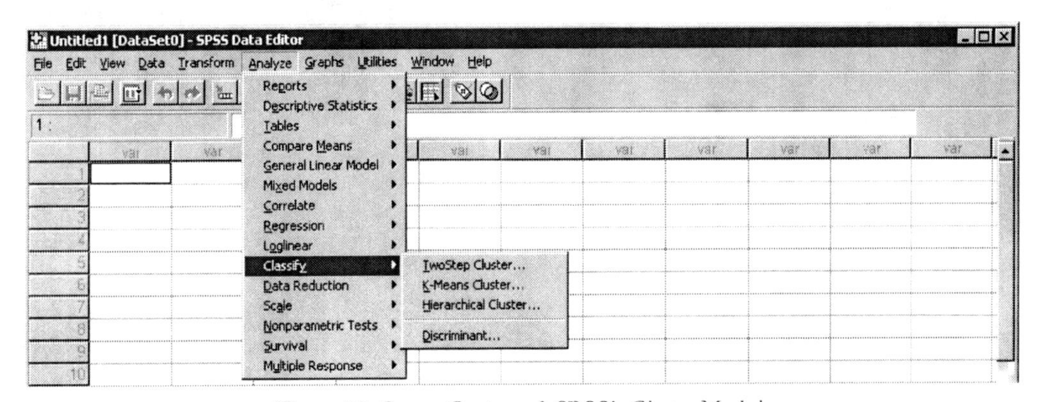

Figure 39. Screen Capture of SPSS's Cluster Module.

There are other approaches to cluster analysis that you may also find useful. K-means cluster analysis allows you to specify starting points for the clusters and the final number of clusters. The method then finds the best allocation of samples (or variables) to the clusters given those constraints. Block clustering clusters samples and variables simultaneously. This method is useful for more explicit explorations of sample-variable associations.

Each of the cluster analysis method has a variety of options for calculating the mathematical distance between clusters and samples (or variables) and for deciding which sample (or variable) to add to which cluster. These are the crux of specifying a cluster analysis that you'll have to learn more about before going further.

Clustering variables is a good way to explore which variables contain the greatest amount of information and which variables may be redundant. Other methods for doing this include factor analysis (including principal components analysis) for variables with continuous scales, correspondence analysis for variables with discrete scales, and multidimensional scaling for any set of variables from which a correlation matrix can be produced. These techniques rearrange the information in the variables so that you have a smaller number of new variables (called factors, components, or dimensions, depending on the type of analysis) that represent about the same amount of information. Because of this increase in information density, analyses can be more efficient. The downside is that the new variables often represent latent, unmeasurable characteristics of your samples, making them hard to interpret.

Figure 40 shows a variety of multivariable statistical techniques on Statistica's multivariate statistics menu.

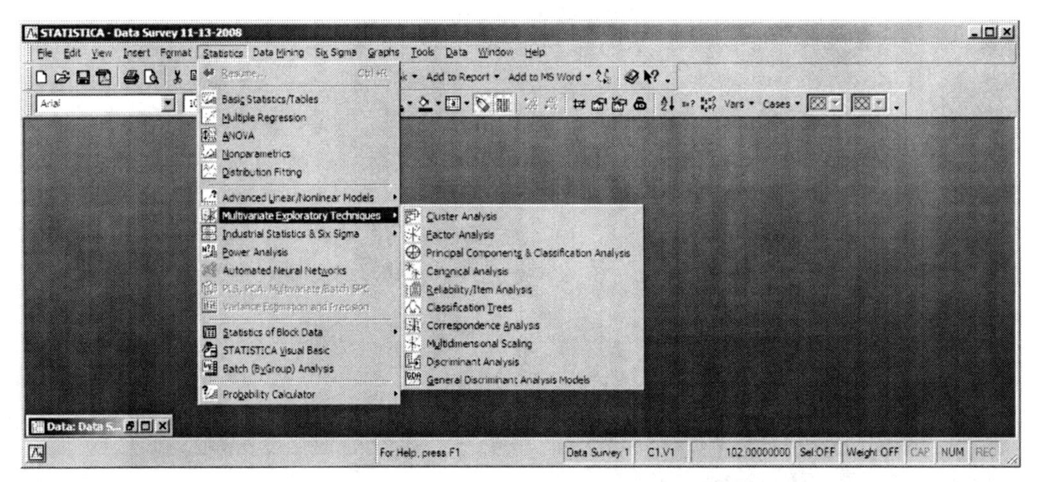

Figure 40. Screen Capture of Statistica's Multivariate Statistics Menu.

Factor Analysis

Factor analysis (FA) is based on the concept that the variation in a set of variables[3] can be rearranged and attributed to new variables, called factors. In a sense, factor analysis is like a complete transformation of all the original variables in which the original information content is homogenized and then resorted into factors having more desirable mathematical properties than the original variables. While a FA produces as many factors as the number of original variables, usually only a few factors account for almost all of the variation in the orginal variables. Using just those few factors instead of all the original variables can produce a much more stable model. As a consequence, FA is often called a data reduction technique because it can dramatically reduce the number of variables needed in a model.

One difference between a cluster analysis of variables and a FA is that the original variables are retained in cluster analysis, albeit reordered, whereas the original variables are replaced by factors in a FA. Like cluster analysis, FA requires some intuition to interpret. Where cluster analysis produces tree diagrams, FA produces equations that define each factor in terms of the original variables:

3 You can also factor samples and even times from a time series but FA is most commonly applied to variables.

$$F_1 = a_{11}x_1 + a_{12}x_2 + a_{13}x_3 \ldots a_{1n}x_n$$
$$F_2 = a_{21}x_1 + a_{22}x_2 + a_{23}x_3 \ldots a_{2n}x_n$$
$$\vdots$$
$$F_m = a_{m1}x_1 + a_{m2}x_2 + a_{m3}x_3 \ldots a_{mn}x_n$$

where:

$\mathbf{F_1}$ through $\mathbf{F_m}$ are the m factors that replace the original n variables
$\mathbf{x_1}$ through $\mathbf{x_n}$ are the original variables
a_1 through a_n are factor analysis weights.
m is always less than or equal to **n**, and if you're lucky, a lot less.

What you have to do is look at the correlations between the original variables and the factors, and then figure out what each factor might mean in terms of the phenomenon you are analyzing. It's like being given a big box of parts—gears, transistors, tires, fabric, motors, pipes, wires, and lumber—and trying to figure out what they're supposed to make. Some parts will be integral and others will be left over.

As with cluster analysis, there are a number of specifications in FA that you have to become familiar with including factoring methods, communalities, and rotation methods. If you explore FA further, you'll find that these specifications become very complicated very quickly. Compared to cluster analysis, FA has much greater utility but at the cost of complexity.

Correspondence Analysis

Correspondence analysis (CA) is like FA in that variance is rearranged to reveal underlying data structures and reduce the dimensionality of a dataset. Also like FA, unfortunately, the data structures have to be interpreted to provide meaning. The difference between FA and CA is that in CA the original data is measured on a categorical scale. CA is applied to tables—cross-tabulation tables, contingency tables, frequency tables, and any other table where there is a numerical correspondence between the columns and rows. CA makes no distinction between dependent and independent variables.

Multidimensional Scaling

Multidimensional scaling (MDS) is also like FA, the difference being that MDS uses any kind of similarity or dissimilarity matrix, not just correlation matrices.

FA, CA, and MDS are like photographs. A photograph conveys a single moment in time, only two of three spatial dimensions, and no information about odors, sounds, temperature, or other circumstances, yet it still presents enough information so that observers can discern what is happening. So the term *data reduction* shouldn't be taken as a pejorative descriptor for these methods. Sometimes simplifying a problem is the best way to solve it; after all, isn't that what modeling is about?

Methods for Dependent and Independent Variables

If data analysis is to be well done, much of it must be a matter of judgment, and "theory," whether statistical or nonstatistical, will have to guide, not command.

John W. Tukey, *The Future of Data Analysis in Annals of Mathematical Statistics,* 1962

If you have more than one variable and you can distinguish between dependent and independent variables, the last major decision point is whether your variables are autocorrelated. If the variables are not autocorrelated, virtually the whole field of classical statistical techniques is open to you. For characterization, you can do the same types of analysis that were mentioned for undifferentiated variables. For detection, you can do the same analyses that were mentioned for undifferentiated variables plus statistical tests involving multiple populations, including the analysis of variance (ANOVA) and the analysis of covariance (ANCOVA).

Analysis of Variance

ANOVA is a statistical procedure for determining if there are differences in groupings of data. In ANOVA, the dependent variable is measured on a continuous scale and the independent variable(s) is measured on a discrete (i.e., nominal or ordinal) scale. When continuous-scale independent variables are included, commonly to control excess variation, the technique is called analysis of covariance (ANCOVA). The ANOVA/ANCOVA model is:

ANOVA

Continuous scale dependent variable ↔ **Categorical** scale independent variables

ANCOVA

Continuous scale dependent variable ↔ **Categorical** scale independent variables + **Continuous** scale covariates

Or in terminology only a mathematician could love:

$$y = a_0 + a_1 x_1 + a_2 x_2 + a_3 x_3 \ldots a_n x_n + e$$

where:

y is a continuous-scale dependent variable.

$\mathbf{x}_1$ through $\mathbf{x}_n$ are the categorical-scale independent variables. In ANCOVA, one or more of the independent variables may be continuous-scale covariates.

a_0 through a_n are constants. If a constant is statistically different from zero, the effect represented by the independent variable associated with it is considered significant.

e is the error term of the model. If the error is small compared to the rest of the model, the model is considered to be significant.

Figures 41 and 42 show the menus for SPSS's general linear models module and Statistica's advanced models module.

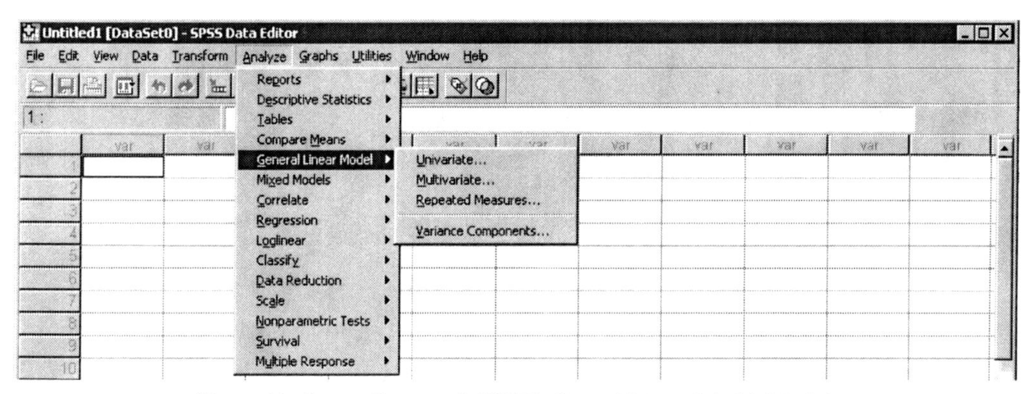

Figure 41. Screen Capture of SPSS's General Linear Models Module.

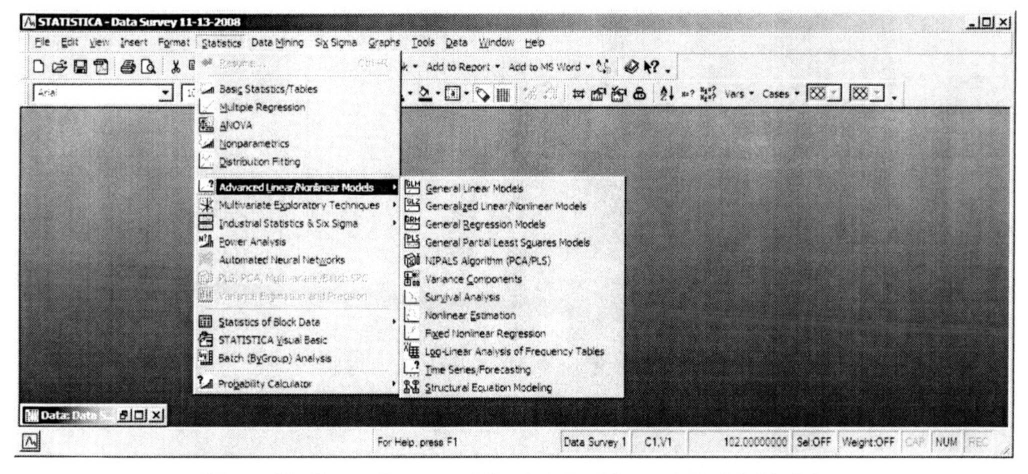

Figure 42. Screen Capture of Statistica's Advanced Models Module.

Regression

If you've taken a course in statistics, you've probably heard of regression analysis. It's a method for finding the equation of a line that fits a set of data points. The equation is developed by minimizing the squares of the errors (i.e., the differences between the observed values of the dependent variable and the values predicted by the model, called residuals). Simple regression involves one dependent variable and one independent variable. Multiple regression involves one dependent variable and more than one independent variable, based on the model:

Continuous scale dependent variable ↔ **Continuous** scale independent variables

or

$$y = a_0 + a_1x_1 + a_2x_2 + a_3x_3 \ldots a_nx_n + e$$

where:

$\mathbf{y}$ is the continuous-scale dependent variable
$\mathbf{x}_1$ through $\mathbf{x}_n$ are the continuous-scale independent variables
a_0 through a_n are constants
e is the uncertainty, the error term of the model.

There are many variations of statistical regression that are useful in certain circumstances. In stepwise regression, independent variables are automatically added to (or removed from) a model, one at a time, by assessing which variable would most improve (or least impair) some diagnostic statistic, usually the coefficient of determination. The stepwise process ends when no other variable can improve the model or when all the variables are included in the model. Nonlinear regression refers to statistical methods for generating models in which the parameters of the model (the a_1 through a_n terms) are not constants. These models are often applied to data representing changing rates of growth or decay. Figure 43 shows the types of regression provided in SPSS's regression module.

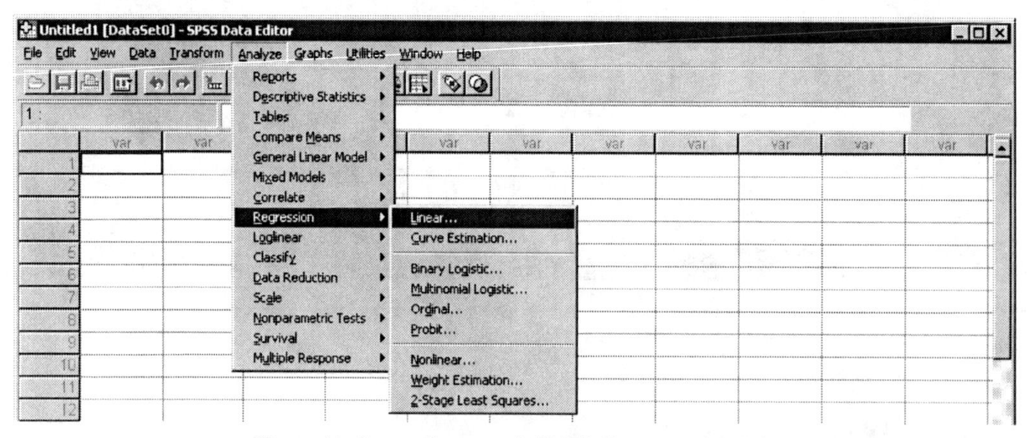

Figure 43. Screen Capture of SPSS's Regression Module.

In situations in which there are many independent variables, it is sometimes useful to replace the independent variables with principal components, a variation of factors in FA. Principal components (PCs) are not correlated, so the absolute contribution of each component to a statistical model can be quantified.

Discriminant Analysis

Discriminant analysis is used in situations in which the dependent variable represents group membership and the goal is to develop a model to evaluate the differences between the groups relative to the differences within the groups. The discriminant analysis model is:

Categorical scale dependent variable ↔ **Continuous** scale independent variables

or

$$y = a_0 + a_1x_1 + a_2x_2 + a_3x_3 \ldots a_nx_n + e$$

where:

y is the categorical-scale dependent variable.

$\mathbf{x}_1$ through $\mathbf{x}_n$ are the independent variables. If **y** has two levels, the *x*s represent the original variables. If **y** has more than two levels, the *x*s are replaced by principal components derived from the original variables.

a_0 through a_n are constants

e is the error term.

Canonical Analysis

Canonical analysis is used in situations in which there are multiple continuous-scale dependent variables as well as multiple continuous-scale independent variables.[4] The aim of canonical analysis is usually to explore underlying relationships between the two sets of variables.

The general canonical analysis model is:

Continuous scale dependent variables ↔ **Continuous** scale independent variables

or

$$b_1y_1 + b_2y_2 + b_3y_3 + \ldots + b_my_m + e = a_1x_1 + a_2x_2 + a_3x_3 + \ldots + a_nx_n + e$$

4 In canonical analysis, the distinction between dependent variables and independent variables isn't that important. The two groups of variables could just as well be called variable set A and variable set B.

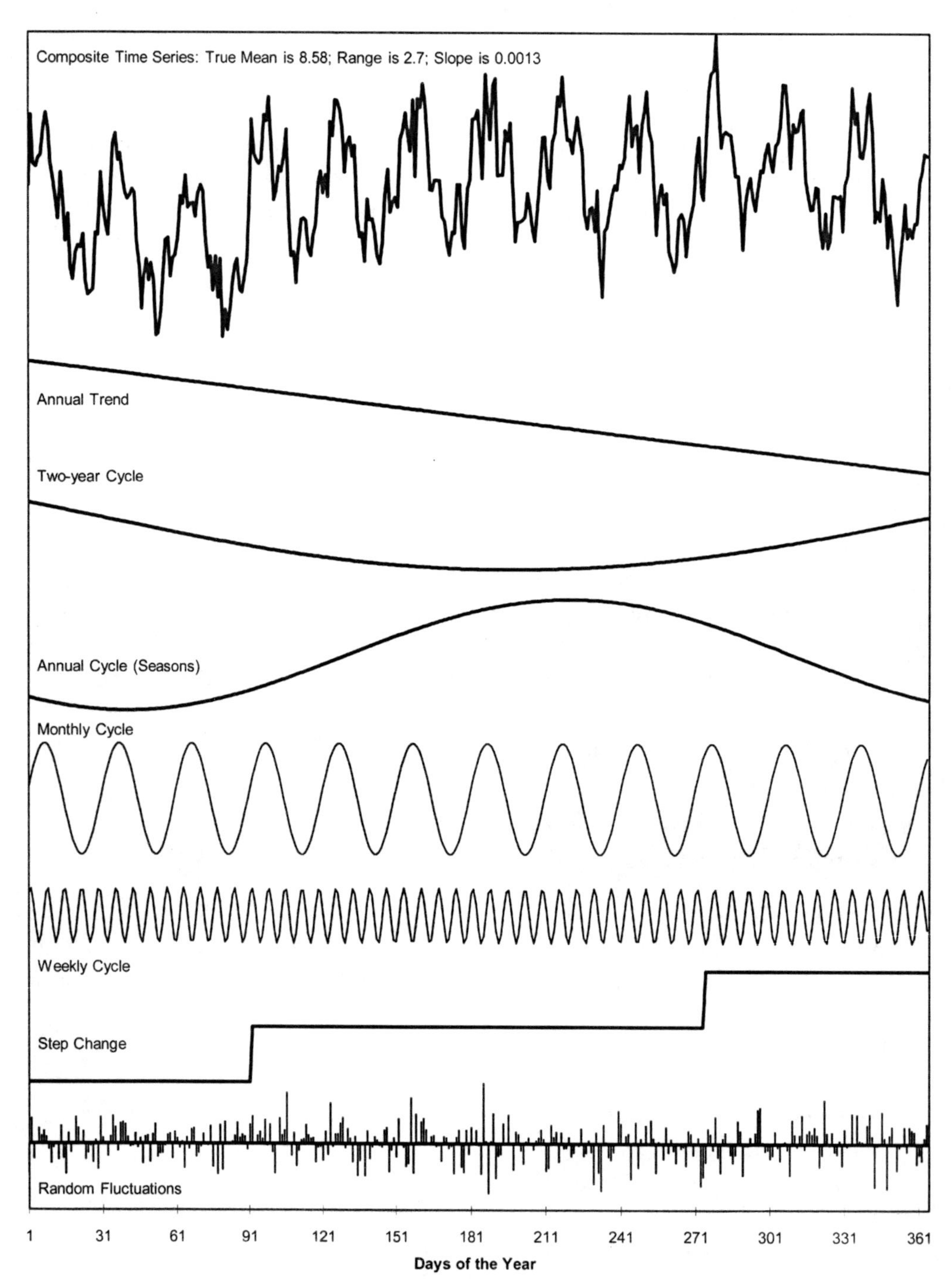

Figure 44. An Example of a Time Series with Seven Components.

where:

$\mathbf{y}_1$ through $\mathbf{y}_m$ are the original dependent variables used to create principal components.

$\mathbf{x}_1$ through $\mathbf{x}_n$ are the original independent variables used to create principal components.

a_1 through a_n and b_1 through b_m are constants

e_y and e_x are error terms.

As with FA, the trick is to interpret the meaning of the new components. With canonical models, there will be either n or m sets of canonical functions, whichever is less. Each dependent component will be associated with only one independent component, which can make interpretation somewhat easier or much more difficult.

You may have noticed by now that many of the models look alike except for the types of measurement scales of the variables used in the models. It's more complicated than that, of course, but this explains why you have to understand scales of measurement before you can decide what statistical methods to use.

Methods for Serially Correlated Variables

There are many methods for predicting the future. For example, you can read horoscopes, tea leaves, tarot cards, or crystal balls. Collectively, these methods are known as "nutty methods." Or you can put well-researched facts into sophisticated computer models, more commonly referred to as "a complete waste of time."

Scott Adams, *The Dilbert Future,* 1997

If your dependent variable is serially correlated, you have to use special techniques to address the dependency. For objectives involving characterization or detection, the most common approach is to use an ordinal-scale variable for some measure of time or location. For example, a time-dependent measure could be characterized by grouping the measurements into seasons or months. If you need to predict or explain a time-dependent variable, you could use a smoothing method, time-series regression, ARIMA (autoregressive, integrated moving-averages), spectral analysis or neural nets. If you need to predict or explain a location-dependent variable, you could use a smoothing interpolation method, trend-surface regression, or geostatistics (variogramming and kriging). If you have no idea what some of these words mean, get some more coffee and keep reading.

Ti i i ime is on my side. Yes it is.

Time Is on My Side

Time-series data are probably the most difficult type of data to analyze. Measurements involving time are usually autocorrelated, so using conventional statistical procedures can produce

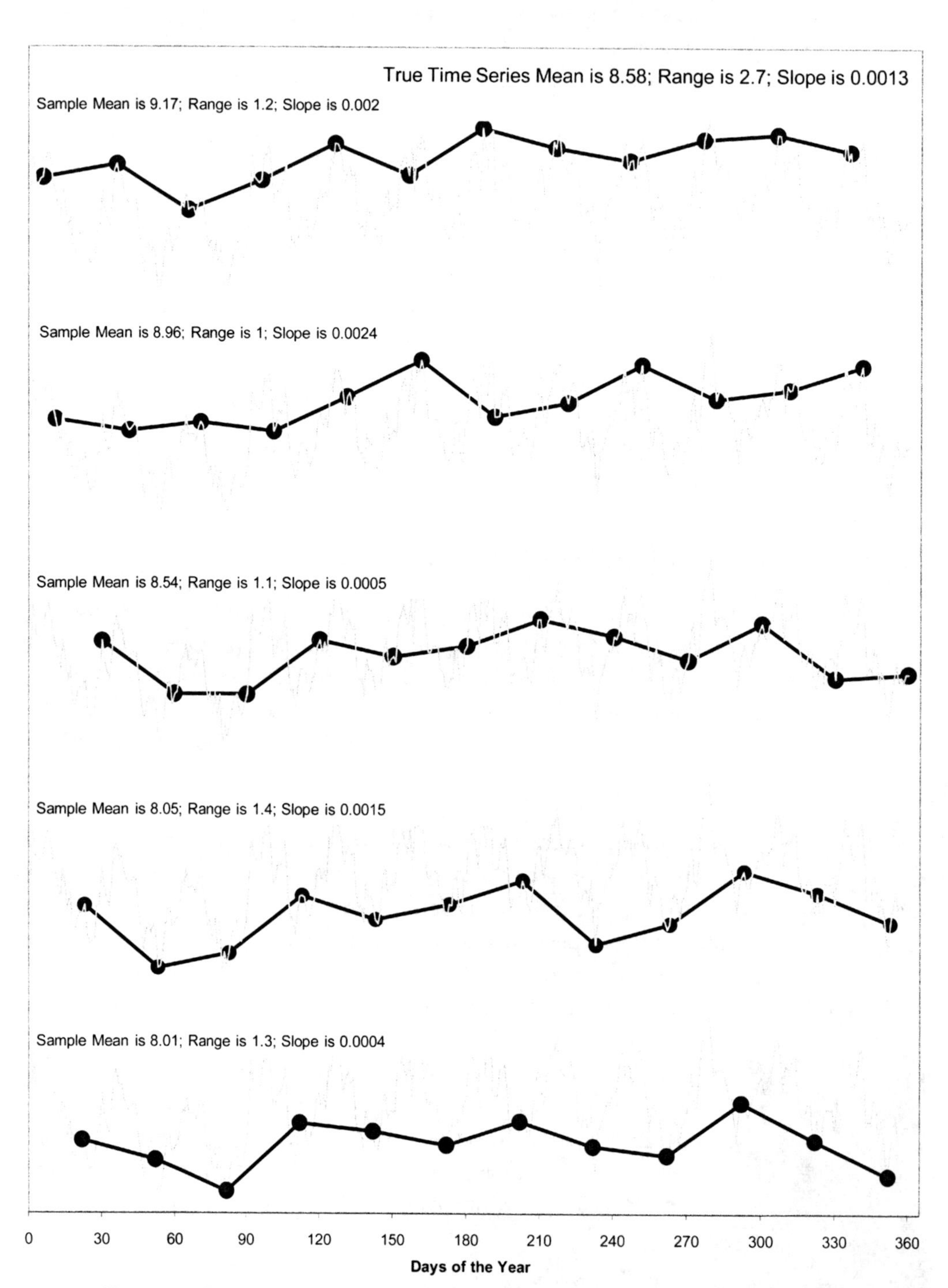

Figure 45. Results from Quarterly, Monthly, and Weekly Sampling of Time-Series.

biased results. But there are several other aspects of temporal variables that add to the confusion.

Pick a Scale. Any Scale.

Measurements of time have some quirky properties. Time might be measured on an interval scale given its capability of being subdivided and its arbitrary starting point. Time measurements follow in a sequence, so they are at least an ordinal scale, but they also repeat. Month 13 is the same as month 1. So time can also be treated as being measured on a nominal scale. Time units used for durations are measured on a ratio scale. Durations can be used in ratios, they have a starting point of zero, and they don't repeat (a duration of thirteen months isn't the same as a duration of one month).

Ch-Ch-Ch-Ch-Changes

Time-series data can exhibit a variety of patterns, including step changes, linear and nonlinear trends, and cyclic fluctuations. The effects may be superimposed on each other within a given time period or spread over many different time periods. For example, a change in the discharge of a river may be attributable to abrupt and ephemeral causes such as failure of a dam or a sudden downpour (shocks), abrupt and long-term causes such as natural changes in a drainage way or a man-made diversion (step changes), long-term causes such as drought or changes in water consumption (trends), repetitive changes such as seasonal cycles related to rainfall or irrigation (cyclic fluctuations) as well as random variations. Confounded effects are often impossible to separate, especially if the data record is short or the sampled intervals are irregular or too far apart.

It's All in Your Imagination

Consider the time series shown in Figure 44. The time-series consists of eight components (shown in the bottom portion of the figure):

- One linear (annual) trend
- Four cycles: two-year, one-year (seasonal), monthly, and weekly
- Two step increases
- Random fluctuations.

These components combine to create a fairly complex time series, shown at the top of the figure. However, the apparent time-series, the pattern you would see from periodic samples or measurements you collect, will depend on how often you collect the samples and where in the time-series you start.

Figure 45 (next page) shows the results of collecting samples at three different intervals from the time-series described in Figure 44. The results of quarterly sampling show little more than a linear trend. Seeing the result of this sampling interval would lead an investigator to believe the time series is simple when in fact it is not. But look again. The apparent trend is in the opposite direction of the real annual trend. What the sampling interval has detected is a smoothed version of the two step increases. The time-series isn't long enough to be sure you're seeing an annual trend. The results of monthly sampling suggest the presence of a seasonal cycle. The weekly sampling shows a 28-day cycle that drowns out other patterns.

Here's a rule of thumb. Collect time-series measurements at intervals at least one time-unit smaller than the interval you are interested in. For example, if you want to evaluate a weekly period, collect data daily. If you want to evaluate seasonal cycles, collect data at least monthly. The duration of sampling should be at least several times the length of the pattern you want to detect.

In the Blink of a While

Time-series measurements are not all collected at a single instant in time. Some measurements are composites over time. For example, a flow measurement (e.g., stream, air) may be an instantaneous discharge or a total discharge over a selected time period. A sample may be collected at one time or be a composite of several samples collected at discrete time intervals and combined into a single sample container. The period over which each measurement is averaged is called the support.[5] Obviously, you can't evaluate a given time interval if your support is the same or larger than the interval.

Not Backwards-Compatible

There is a dilemma involving time-series that are measured over many years. It goes like this. As knowledge and technology improve, the greater the chance that there will be improvements in sampling and analysis procedures that will reduce the overall variability of more recent measurements. That leads to violations of another assumption of most statistical procedures, homoscedasticity, or equality of variances. Sometimes, you just can't win.

In the Year 3000...

With most types of analysis, both statistical and deterministic, the range of the data covers the area of interest. If you want to analyze a chemical reaction at 100 degrees, you might analyze the reaction at temperatures between 80 degrees and 120 degrees. But you wouldn't test the reaction at 40 to 80 degrees and extrapolate to what might happen at 100 degrees. In fact, scientists are taught never to extrapolate outside the range of their data. But you have to extrapolate with time-series data because you almost always want to know what will happen in the future. If you wait to see what actually happens, then it's no longer interesting because it's the past. And in the ultimate of ironies, you often can extrapolate time-series data because they are ... autocorrelated. So the same property that makes time-series data difficult to analyze is what allows them to be extrapolated to future times, a process called forecasting.

Furthermore, with other types of data, even autocorrelated spatial data, you can

---

5 Support applies to spatial samples, too.

verify predictions whenever the need arises. With forecasts for a time-series, you have to wait until the time in question arrives. Then you have just one chance. You can't go back if something goes wrong and you miss collecting the verification data. Hence, you can't control verification.[6]

The Future Is How You Look at It

How to analyze time-series data depends to a great degree on:

- What question you want to answer
- How many data points you have
- The interval at which the data points were collected
- How you want to apply the results.

Consider three fundamental questions that you might ask about time series data. First, did something happen? In other words, is there a difference between some baseline period and some period after an event? Second, is there a trend? Has a slow change occurred over time? Third, what does the future hold? Is it possible to predict when an event will occur? Here are some ways to analyze time series data to answer these questions.

Did Something Happen?

Sometimes, measurements are made over time to see if some condition has changed. In these cases, the change in condition is instantaneous, or at least, instantaneous relative to the measurements taken. Typically, there may be only a few samples collected before and after the event to make the statistical determination.

What happened?

This type of situation is often dealt with using a statistical test in what is called an intervention approach. The concept is that measurements taken before and after some intervention or event will reveal the effect of the intervention. Say, for example, you run your company's billing department and want to test the effect of a new procedure on the accuracy of customer invoices. By comparing customer complaints before and after the procedure is implemented (the intervention) it may be possible to determine if something statistically significant happened. Events or treatments sandwiched between a baseline data period and a post event period can be analyzed this way.

---

6 Time travel did not exist when I started writing this book. That may no longer be true by the time I finish.

But what happens if the change is gradual and extends across many sampling rounds? In this situation the question is not "did something happen?" but rather "is there a trend?"

Is There a Trend?

The question of whether a set of time-series data displays a trend is very common. Is business improving? Is a patient's condition improving? Is environmental quality getting worse? There are several approaches to detecting trends.

Sen's test is a simple nonparametric test to determine if there is a linear trend in the data. The test involves calculating the slope between consecutive data points. The slopes are then ranked from largest to smallest. The farther away the median slope is from zero, the greater the likelihood of their being a significant trend. Sen's test can identify only linear trends. Time-series cycles will appear as if there is no trend.

In the Mann-Kendall test, the data are ranked from largest to smallest. Then all possible differences between data points are calculated. Positive differences are recoded as +1 and negative differences are recoded as -1. If the sum of the values is a large positive number, it is likely that there is an upward trend. If the sum is a large negative number, it is likely that there is a downward trend. The Mann-Kendall test can be modified to account for cyclic changes, such as seasonal fluctuations, by calculating the test separately for each part of the cycle.

Another method of testing for trends in time-series data is to fit a regression equation to the data and then examine the t–test for the regression coefficient to see whether the slope of the equation is significantly different from zero. There are several advantages to this approach. Missing and censored data can be accommodated, and many types of intrinsically linear and cyclic trends can be assessed. Furthermore, regression statistics are commonly provided in spreadsheet software (as opposed to Sen's test and the Mann-Kendall test which are provided only in specialized statistical software). On the other hand, this approach usually requires more data and assumes that residuals from the regression analysis are Normally distributed.

In repeated-measures ANOVA, changes over time can be detected with increased resolution because each sampling point serves as its own control. Natural variability and sampling-and-analysis variability are separated from changes over time, so the sensitivity of the statistical test is increased. The price of this enhanced detection capability is an increased level of difficulty in applying the technique. A well-planned statistical design, specialized software, and an evaluation of test assumptions are required.

How Long Until ...?

The ultimate question in analyzing time-series data is "How long until something happens?" the classic case of forecasting. It is the difference between the relatively simple statistical tests for identifying changes and trends, and advanced statistical modeling. So why isn't statistical time-series modeling done more often? Well, for one, this type of modeling commonly requires a considerable amount of time-series data. Some temporal modeling techniques require data points to be collected over regular intervals from a single sampling point. Specialized and relatively sophisticated statistical expertise is also required. Consider, for example, three types of statistical time-series models—smoothing models, regression models, and ARIMA models. Examples of these models are shown in Figure 46.

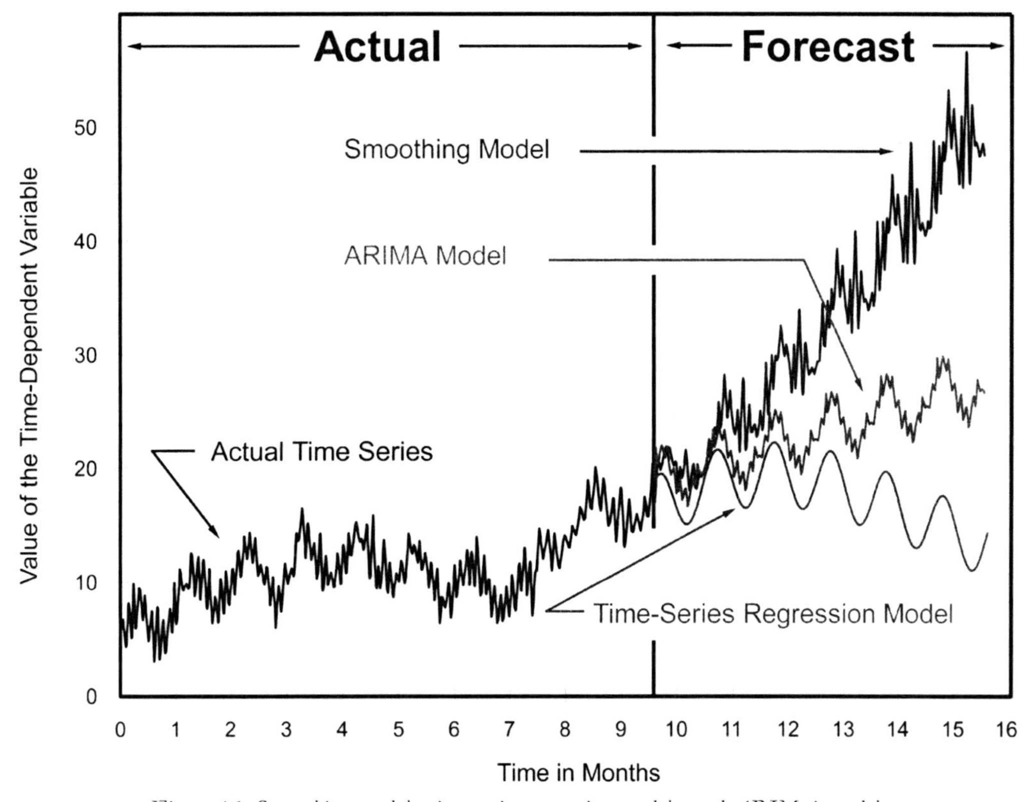

Figure 46. Smoothing models, time-series regression models, and ARIMA models.

Smoothing models use prior time-series measurements to produce estimates of later time-series measurements. A moving-average model is a simple time-series model in which a certain number of prior measurements are averaged together to produce a later measurement. The number of measurements in a moving-average model can be limited to accentuate variability or increased to eliminate perturbations. Weighted averages can be used to emphasize the contribution of more recent data. There are many types of smoothing models. Some models include linear, exponential, or seasonal trends. Some models are additive, and some are multiplicative. The possible combinations available for smoothing models are limited only by the specialized software being used for the analysis. Although every smoothing model has an equation underlying its forecasts, they're mostly used for graphical presentations. A common application is in the display of stock market prices over time.

Linear regression can also be used to develop time-series models. The advantage of this method is that the software is readily available. The disadvantage is that the independent variables in the regression model must usually be selected on the basis of trial-and-error and experience.

There are five types of functions that are used in time-series regression models—trends, cycles, lags, differences, and nontemporal predictors.

- **Linear trends** can be represented by the number of days after some selected starting date. Such a trend could also be represented by any intrinsically linear function, such as logarithms or reciprocals.
- **Cyclic trends**, trends that oscillate up and down in a regular fashion, can be represented by trigonometric functions. However, the period and the starting point of the cycle may be quite difficult to identify. Usually, meaningful periods (such as days, months, quarters, or years) are assessed to see if they produce significant autocorrelations (that is, correlations between a given time-series data point and prior time-series data points).
- **Lags** assess the autocorrelation between time-series data points and points from prior periods. A first-order lag assesses the autocorrelation between a given data point and the data point measured in the previous time period. A second-order lag assesses the autocorrelation between a given data point and the data point measured two periods previously. And an nth-order lag assesses the autocorrelation between a given data point and data point majored n periods previously.
- **Differences** are simply the result of subtracting the value of a time series at one time from it's value at another. As with lags, an nth-order difference is the arithmetic difference between a given data point and the data point measured n periods previously. Thus, rather than looking at similarities (as do lags), differences look at perturbations from the trend.

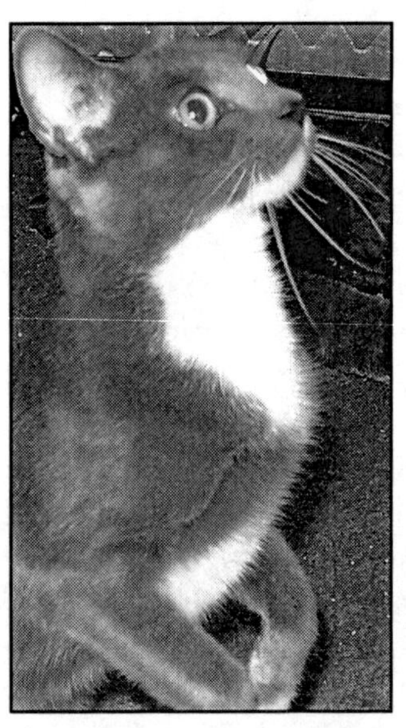

Are you going to feed me soon?

Time-series regression models can also include any nontemporal predictor term that contributes to successful forecast.

Like smoothing models, time-series regression models can be arduous to fit properly. Their advantage however is that more people are familiar with the concepts of regression analysis than with other types of time-series modeling.

Autoregressive integrated moving average (ARIMA) models are highly effective models for time-series data. An autoregressive term in an ARIMA model refers to a prior measurement in a time series. A moving-average term refers to an average of some number of prior measurements. An integrated term refers to the difference between two prior measurements in a time series. ARIMA models can contain one or more of these terms. They can incorporate all the types of functions that regression models can and are usually easier to fit because there are statistical procedures for identifying the most appropriate terms to include in the model. The difference between the two types of models is that regression models are based on the number of time periods after an arbitrary start date whereas ARIMA models are based on the time difference between data points. Regression models do not require that all data be collected at regular intervals whereas ARIMA models do. Finally, ARIMA models require specialized software. While they are more difficult to develop, ARIMA models do tend to produce better forecasts than other types of time-series models.

In addition to smoothing models, regression models, and ARIMA models, there are several other types of time-series models. Decomposition models attempt to separate out time-series trends, cycles, lags, and differences, analyze each component separately, and then combined the results into a composite model. The Census Bureau and the stock market use this approach frequently. Spectral models attempt to fit cyclic functions, particularly sine and cosine functions, to time-series data. This type of model usually requires hundreds of observations, and is used often in physics and astronomy.

Methods for Location-Dependent Variables

If the terrain and the map do not agree, follow the terrain.

Swedish army manual

Autocorrelated time measurements are certainly a challenge to deal with, but in some respects, autocorrelated location measurements are even worse. Time measurements may have many perplexing quirks, but at least time represents a single dimension.

Location. Location. Location.

This is My Space. You can use Facebook.

Location measurements can represent one, two, or even three dimensions, such as length, width, and height or depth.

In one dimension, location measurements can be analyzed like time measurements with one big difference. With time measurements, a point in time can only be influenced by earlier points in time. With location measurements, a point in space can be influenced by points anywhere near it.

In two dimensions, location measurements may be represented by maps or cross-sections. Maps are used commonly to illustrate changes in spatially dependent measurements over an area. Both maps and cross-sections can be contoured, which involves placing lines on a representation of the area to show where equal values of the measurement might be.

Did I hear the can opener? Way ahead of ya, Sid.

In three dimensions, location measurements may be represented by block diagrams, consisting of a map and one or two cross-sectional views.

Measurements that change at every location they are measured at are called regionalized variables. A regionalized variable is contoured either to depict spatial variations in the variable or to predict values of the variable at locations where it has not yet been measured. Contour maps can focus on either trends in the data or anomalies that deviate from the trends.

Competitive Gridding

The first step in the process of contouring is to create a grid of estimated values derived from the original data. This process is called gridding or interpolation. Manual interpolation involves estimating a few strategically placed values by taking a simple average between two adjacent data points. Computers, on the other hand, create grids of interpolated values using many data points. These grids are then contoured by connecting points of equal values at some user-defined interval.

Commercially available contouring programs employ a variety of algorithms for producing interpolated grid values. An algorithm is simply a mathematical process that is repeated to produce a set of values. Commonly used interpolation algorithms include: moving averages, inverse distance averages, splines, and geostatistics. Table 22 summarizes aspects of these algorithms.

The methods listed in Table 22 address spatial correlation by incorporating information about the distance between samples. The most sophisticated of the techniques, geostatistics, involves considerable effort to fit a theoretical mathematical model to data variograms, plots of the spatial variance on the y-axis versus the distance between the samples on the x-axis. Modeling a variogram first involves identifying a theoretical model shape that describes the behavior of the spatial variance. The model used most commonly in geostatistics is the spherical model. In a spherical model, spatial variance increases as the distances between samples increases up to a point at which the spatial correlation approaches zero. Fitting a spherical model to spatial data involves estimating three parameters:

- **Sill**—the spatial variance between locations that are statistically independent.
- **Range**—the distance at which locations become statistically independent.
- **Nugget**—the nonspatial error inherent in the measurement being taken.

Moving Average Interpolation	A *moving average* value for a grid node is produced by taking a simple average of a predetermined number of data values located near the grid node. Moving average surfaces tend to be very smooth and will emphasize trends. They will not produce a surface that extends outside the range of the original data.	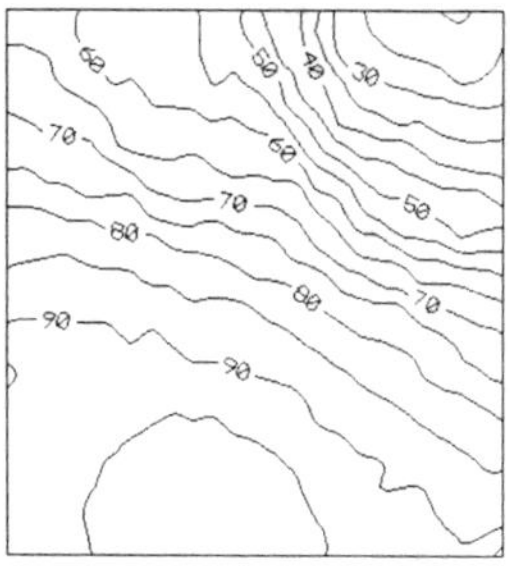
Inverse Distance Interpolation	*Inverse-distance* interpolation produces grid values by weighting the data values by the reciprocal of the distance from the data point to the grid node. The farther away the data point is from the grid node the less it contributes to the node's calculated value. Inverse distance surfaces tend to be less smooth than moving average surfaces, will emphasize both trends and anomalies, and will not produce a surface that extends beyond the original data.	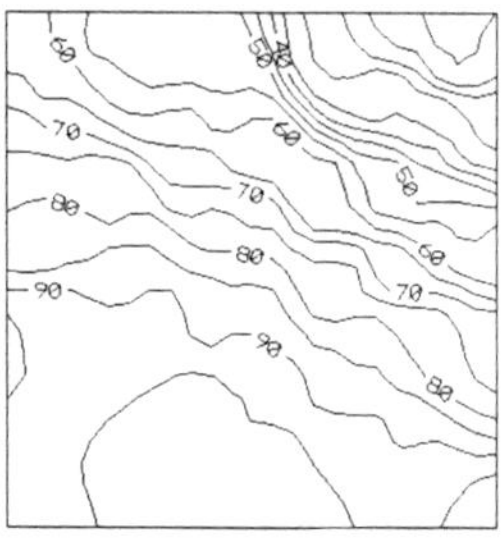
Geostatistical Interpolation	*Geostatistical* interpolation is like inverse-distance interpolation except that the gridding process takes into account the spatial autocorrelation of the data. To produce a geostatistical surface, first the autocorrelation structure of the data is characterized in a process called *variogramming* (or semivariogramming to geostatistical purists). Grid values are interpolated based on the correlation structure in a process called *kriging*. Geostatistical surfaces will emphasize both trends and anomalies.	
Spline Interpolation	*Splines* are mathematical functions that minimize the change in a surface's slope. Spline surfaces tend to be smoother than inverse distance surfaces and will emphasize trends rather than anomalies. Splines can also produce surfaces that extend outside the range of the original data.	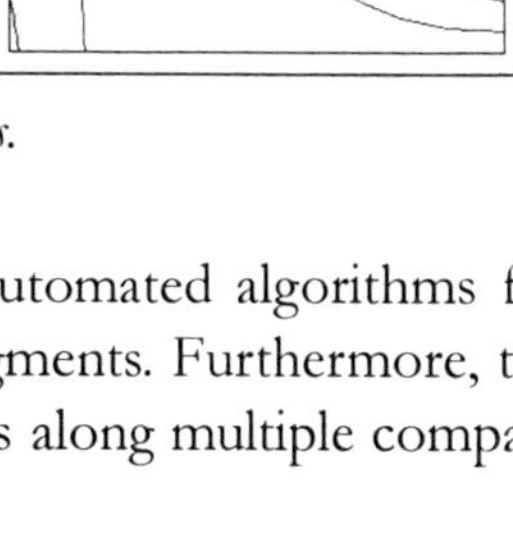

Table 22. Four Spatial Interpolation Algorithms.

This is not a trivial process because, though there are automated algorithms for fitting the model, the data analyst still has to make many judgments. Furthermore, the model has to be fit many times to assess correlation structures along multiple compass directions, termed anisotropy.

The other method of addressing spatial correlation is to incorporate information about the location coordinates of the samples. This is the strategy behind trend surfaces, which are produced by using statistical regression of location coordinates (the independent variables) to generate grid values (of the dependent variable). For example, a simple, flat trend-surface model would be:

$$\mathbf{y} = a_0 + a_1(\textbf{North coordinate}) + a_2(\textbf{East coordinate}) + e$$

where **y** is the spatially autocorrelated dependent variable. Commonly, trend surfaces use higher-order polynomials, for example:

$$\mathbf{y} = a_0 + a(\mathbf{N} + \mathbf{E})^2 + e$$

Where N is the north coordinate and E is the east coordinate. This equation is equivalent to:

$$\mathbf{y} = a_0 + a_1\mathbf{N}^2 + a_2\mathbf{NE} + a_3\mathbf{E}^2 + e$$

Trend surfaces have the advantage of being able to use rotated and translated coordinates, other types of mathematical functions of the geographic coordinates, and non-location variables to achieve a better fit. This better fit comes at the cost of a lot more effort. In general, trend surfaces tend to be very smooth and will emphasize trends even well outside the range of the original data.

Controlling Perceptions

Even with an interpolation algorithm selected, there are still many ways to control the appearance of a contour map. For example, there are several options that contouring software provide to selects data points to use to interpolate a grid node. You might specify the number of points to use and the software will find the closest data. Or you can also specify a patterned search in which the software will look for the specified number of points in quadrants, octants, or other pattern. Specify a lot of points and search sectors, and you'll make your surface look much smoother than it might really be. That's why it's important to determine the geostatistical range.

If the contouring algorithm adjusts the interpolated grid so that the contoured surface agrees with the original data values, it is called an exact interpolator. If the algorithm does not force the grid to honor the original data points, it is called an inexact interpolator. Most algorithms are inexact interpolators except for some splines and geostatistics, which can be either. Choosing an exact interpolator over an inexact interpolator is a matter of philosophy more than mathematics. If you believe there may be random error or uncertainty in your measurements,[7] use an inexact interpolator. If you believe your measurements are accurate and precise, and would be identical upon replication, use an exact interpolator. In a sense, using inexact interpolation shows a mistrust of the data; using exact interpolation shows a mistrust of the algorithm. Take your pick. I've found that most clients prefer exact interpolation. For some reason, they don't like the idea that they paid a lot of money for data that doesn't appear to be on the map you make.

Even before the days of computer contouring software, there were plenty of ways to control the appearance of a data surface. Picking an appropriate contour interval was, and still is, an art. Too large an interval and you won't see much. Too small an interval,

7 And what statistician doesn't?

and you'll see stuff that's not really there. And with modern contouring software that can shade, colorize, and even produce three-dimensional images, it's too easy to lose the data behind the flashy façade.

There are many other interpolation algorithms, data search controls, and other ways to customize a contoured surface. Your options will depend on the contouring software you plan to use.[8] Think carefully about the choice of interpolation controls and contour intervals. How you contour your data will have a profound effect on how you and your readers interpret reality.

8 You can always contour your data by hand, which is how it was usually done before the 1990s. This is still a useful approach if you have only a few data points or you need to incorporate the effects of some qualitative feature into your interpretation of the spatial variable. Contouring software has become like indoor plumbing; you only go back to the old ways in an emergency.

23 Models and Sausages

Models are like sausages. They look good on the outside, but there are a lot of things that go into them that you wouldn't necessarily expect. It's usually better to know the truth, however gruesome it may be. So let's try making some statistical sausage.

The Jungle That Is Modeling

Nothing in these abstract economic models actually works in the real world. It doesn't matter how many footnotes they put in, or how many ways they tinker around the edges. The whole enterprise is totally rotten at the core: it has no relation to reality anymore—and furthermore, it never did.

Noam Chomsky, "The Fraud of Modern Economics" in *Understanding Power,* 2002, p.251

Imagine this. Your boss asks you to analyze the satisfaction of the customers of the business for which you work. He says he wants to know what the most important factors are to satisfied customers so that he can focus on doing those things.

The Phenomenon

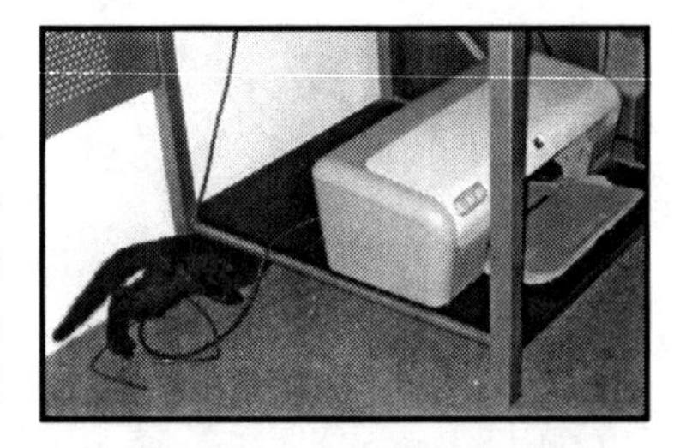

I see the problem. Something's been chewing on your cables. Nom. Nom. Nom.

Your boss wants you to analyze customer satisfaction, so first you have to understand what it is. What do you do? You thank Gore[1] for the Internet and Google "customer satisfaction." From the Internet, you learn that researchers believe there are ten aspects (called domains) of customer satisfaction: quality, value, timeliness, efficiency, access, environmental, teamwork, service, commitment, and innovation.[2] Furthermore,

1 "Former United States Vice President Al Gore did not invent the Internet. What's more, he never said he did! However, Gore's contributions to the development of the Internet as we know it were quite significant." From: www.boutell.com/newfaq/history/gore.html.

2 Berry, Bart Allen. 2002. *What Customers Want!* Aquarius Publishing San Diego California.

satisfaction is regarded as the disparity between the customer's perception of a business' performance and the customer's expectation of a business' performance. After a few more hours of research, you decide it's time to talk to your boss.

Model Use and Specifications

You again ask your boss what he wants the model to do. He then gives you a twenty-minute lecture on the importance of customer satisfaction. No help there, so you try a different approach. You ask what decisions he expects to make based on the analysis results. He reiterates that he wants to know what the most important factors are to satisfied customers so that he can focus on doing those things. There are two things troubling with this statement. First, when people repeat what they've already told you to different questions, it often means they haven't thought out the next step.[3] The boss hasn't thought about what decisions he might have to make. He hasn't linked your analysis with something he must do, so he is not invested in the analysis. This is a very bad sign. Second, he said he wants to focus on what works, not what's broken. Now it could be argued that if something makes you successful, you should keep doing it. But at the same time, you can't ignore what isn't working. Your car's engine may be running great, but you'll still have to stop if you have a flat tire.

This is where you have to be more than just a number cruncher. You talk about the ten elements of customer satisfaction and how you might investigate them. Then you mention a few scenarios for what could be done in response to the study's findings, like employee training and team-building exercises, a new PR campaign, and investing in new technologies to improve communications. These could be very expensive responses, so it's important to do the analysis right. You hope that some of the information sinks in.

So it's now clear to you that this would be a one-time, single-purpose model for explaining an abstract condition (i.e., satisfaction). Obviously, you won't be building a physical model. You'll be representing customer satisfaction as numbers, a graphic like a flow chart, or text. Now you just have to figure out how to do it.

Variables, Samples, and Data

You got lucky selecting concepts for the model. Other people did the work for you and put it on the Internet. So all you have to do is figure out how to measure quality, value, timeliness, efficiency, access, environmental, teamwork, service, commitment, and innovation.

This is where it gets challenging. You start by identifying candidate variables that would measure the customers' satisfaction. You might:

- Ask the customers directly if they are satisfied with your company's product.
- Find out how much of a certain type of product that a customer buys is from your company.

3 Or, they're a White House press secretary.

- Determine how long customers have purchased products from your company.
- Determine how frequently customers have purchased products from your company.
- Research how many complaints customers have filed.
- Analyze how quickly invoices are paid.
- Interview your sales force to obtain their view of customers' satisfaction.

After some reflection, you decide you can forego a survey of your own sales force. They're going to give you their perceptions, which may not actually reflect customers' feelings. The customers' payment history also isn't a good criterion measure, at least for the purpose of this model, because payment delays may be attributable to administrative procedures or cash-flow problems rather than satisfaction. Frequency of orders may be misleading as a measure of satisfaction because all customers won't need your product on the same schedule. The percent of a product a customer buys from your company may also be problematical because you would need your customers to provide information they might consider to be confidential.

Direct observation is always better than opinion surveys.

That leaves a survey of customers' satisfaction, the durations customers have done business with your company, and the numbers of customer complaints. The lengths of time customers have purchased your products and the numbers of complaints have the advantage of being hard data. These measures reflect information on behaviors that were demonstrated in the past, not just intangible feelings. However, they are indirect measures of satisfaction. They would also have to be standardized to account for the ages and sizes of the customer businesses. There's also no way to compensate for corporate culture; some customers complain while others don't. They may also complain verbally, so there is no data trail to follow up on.

A direct survey of customer satisfaction, on the other hand, would reflect a single point in time, is intrusive on the customer, and may be biased by respondents who alter their responses to fit some specific agenda rather than their true feelings. But it is really, at least on the face, the only direct way to measure customers' satisfaction. Table 23 summarizes the candidate dependent variables.

Table 23. Relationships of Candidate Dependent Variables to Example Phenomenon

	Hard Data (tangible, observable)	**Soft Data (intangible, opinions)**
Directly Related to the Phenomenon	?	Survey of Customer Opinions
Indirectly Related to the Phenomenon	**History of Complaints** **Time as Customer** *Percentage of Purchases* **Frequency of Purchases** **Invoice Payment History**	Interviews with Sales Force
Ease of Data Collection	**Bold Font = EASY**; *Italics Font =DIFFICULT*	
Anticipated Data Quality	Large Font = GOOD; Small Font = FAIR	

So you decide to conduct a survey of your customers in which you'll ask a variety of questions to determine how satisfied they are with your company's product. You'll also try to track down the ordering and complaint data to see if it might be useful.

Now you have to decide specifically what questions to ask and what scale to use. Everyone has taken an opinion survey at one time or another. Unfortunately, some people think this is all the training they need to conduct their own surveys. It's OK to create a whimsical poll to augment a blog, jazz up a website, or irritate the liberals on Facebook, but if you want meaningful data, you have to do it right. There are scores of books and thousands of Internet resources on survey design that are readily available. You don't have to become an expert, just be sure to learn enough to generate good data. Here are a few things you consider:

- **Manage text or numbers**—Before you write your questions, decide what form you want the responses to take. If you just want numbers that you can count and average, don't ask any *open-ended* questions. Results expressed in numbers are far easier to analyze than text responses. On the other hand, scaled questions are constrained to what you ask about while open-response questions can elicit observations on things you may not have thought of. You can use both in the same survey. Keep in mind when you analyze the data though, that you can't weight the importance of open and closed-ended responses the same. A free-form text comment is an anecdote. It is qualitative. It may represent only the respondent who made the comment or it may represent many respondents who did not think of making the same comment. You won't know.

- **Define your scales.** What degree of precision do you think you'll need? Preference scales can have any number of levels. The most commonly used scales have two levels (yes, no), three levels, four levels, five levels, and ten levels.

  » Use a scale that lends itself to the question. For example, if you want an either/or dichotomy, use a two-level scale. If you want a proportion, use a ten level scale.

» Respondents tend to have a harder time answering questions when the scales have "too many" levels. So, don't use a five-level scale if you could make your decision based on a three-level scale.

» Define all the levels of the scale, especially if you use a scale with an odd number of levels. Scale midpoints confuse some people. Others like to sit on the fence between the extremes of a scale and not commit one way or the other. For these reasons, some pollsters prefer using scales with an even number of levels. That way they don't have to deal with trying to interpret the middle ground.

» Try to use as few different scales as possible. Also, try not to mix scales in a block of questions. If you do use multiple scales, don't reverse the directions of the scales. For example, don't make a response of 1 mean best on one scale and have the same scale mean least on another. Be consistent in having the 1 response always mean least , worst, disagree, ineffective, unimportant, small, dissatisfied, bad, slow, poor, unlikely, and so on.

- **Ask only for what you'll use.** Don't waste respondent's time on questions that won't generate a decision. For example, if you are constrained from changing the color of your product, don't ask if another color might be preferable. If you won't change your product, an overall satisfaction question might be fine. But if you want to improve your current product, several questions on different aspects of the product would be necessary.

- **Ask the right questions the right way.** This is by far the hardest part of survey design. Try to avoid introducing any bias in your questions. Bias is a particular problem in surveys because respondents sometimes provide the answer they think you want instead of their true opinions. If a certain topic is critical to a decision you have to make, ask about the topic more than once in different ways. Try eliciting opinions both directly and indirectly. For example, in addition to asking if they like your product, ask if they would recommend the product to others, or if they use your product routinely in their work, or if they would try a competitor's products if given the chance. Sometimes it's good to phrase the question in terms of their expectations. A survey respondent might say that they like your product but admit it didn't meet their expectations.[4] Also, be as specific as possible. Specific questions tend to generate better results.

---

4 Have you ever purchased a product or service without actually seeing the goods? Of course you have. You might have gone to a movie or a restaurant or purchased something through the mail or over the Internet. What you got might have been of fine quality, but did it meet your expectation? Read some of the reviews of products at Amazon.com or comparable websites. It's not hard to find reviews of products that award four or even five-stars yet still lament about what the reviewer wishes the product could do. Have you ever received a performance review from your boss in which he or she said your work was satisfactory but below what they expected from you? Sometimes there is a huge gap between satisfaction and expectation that can be advantageous to explore in a survey.

- **Consider the Delivery System.** There are a number of ways to conduct surveys—in person, by phone, by mail, or by the Internet. Each method has advantages and disadvantages. If you (or your designees) will conduct the survey in person or by phone, you'll need to have a script, a plan for how to handle questions and unanticipated circumstances, and training to be sure all the surveyors are consistent. Personal surveys tend to cost much more than self-administered surveys because of the manpower and scheduling aspects of the work but you will have more control over data quality and response rates. If you will conduct the survey by mail or by the Internet, you'll have to ensure that the survey instructions and the questions are easy to understand as you won't be there to provide support. Internet surveys, in particular, are very inexpensive, can be directed to specific individuals or a broad group of individuals, and can accumulate responses without additional data entry. However, not everyone has (or uses) Internet access and occasionally there are technological glitches that causes problems.

- **Keep it short and simple.** Long surveys frustrate respondents, so try to keep the survey short and simple to complete. Always provide an estimate of how long the survey will take to complete. Shoot for less than five minutes and don't exceed ten minutes without some incentive. Don't be afraid to add enough text to the survey so that respondents understand the questions and the reasons why their careful responses are important.

- **Give them time.** Surveys that are not time-sensitive usually run two to four weeks. Spanning at least one weekend allows workers who are traveling some time to get back into the office. Try to avoid holidays that might cause the survey to be put on the to-do pile and forgotten. Some statisticians like to deploy their surveys on Tuesday mornings because many workers now work at home on Mondays and Fridays or take those days off to extend vacations.

- **Know your respondents.** Surveys are almost always either anonymous (i.e., the researcher does not know who provided the response) or confidential (i.e., the researcher knows who provided the response but does not divulge their identity). Anonymity is often used to reduce the possibility of researcher bias and to assure respondents of their privacy. Still, it is usually essential to know something about each respondent. You might design a few questions to categorize how long they've been a customer and how much business they do with your company, for instance. Then you can use those categories to compare their perceptions of satisfaction. Put these questions at the end of the survey so that your respondents don't become hesitant to provide honest responses.

- **Remind but don't nag.** Most of the responses you will receive to a voluntary, non-interview survey will be completed within the first twenty-four hours of respondents receiving the notice. This can be as much as a third to a half of the people who will respond. If the number of responses in the first twenty-four hours (not on a weekend) is not about half of the number you want, you'll have to sent one or two reminder messages. After each reminder, you'll get a spike of returns that will diminish to nothing over a few days. Manage your surveys so that you don't send reminders to people who have already responded.

- **Think about the future.** If you think you might repeat the survey in the future, try to ask the questions in ways that won't make them dated so that the data can be compared. Try not to drop, add, or reword questions unnecessarily.

So you develop a survey with twenty questions. Twelve questions will use a four-level scale, including two questions on overall satisfaction, and ten questions on the domains of satisfaction you identified in the literature. One question will employ a ranking scheme to identify the relative importance of the ten domains of satisfaction. One open-ended question will solicit any other comments customers may have. Six additional questions will categorize the customers' use of your products and the individual completing the survey for the customer. The whole survey should take no longer than five minutes to complete. You also plan to put a code on each survey so that you can later merge the survey data with business data your company maintains on the customer. This makes the survey confidential but not anonymous so you'll have to have a system to keep the information secure.

Statistical Methods

There are a lot of things you could do with these data. Starting with what your boss actually asked for: What is it we do well that our customers are most satisfied with? So the first thing you might do is calculate descriptive statistics for all the questions.

Now is a good time to think about how you view the scales for the survey questions. Are the scales ordinal or interval scales? Take, for instance, a question like:

Are you satisfied with the services (or products) you purchase from the XYZ Company?

» Yes, even more satisfied than I expected
» Yes, I am satisfied
» No, I am somewhat disappointed
» No, I am very dissatisfied.

Do you believe that respondents could subdivide these levels further if given the chance? Do you believe there are many shades of gray in the answers? For example, could there be many steps between disappointment and downright dissatisfaction? If you answered these questions no, you are probably thinking of the scale as ordinal. Ordinal scales are like stairs, discrete positions that are arranged in a sequence, where you can only be on a step, not between steps. If you answered these questions yes, you are probably thinking of the scale as interval. Interval scales are more like a ramp, a continuous area where you can be anyplace your stride takes you.

If you are using an ordinal scale, you are pretty much limited to percentages. So to answer your boss's question, you might calculate:

- Percents of the respondents who selected each of the response levels. For a four-level scale, that will produce four percentages for each question.

- Percent of the respondents who said they were satisfied, (i.e., the top two response levels).
- Percent of the respondents who said they were satisfied minus the percent of the respondents who said they were not satisfied. This could be calculated as either the percentage of responses of the top level minus the percentage of responses of the bottom level, or as the percentage of responses of the top two levels minus the percentage of responses of the bottom two levels.

If you are using an interval scale, you can and should calculate percentages but you also have other options. First, you recode the word responses into numbers (e.g., very satisfied = 4; satisfied = 3, dissatisfied = 2; and very dissatisfied = 1). Then you can calculate descriptive statistics, such as the average of the responses for each question. If you find it difficult to work with a scale that is not based on a multiple of ten, you can convert the average response from a ***k***-level scale to a scale of 10 by multiplying by 10/***k*** (or to a scale of 100 by multiplying by 100/***k***).

So to answer the boss's question about what your customers are most satisfied with, you calculate percentages or average scores for the responses. If you are using an interval scale, you could also use regression analysis to evaluate the influence of the ten domains of satisfaction on overall satisfaction using the model:

$$
\begin{aligned}
\mathbf{y}_{\text{overall satisfaction}} = \mathbf{b}_0 &+ \\
&\mathbf{b}_1\mathbf{x}_{\text{satisfaction with quality}} + \\
&\mathbf{b}_2\mathbf{x}_{\text{satisfaction with value}} + \\
&\mathbf{b}_3\mathbf{x}_{\text{satisfaction with timeliness}} + \\
&\mathbf{b}_4\mathbf{x}_{\text{satisfaction with efficiency}} + \\
&\mathbf{b}_5\mathbf{x}_{\text{satisfaction with ease of access}} + \\
&\mathbf{b}_6\mathbf{x}_{\text{satisfaction with environment}} + \\
&\mathbf{b}_7\mathbf{x}_{\text{satisfaction with inter-departmental teamwork}} + \\
&\mathbf{b}_8\mathbf{x}_{\text{satisfaction with front line service behaviors}} + \\
&\mathbf{b}_9\mathbf{x}_{\text{satisfaction with commitment to the customer}} + \\
&\mathbf{b}_{10}\mathbf{x}_{\text{satisfaction with innovation}} + \\
&e
\end{aligned}
$$

Then you could compare the regression results with the results of the question in which respondents ranked the importance of the domains of satisfaction they believed were most important to their overall satisfaction. If the results aren't similar, it might mean there is some issue with the way you asked the questions, there is substantial multicollinearity between the responses, there are outliers or errors, or there is a disconnect between the way customers actually perceive satisfaction and how they responded (e.g., they may be satisfied but their expectations were not met).

You could also use regression analysis to evaluate the influence an *individual's* background has on overall satisfaction using the model:

$$\begin{aligned}
\mathbf{y}_{\text{overall satisfaction}} = \mathbf{b}_0 &+ \\
&\mathbf{b}_1\mathbf{x}_{\text{years purchasing from XYZ Company}} + \\
&\mathbf{b}_2\mathbf{x}_{\text{years with current employer}} + \\
&\mathbf{b}_3\mathbf{x}_{\text{percent of time spent purchasing}} + \\
&\mathbf{b}_4\mathbf{x}_{\text{number of complaints}} + \\
&e
\end{aligned}$$

Or you could use regression analysis to evaluate the influence a *company's* background has on overall satisfaction using the model:

$$\begin{aligned}
\mathbf{y}_{\text{overall satisfaction}} = \mathbf{b}_0 &+ \\
&\mathbf{b}_1\mathbf{x}_{\text{years purchasing from XYZ Company}} + \\
&\mathbf{b}_2\mathbf{x}_{\text{amount of business done with XYZ Company}} + \\
&\mathbf{b}_3\mathbf{x}_{\text{average days receivable}} + \\
&\mathbf{b}_4\mathbf{x}_{\text{number of complaints}} + \\
&e
\end{aligned}$$

The terms in these models would depend on what questions you asked at the end of the survey to characterize the respondents and their company, and what business information you might have available in your company's databases.

You could use analysis of variance (ANOVA) to evaluate the influence of factors that categorize respondents, such as the model:

$$\begin{aligned}
\mathbf{y}_{\text{overall satisfaction}} = \mathbf{b}_0 &+ \\
&\mathbf{b}_1\mathbf{x}_{\text{company size category}} + \\
&\mathbf{b}_2\mathbf{x}_{\text{geographic region}} + \\
&\mathbf{b}_3\mathbf{x}_{\text{major NAICS or SIC category}} + \\
&e
\end{aligned}$$

The independent variables in this ANOVA model would be measured on either nominal or ordinal scales.

You could use cluster analysis or correspondence analysis to group your customers into types based on their responses. Then use the categorization to fine-tune how you interact with each client type.

So there are many kinds of knowledge you can extract from the survey data besides the answer to your boss's simple question. Performing more than one type of data analysis is often a good way to confirm findings from different analyses. You may also develop new business intelligence and identify data patterns that should be explored further. This is important when you consider the effort and cost of generating good data.

The Modeling Process

You already have all the software and reference materials you'll need to analyze the survey results, so the important thing you'll have to do at this point is collect the data. In this case, that means actually conducting the survey. So first, you have to put together a list of clients to send the survey to. After getting approvals from seven organizational levels of your company, you are granted an audience with the high priest who maintains the customer database. He tells you that the company has 12,810 unique customers, 5,863 of which have purchased services within the past three years. The database has business data and contact information on 2,781 of those customers. So you have the cleric create you a database of that information and thank him for his benevolence.

Now you have to fine tune your sample size. You figure a margin of error of ±5 percent will be adequate for the survey, so you calculate you'll need about 400 responses.[5] You've never surveyed these customers before, so you think you might only get a 25 percent response rate. That means that you'll have to survey 1,600 customers to get 400 responses. Ouch! So you figure you might as well just survey every company since you'll probably get unusable responses. The additional responses also couldn't hurt if you decide to group the companies or do regression analyses. You consider using incentives, such as small gifts or honoraria, to improve the response rate but your boss nixes the idea.[6]

You'll need to use one of these if you want me to answer your survey.

Your next task is to decide how to deliver the survey. To conduct the survey by personal interviews or even by telephone, you would have to hire

---

5 The number of samples needed to have about a 5 percent margin of error with 95 percent confidence is equal to $1/\sqrt{0.05}$. In general, the number of samples (n) needed to have confidence (α) that a margin of error (moe) will include a population proportion estimated from a sample of the population (p), is equal to $z_\alpha^2 * p * (1\text{-}p) / moe^2$. The quantity z_α is the value of the standard normal distribution where the area in each tail is $(1\text{-}\alpha)/2$). If the number of samples is close to the number of individuals in the population (N), multiply the moe by (1- (n/N)). These formulas only apply to proportions with a restricted range of 0 percent to 100 percent.

6 Including gifts to encourage survey recipients to respond is common when candidate respondents are less likely to answer surveys, such as doctors, business executives, and statisticians. Once I mailed a survey with an inexpensive gift pen to an employee of the U.S. government only to have the enveloped returned. It was stamped "unable to deliver" and "return to sender," and the new address was printed on a label and affixed to the envelope. The label had the same name, the same city, state, and zip, the same government agency and department, the same street, the same building number, but a different room number. It gets better. The envelope was opened and the pen was removed, then taped closed again. So don't believe any claim you hear that government employees aren't trying to minimize expenditures for office supplies.

a contractor or train temporary staff. That would take too long for your boss, so you decide to use a mail survey. If you had programming support in your company, you would have considered an Internet survey, but that isn't an option. So snail mail it is. You estimate that it will take less than ten minutes for a respondent to complete the survey, even while drinking coffee, talking on their cell, and/or using the rest room. You also decide a six-week deadline for receipt of the surveys is a bit long; a month ought to do it.

So you have the surveys printed, you code the forms, and you write a cover letter with a description of the purpose of the survey, an estimate of the time to complete it, and a reply-by date. Then you stuff the envelopes, include a SASE[7], and mail them. In a few days, you start receiving undeliverable surveys back from the post office. You fix what you can and send a list of the remaining bad addresses to the high priest of the customer database so that he can correct his files. Two weeks after you mail the surveys, you send a reminder email to all your candidate respondents. Once again, a few of the addresses bounce back to you, so you send the list of the bad email addresses to the high priest. You might have considered also making follow up calls to remind respondents about the reply by date, but by then, you have a 30 percent response rate. Success!

The next month is lost as the data is entered and verified and arranged into a structure amenable to statistical analysis (see Chapter 16). You then spend another few weeks scrubbing the data (see Chapters 17 and 18), do some exploratory analyses (see Chapter 19), and beef up the dataset (see Chapter 20). The data analysis and reporting takes another few weeks. But in the end, your boss loves the results and praises your work to the CEO who then gives you a big raise, a private office, and a company car.

You Can Lead a Boss to Data but You Can't Make Him Think

It ain't so much the things we don't know that get us into trouble, it's the things we do know that just ain't so.

Artemus Ward (Charles Farrar Browne), U.S., writer

The most carefully planned data analysis may not survive the intervention of a boss (or a client or other reviewer), whether well intentioned or not. Your aim may be to generate sound data and conduct a thorough and valid analysis, but your boss may have different motives and concerns. He or she may have budget or schedule constraints, not to mention business vulnerabilities and office politics to contend with. So be prepared when the unthinkable happens.

A few days before you planned to send out the survey, your boss calls you into his office and speaks some disconcerting words…

Change This Question Before You Send It

Add a question, reword a question, drop a question, and other requests that will defile the perfect survey you spent so much time creating. What do you do?

7 Self Addressed Stamped Envelope.

Adding a question (or other measurement) may not be a problem so long as you haven't already printed and mailed the questionnaires. Add the question and don't worry about it too much.

If I weren't so full of tuna, I'd make him pay for that.

Rewording questions may or may not be a problem. Keep an open mind for better wording. It's when the meaning of the questions is changed that you have to be careful. Sometimes the changes may seem subtle. For example, asking respondents if they do something once a week is not the same as asking if they do something often. Asking if something is good is not the same as asking if something is better than expected. If you have to change a question, don't compromise quality by making it vague.

Dropping a question is problematical. If you didn't need it, you wouldn't have included it in the first place, but your boss won't see it that way. Questions that are especially susceptible are the ones at the end of the survey that characterize the respondents. They're tough to lose but it's better to have something over nothing. Make sure the boss knows what he'll be losing and go with what you can.

We Can't Afford This

That's more than I expected to spend is a mantra you'll hear often. You may even have said it yourself to the dealer of the hot red convertible everybody wants, or to the plumber who came to fix your leaky toilet, or to the tax preparer you finally cornered at 5:00 PM on April 15. Did you pay up, pass on the offer, or negotiate something in between? Likewise, there are three things that your boss can do in this situation:

- Relent and pay for the study
- Negotiate a reduced price
- Cancel the study (or get someone cheaper to do it).

You want the first thing to happen and can live with the second, but the third would be a disaster.

So put yourself in your boss's position. Is the analysis something he has to do? If so, gently review the consequences of not doing it. Will his boss be upset? Will a regulatory agency come calling? Paint a picture of the cost of the consequences compared to the cost of the study. If the analysis is something he doesn't have to do, you'll have to convince him that it is something he needs to do it, or even better, he wants to do. Consider what the boss is looking to do with the results. Show him how your data analysis will add value to his operation. Give him a clear sense of what the payback will be.

If that doesn't work, try working out a compromise. Just remember, a cut in price has to be balanced by a cut in scope, otherwise you'll have no credibility. If the cuts will impair the analysis so much that little will be gained, it's better to pass on the study and survive to analyze another day.

That's Too Many Samples

Having your boss (or client) ask if you can get by with fewer samples is almost a given. Unless the boss understands statistics and variability, it's unlikely he'll see the need for as many samples as you planned to collect. Agreeing to reduce the number of samples is a self-inflicted but nonfatal wound. The margin of error will be bigger but quantifiable, and you may not be able to do the modeling you hoped to do. Let the boss know what he'll be getting. He should appreciate the concept of trade-offs since he has to deal with them all the time in his own work. This assumes, of course, that the boss is at least in the same ballpark as you. If you want 800 samples, and he was thinking of just asking a few customers some questions over lunch, you're toast.

Here's the List of People to Survey

After spending a lot of time ensuring that your sample will be truly representative of your population, your boss gives you his list of who to survey. The list may focus on the biggest or oldest customers, or worse, the customers who will give the best reviews. It's clear that the list from the boss won't fairly represent all customers. What do you do?

This can be a deal breaker from your perspective. There's no reason to do a survey, or any statistical analysis for that matter, with a judgment sample let alone a highly biased judgment sample. You'll only be deluding your boss and yourself. Try to convince your boss that the directed samples will invalidate the survey. If you can't, take a list as a window onto your boss's real reason for conducting the survey. He may just want something flattering to show his boss. If you still have to analyze the results even with the biased list, be sure to caveat your findings.

We're Not Going to Do This After All

There are two common reasons the boss will pull the plug on a survey he asked for. The first is that he didn't get approvals from his bosses. If you raise the level of attention of the survey in the organization early in the process, like when you go in search of data, this shouldn't be a problem because you'll be shot down before you do much work.

The second reason is funding. Some bosses relax during the first two quarters of the fiscal year then suddenly realize in the third quarter that they aren't going to make their business plan unless they take drastic measures. Your survey will be the first thing to go, deferred if you're lucky, cancelled if you're not. Try to schedule your survey for early in the fiscal year.

The Results Aren't What We Expected

If your boss says he is surprised with your results, it could mean a couple of different things.

If the boss says either that the results were too general to tell anybody anything or that the results were too detailed for anybody to figure out, it may be a presentation problem. That's actually a good thing. It can be fixed. Don't be reluctant to get a communications specialist to help translate your work into a better presentation. You will benefit from more people being able to understand what you did.

If the boss is surprised by your results, and it's not the presentation, it's a virtual certainty that the results are unfavorable. There are at least four possible scenarios to consider:

- The boss really is surprised by the results because he is out of touch with his customers. In this case, watch how he reacts as you explain some of the intricacies of the findings. As long as he doesn't become defensive, it could be a great opportunity for both of you, not to mention the company and the customers. If he does become defensive, try stressing that the results are the perceptions of customers who don't appreciate all the constraints he is under. Protect his fragile ego. You need him to take the next step. It's irrelevant whether the boss believes he is addressing reality or the fantasies of misinformed customers as long as he does something positive.
- The boss is really surprised by the results because he thinks there may be a problem with the data or the analysis. Double-check your work to be sure nothing is amiss statistically, and keep an open mind. The boss may be aware of customer relationships or other business factors that may have skewed your results.
- The boss is not surprised by the results but doesn't want to admit it because he was hoping for something different. He's playing dumb. This is his way of trying to deflect some responsibility away from himself. Play along. Remember, the only important thing is that he takes some positive action based on your results.
- The boss was not surprised by the results but doesn't want to admit it because the results don't match what *his* boss wants to do. There's not much you can do about this. The boss will probably take the results, and you'll never hear of it again. Consider it a valuable life experience.

Just Give Me the Results; I'll Take Care of the Rest

Everybody who has a boss has lived this alternative reality, whether you're analyzing customer satisfaction data for the CEO or wrapping burgers for hungry customers. You give your work to the boss who becomes the face of the product. The boss gets all the accolades, and perhaps, an occasional complaint, while you get little if any recognition.

True, some bosses do this to try to claim credit for work they didn't do, but this is usually transparent to anyone who matters. The CEO knows your boss neither conducted the survey nor has the knowledge or the time to have done it. He is responsible, though, for the work *you* do. You know when you go to a restaurant that the host didn't prepare the food. You know when the politician cuts the ribbon at the opening of the

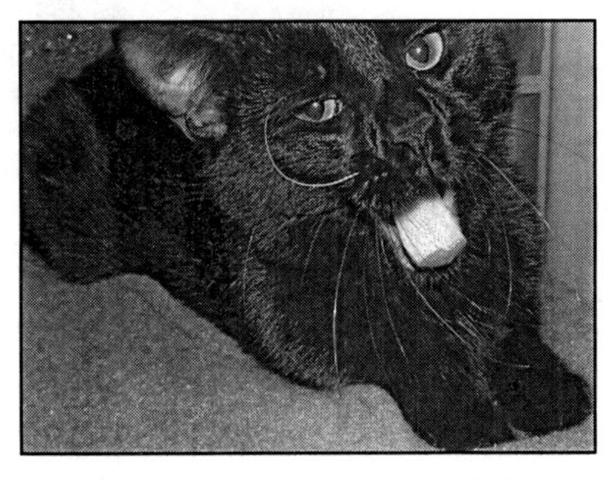

Just give me the tuna. I'll take care of the rest.

new library that he didn't build the building or shelve the books. But those people represent the work of many.

But a boss may want to be the spokesperson for your work for other reasons. He may want to bury unfavorable results. He may want to control access to the results because knowledge is power (after all, that's why you had such a hard time getting the customer information from the high priest of the database). Even more important, he may want to control access to *you*. Your newly demonstrated skills make your boss even more powerful. Finally, it's possible that your boss doesn't want you involved in decisions made as a result of your work. While this judgment is usually blamed on ego flexing, it may also be attributable to self-loathing over the decision-making process. If you think data analysis is messy, you should see what decision makers do. They are often illogical and inconsistent. They choose paths of least resistance over avenues of greatest effectiveness. They pay more attention to anecdotes than data. They worry about obscure possibilities and ignore likely improvements. They have a penchant for inaction and a desire to preserve the status quo. Unless you have a frustration deficit in your life, it might be better to let this one go.

One big issue with this response is that you don't get any feedback. It's important to get recognition for your work, but it's absolutely essential that you get feedback. That's how you grow as a professional and as an individual. If you can't get any feedback from your boss, look elsewhere. How did the high priest of the database like working with you? Do you know any customer contacts well enough to ask them their opinions of the survey? Do you ever run into the CEO or other managers in the hallway who might be familiar with what you did? Take whatever you can get (without making your boss paranoid) and put the knowledge to good use.

Ockham's Spatula

We have a habit in writing articles published in scientific journals to make the work as finished as possible, to cover up all the tracks, to not worry about the blind alleys or describe how you had the wrong idea first, and so on. So there isn't any place to publish, in a dignified manner, what you actually did in order to get to do the work.

Richard Feynman, Nobel Lecture, 1966

I had a client, a very skilled engineer, who wanted a model to predict how many workers he would need to hire during the year. His company produced three lines of products, most of which were customized for individual customers. A few years earlier, he had gone to great effort and expense to develop a model to predict his man-power needs. He collected data on how many of each type of product he had produced over the past five years and from that data had his managers estimate how long it took to make each product and complete the most common customizations. Then he had his sales force estimate the number of orders they expected the following year. He reasoned that adding up the time it took to produce a product multiplied by the number of expected orders would give him the number of man hours he would need. The only problem was that it didn't work. Even after tinkering with the manufacturing times and correcting for employee leave, administrative functions, and inefficiency, the model still wasn't very accurate. Moreover, it took his administrative assistant several weeks each year to collect the projected sales data to input into the model. Some of the sales force estimated more sales than they expected to try to impress the boss. Others estimated fewer sales so that they would have a better chance of making whatever goal might be given to them. A few avoided giving the administrative assistant any forecasts at all, so she just used numbers from the previous year.

Using a top-down, statistical modeling approach, I found that his historical staffing was highly correlated to just one factor, the number of units of one of the products he produced in the prior year. It made sense to me. His historical staffing levels were appropriate because he had hired staff as he needed them. His business had also been growing at a fairly steady rate. So long as conditions in his market did not change, predicting future staffing needs was straightforward. He didn't need to rely on projections from his psychologically fragile sales force.

But my model proved to be quite unsettling to many. The manager of the product line that was used as the basis of the model claimed the model proved his division merited a greater share of corporate resources, and bigger bonuses for him and his staff. Managers of the two product lines that were not included in the model claimed the model was too simplistic because it ignored their contributions to the workload.

At that point, the client had a complex model that he liked but didn't work and a simple model that worked but nobody liked. He probably would have continued to use the complex model if it didn't take so much work to gather the input data. Valid or not, the simple model had no credibility with his managers. He could calculate a forecast with it but was reluctant to favor the model over their intuitions. So given his two flawed alternatives, the client moved man power forecasting to the back burner until the next crisis again brought it to a boil.

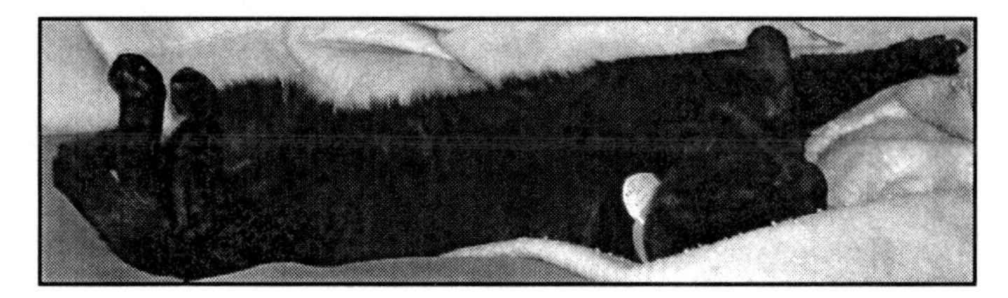

Statisticians, like cats, have to be flexible.

I wish I could say that this was an isolated case, but it's more of a rule than an exception especially with technically oriented clients. I once developed a model for a client to predict the relative risks associated with real estate they managed. The managers wanted a quick-and-dirty way to set priorities for conducting more thorough risk evaluations of the properties. I based my model on information that would be readily available to the client. They could evaluate a property for a few hundred dollars and decide in a day or two whether further evaluation was needed immediately or whether it could be deferred. When the model-development project was done, the model was turned over to the operations group for implementation. The first thing the operations manager did was invite "experts" he worked with to refine the model. Very quickly, the refinements became expansions. The model went from quick and dirty to comprehensive and protracted. It took the operations group on average $50,000 over six-months to evaluate each property. The priorities set by the refined model set were virtually identical to priorities set by the quick-and-dirty model.

Was one of these models good and the other bad? Not exactly, there's an important distinction to be made. Statisticians, and for that matter, scientists and engineers and many other professionals, are taught that, all else being equal, simple is best. It's Ockham's razor. A simple model that predicts the same answers as a more complicated model should be considered to be better. It's more efficient. But sometimes you, as the statistician, have to be more flexible.

The operations manager wasn't comfortable with a simple model. He needed to be confident in the results, which, for him, required adding every theoretical possibility his experts could think of. He didn't want to ignore any sources of risk, even if they were rare or unlikely. That made for a very inefficient model, but if you don't have confidence in a model and don't use it, it's not the tool you need.

These cases illustrate how there's more to modeling than just the technical details. There are also artistic and psychological aspects to be mastered. Textbooks describe statistical methods to find the best model components but not necessarily the ones that will work for the model's users. Sometimes you have to think of Ockham razor as more of a spatula than a cleaver.

Model building is like climbing a mountain. It's what you spend so much time planning for. It's what everybody wants to talk about. It's what gives you that euphoric feeling of accomplishment when you're finished. But just as mountain climbers have to descend, model builders have to deploy. You have to put your model in a form that will be palatable to users.

"... the best material model of a cat is another, or preferably the same, cat." From The Role of Models in Science by A. Rosenblueth and N. Wiener in Philosophy of Science, v. 12, no. 4, p. 320, 1945.

Like sausages, models need to look good on the outside especially if there are things on the inside that might make most users choke. You have to package the model. First, it can't look so intimidating that users break out in a sweat when they see it. Leave the equations to the technical reviewers; hide them from the naive users. USDA inspectors have to look inside sausages, but you don't. Second, put the model in a form that can be used easily. That inch-thick report may be great documentation, but it'll garner more dust than users. If your users are familiar with Excel, program the model as a spreadsheet. If you know a computer language, put the model in a standalone application. A model is only as successful as the use to which it is put.

Yes, I admit that a lot of this discussion is negative, pessimistic, and even cynical. Most textbooks in statistics aren't so gloomy. I just thought it might be worthwhile to strike a balance.

PART 6

Saving the World One Analysis at a Time

If your head is spinning after all those models in Part V, take a breath but don't get too relaxed. This is game day. Part VI is where you graduate onto your own analyses.

Chapter 24—Grasping at Flaws is about how to review data analyses done by other people. If you are a statistical novice, you might think this will never happen. I can tell you from experience that it happens all the time. In the office, the boss will target you for this chore if you're a recent college graduate, an engineer, or just standing around kibitzing at the water cooler. You'll find you can use some of these tips in your personal life too, especially if you like to debate on political forums. This chapter will tell you how to make constructive, thought-provoking comments without knowing all that much about statistics.

Chapter 25—The TerraByte Zone is where you get to apply some of the skills you might have picked up from this book, statistics classes, and your own experiences analyzing data. There are no calculations to do. There are no unique correct answers. All that will come in time with practice. This chapter aims to get you over your inertia and insecurity by providing data analysis scenarios for you to think about, and perhaps, discuss with your cohorts.

The ball is now in your court. Good luck.

24 Grasping at Flaws

Even if you're not a statistician, you may one day find yourself in the position of reviewing statistical analyses that were done by someone else. It may be an associate, someone who works for you, or even a competitor. Don't panic. Critiquing someone else's work has got to be one of the easiest jobs in the world. Doing it in a constructive manner is another story. Like compassionate conservatism, compassionate criticism is feasible but regrettably too rare.

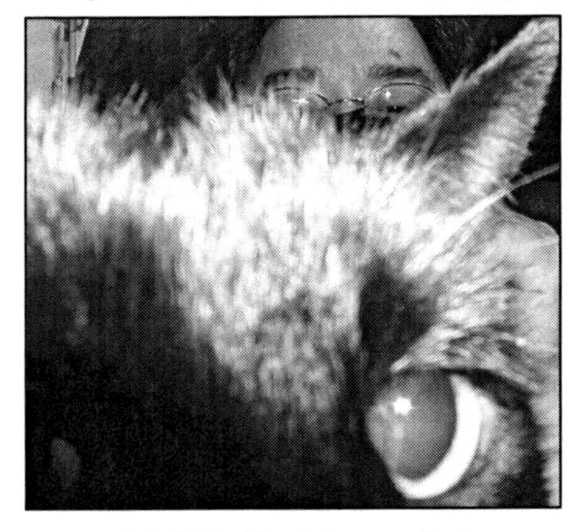

I feel like I'm being watched.

So here's the deal. I'll give you some bulletproof leads on what criticisms to level on that statistical report you're reading. In exchange, you must promise to be gracious, forgiving, understanding, and, above all, constructive in your remarks. If you don't, you will be forever cursed to receive the same manner of comments that you dish out.

Don't expect to find a major flaw in a multivariate analysis of variance, or a neural network, or a factor analysis. Look for the simple and fundamental errors of logic and performance. It's probably what you are best suited for and will be most useful to the report writer who can no longer see the statistical forest through the numerical trees.

With that said, here are some things to look for.

The Red-Face Test

A "critic" is a man who creates nothing and thereby feels qualified to judge the work of creative men. There is logic in this; he is unbiased—he hates all creative people equally.

Robert A. Heinlein, *The Notebooks of Lazarus Long,* 1978

Start with an overall look at the calculations and findings. Not infrequently, there is a glaring error that is invisible to all the poor folks who have been living with the analysis 24/7 for the last month. The error is usually simple, obvious once detected, very embarrassing, and enough to send them back to their computers. Look for:

- **Wrong number of samples.** Either samples were unintentionally omitted or replicates were included when they shouldn't have been.
- **Unreasonable means.** Calculated means look too high or low, sometimes by a lot. For example, an average pH of 15 would definitely be a red flag that something is amiss. A mean of 8 or 9 for natural waters may be something worth looking into. The cause may be a mistaken data entry, an incorrect calculation, an untreated outlier, or really bad tasting water.
- **Nonsensical conclusions.** A stated conclusion seems counterintuitive or unlikely given known conditions. This may be caused by a lost sign on a correlation or regression coefficient, a misinterpreted test probability, or an inappropriate statistical design or analysis. Don't worry about the cause unless you can easily spot the error. Just explain your rationale for why the results are not sensible. But beware, just because a conclusion doesn't match your expectation (your own bias) doesn't mean that it's wrong.

You'll save the writer considerable embarrassment if you can find any of these flaws before the report is released. If you say it nicely, he or she might even be grateful.

Nobody Expects the Sample Inquisition

A person's opinion of an institution that conducts thousands of transactions every day is often determined by the one or two encounters which he has had with the institution in the course of several years.

William G. Cochran, *Sampling Techniques,* 1952

Start with the sample. If you can cast doubt on the representativeness of the sample, everything else done after that doesn't matter. How could it? Garbage in; garbage out. Think of the Leaning Tower of Pisa. If the foundation is flawed, there's no way to right what's above it.

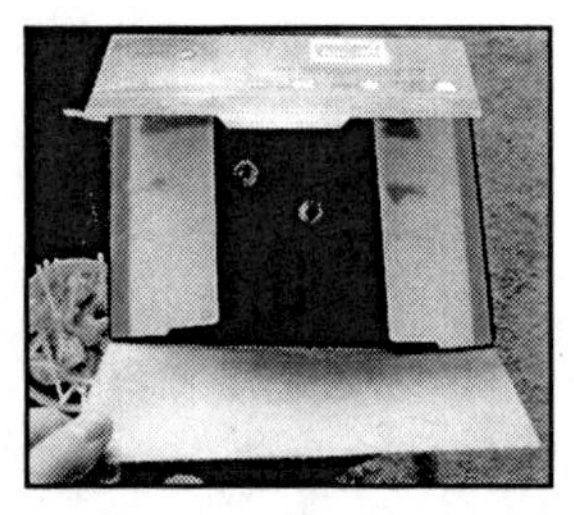

It helps to know where to look for hidden trouble.

If you are reviewing a product from a mathematically-trained statistician,[1] probably the only place to look for difficulties is in the samples. There are a few reasons for this. First, a statistician may not be familiar with some of the technical complexities of sampling the medium or population being investigated. Second, he or she may have been handed the dataset with little or no explanation of the methods used to generate the data. This situation is not uncommon. Third, he or she will probably get everything else right. Focus on what the data analyst knows the least about.

1 As opposed to a professional educated in a non-mathematical discipline who has statistical training.

For example, say you live near a manufacturer that uses a variety of chemicals. The discovery of several animal carcasses prompted an investigation of whether chemical contamination might be entering the food chain and poisoning the animals. The manufacturer hired a consultant who conducted a study of pollutant concentrations in the tissues of field mice. To use statistics to evaluate the levels of pollutants in the mice, the samples would have to be representative of the distribution of mice around the facility, especially in relation to pathways by which the mice might have come in contact with the chemicals. A statistician analyzing the data would have to know something about where the mice were trapped relative to the source of the pollutants, how the pollutants behave in a mouse's metabolism, which tissues were extracted for analysis, and what the detection limits and precisions are of the analyses. The statistician wouldn't necessarily have to be a multidisciplinary genius, just have access to experts or information sources that would provide the information. That may be a lot to ask of someone on a limited budget and schedule, though. That's why even experienced statisticians might overlook some nonrepresentativeness of samples.

Examples of fundamental aspects of sampling that should be examined include:

- **Randomization**. There is no statistical requirement that samples be selected randomly. However, they must be representative of the population of possible samples being analyzed. Randomization is a good way to help ensure representativeness. So start by looking for some bias in the way the samples were selected. Surveys of the general population in which respondents are self-selected instead of receiving a randomized invitation can be biased toward those holding strong opinions. For example, Internet polls open to all website visitors, especially political polls, are often "freeped."[2] Another example that occurs with some frequency involves environmental technicians who prefer to collect samples of the green crud bubbling up around the pile of dead rodents and not the grassy area where the survey flags were placed. Studies involving field sampling also succumb to the bias of moving randomly selected sampling locations so that there is an appearance of a good spatial distribution. If such biased sampling happens once, you might have an outlier. If it happens a lot, you have a biased, misleading dataset. So look for the samples (or at least a method for selecting the samples) to have been specified *before* they were collected. Be skeptical, for example, if the sampling locations for a field study correspond to the only roads in an undeveloped area. If a random sampling scheme wasn't used, look for the samples to have been collected systematically (i.e., on a grid) with some expectation that the process responsible for the distribution of the characteristic being measured was random over the area sampled. If neither of these approaches were used, be suspicious of sampling bias.

- **Independence**. A fundamental assumption of statistics is that samples are not correlated with each other. This is almost never the case with environmental

2 Freeping a poll involves voting repeatedly on an online survey or encouraging others to respond to the poll to bias the results in a particular direction. Freep can also refer to *The Detroit Free Press* or *The New Republic*.

data, which are usually correlated spatially, temporally, or both. Look for some analysis and/or correction for sample dependence to have been made. This usually takes the form of stratification, in which samples are collected in proportion to some measurable condition of the candidate samples. After sample collection, independence can be assessed by plotting sample variance versus distance between samples, called correlelograms for temporal data and variograms for spatial data.

- **Comparability**—Beware of Frankendata. There may be data comparability problems associated with sample collection, especially if the data were collected over a long time span. This is true because preferred sampling protocols change over the years as well as from consultant to consultant. For example, groundwater samples collected with different types of pumps may not be comparable. Time-weighted samples of air or soil-gas may not have been standardized for pump calibration. Sample preservatives may be of inappropriately low purity grades and may contain the very constituents the sample is being tested for. Look for sample variances to be about the same from one sampling round to the next.

Data Alone Do Not an Analysis Make

I have made all the calculations; fate will do the rest.

Napoléon Bonaparte, 1813, in Herold, J. C., *The Mind of Napoleon*, 1955, p. 45

Calculations

Unless you see the report writer counting on his or her fingers, don't worry about the calculations being correct. There's so much cheap and good statistical software available that getting the calculations right shouldn't be a problem. It should be sufficient to simply verify that he or she used commercially available statistical software. Likewise, don't bother asking for the data unless you plan to redo the analysis. You won't be able to get much out of a quick look at a database, especially if it is large. Even if you redo the analysis, you may not make the same decisions about outliers and other data issues that will lead to slightly different results. Waste your time on other things.

Descriptive Statistics

Descriptive statistics are usually the first place you might notice something is amiss in a dataset. Be sure the report provides means, variances, minimums, and maximums, and numbers of samples. Anything else is gravy. Look for obvious data problems like a minimum that's way too low or a maximum that's way too high. Be sure the sample sizes are correct. Watch out for the analysis that claims to have a large number of samples but also a large number of grouping factors. The total number of samples might be sufficient, but the number in each group may be too small to be analyzed reliably.

Correlations

You might be provided a matrix with dozens of correlation coefficients. For any correlation that is important to the analysis in the report, be sure you get two things:

- **Statistical Test**—A t-test is used to determine whether the correlation coefficient is different from zero. If the test is not significant, either the true population correlation is too small to be measured reliably or the sample size is too small to make a meaningful determination.
- **Scatter Plot**—A plot of the two correlated variables is used to verify that the relationship between the two variables is linear and there are no outliers that might enhance or degrade the correlation.

Without these, you can't tell if a correlation really means anything. Don't necessarily expect to get plots for every combination of variables unless you have a forklift to move the report. After that, just remember that correlation might imply causation and it might not. Don't assume one or the other.

Regression

Regression models are one of the most popular types of statistical analyses conducted by non-statisticians. Needless to say, there are usually quite a few areas that can be criticized. Here are probably the most common errors.

- **Number of Samples**—If the ratio of data points to predictor variables isn't at least 10 to 1, the model will be unstable. This means that the regression equation is liable to change substantially if the analysis were repeated with new data. The model may be a good start but it's not the end of the story. Model stability starts at a ratio of about 50 data points per predictor variable.
- **Intercept**—There must be an intercept term in the model unless there is a compelling theoretical reason not include it. When an intercept is omitted, the coefficient of determination, R^2, is artificially inflated and the model will look better than it really is.
- **Variation**—Look at the variation of the predictions, usually expressed as the standard error of estimate. Many models tend to have large variances even when the R^2 value is large. This means that you might have an accurate predictive model that lacks enough precision to be useful. For example, if the confidence band around a pH prediction is ±2 units, you have a problem.

Statistical Tests

With all the cheap statistical software available, statistical tests are done routinely by report writers with no notion of what they mean. Look for some description of the null hypothesis (the assumption the test is trying to disprove) for the test. It doesn't matter if it is in words or mathematical shorthand. Does it make sense? For example, if the analysis is trying to prove that a pharmaceutical is effective, the null hypothesis should be that the pharmaceutical is not effective. It wouldn't make sense to set up the null hypothesis to assume that the pharmaceutical is effective and try to disprove its effectiveness. After that, look for the test statistics and probabilities. If you don't understand what they mean, just be sure they were reported. If you want to take it to the next step, look for violations of statistical assumptions.

Analysis of Variance

An ANOVA is like a testosterone-induced, steroid-driven, rampaging horde of statistical tests. There are many many ways the analysis can be misspecified, miscalculated, misinterpreted, and misapplied. You'll probably never find most kinds of ANOVA flaws unless you're a professional statistician, so stick with the simple stuff. A good ANOVA will include the traditional ANOVA summary table, an analysis of deviations from assumptions, and a power analysis. You hardly ever get the last two items. Not getting the ANOVA table in one form or another is cause for suspicion. It might be that there was something in the analysis, or the data analyst didn't know it should be included. If the ANOVA design doesn't have the same number of samples in each cell, the design is termed unbalanced. Violations of assumptions are more serious for unbalanced designs. If the sample sizes are very small, only large difference can be detected in the means of the parameter being investigated. In this case, be suspicious of finding no significant differences when there should be some. Also, check the formulation of the null hypothesis to make sure it is appropriate for the analysis.

Assumptions Giveth and Assumptions Taketh Away

Any author who uses mathematics should always express in ordinary language the meaning of the assumptions he admits, as well as the significance of the results obtained.

Maurice Allais in "La formation scientifique, Une communication du Prix Nobel d'économie," address to the Académie des Sciences Morales et Politiques, 1997.

Statistical models usually make at least four assumptions: the model is linear; the errors (residuals) from the model are independent; Normally distributed; and have the same variance for all groups. A first-class analysis will include some mention of violations of assumptions. Violating an assumption does not necessarily invalidate a model but may require that some caveats be placed on the results.

The independence assumption is the most critical. This is usually addressed by using some form of randomization to select samples. If you're dealing with spatial or temporal data, you probably have a problem unless some additional steps were taken to compensate for autocorrelation.

Equality of variances is a bit more tricky. There are tests to evaluate this assumption, but they may not be cited by the report writer. Here's a rule of thumb. If the largest variance in an ANOVA group or regression level is twice as big as the smallest variance, you might have a problem. If the difference is a factor of five or more, you definitely have a problem.

The Normality of the residuals may be important although it is sometimes afforded too much attention. The most serious problems are associated with sample distributions that are truncated on one side. If the analysis used a one-sided statistical test on the same side as the truncated end of the distribution, you have a problem. Distributions that are too peaked or flat can result in slightly higher rates of false negative or false positive tests but it would be hard to tell without a closer look than just a review.

Look at a few scatter plots of correlations with the dependent variable, then forget the linearity assumption. It's most likely not an issue. If the report goes into nonlinear models, you're probably in over your head.

We're Gonna Need a Bigger Report

What comes full of virtue from the statistician's desk may find itself twisted, exaggerated, oversimplified, and distorted-through-selection by a salesman, public-relations expert, journalist, or advertising copywriter. [...] As long as the errors remain one-sided, it is not easy to attribute them to bungling and accident.

Darrell Huff, *How to Lie with Statistics,* 1954, p. 101

Statistical Graphics

There are scores of ways that data analysts mislead their readers and themselves with graphs.[3] Here's the first hint. If most of the results appear as pie charts or bar graphs, you're probably dealing with a statistical novice. These charts are simple (what I learned in high school is now taught in primary school) and used commonly (in virtually every issue of *USA Today*), but they are notorious for distorting reality. This is because they present data in one-dimension (along a line) but add a second dimension for appearance. In a pie chart, for instance, a radius based on data is expanded into a slice. Rather than interpreting the radius, the mind perceives the two-dimensional area of the slice. The same is true of bar charts. In fact, some software (e.g., Microsoft Excel) now offers the

3 For more information on common errors in graphs, see G. E. Jones, 2007. *How to Lie with Charts* (2nd ed.). Santa Monica, CA: La Puerta Productions.

option of showing data as either a radius or a slice area. In general, never use a pie or bar chart if you have more than a few percentages to display. A pie chart with thirty slices is about as satisfying as a one-thirtieth slice of a blueberry pie. Use a scatter plot or line graph instead.

Also, be sure to check the scales of the axes to be sure they're reasonable for displaying the data across the graphic. If comparisons are being made between graphics, the scales of the graphics should be the same. Make sure everything is labeled appropriately.

Maps

As with graphs, there are so many things that can make a map invalid that critiquing them is almost no challenge at all.[4] Start by making sure the basics—north arrow, coordinates, scale, contours, and legend—are correct and appropriate for the information being depicted. Compare extreme data points with their depiction. Most interpolation algorithms smooth the data, so the contours won't necessarily honor individual points. But if the contour and a nearby datum are too different, some correction may be needed. Check the actual locations of data points to ensure that contours don't extend (too far) into areas with no samples. Be sure the northing and easting scales are identical, easily done if there is an overlay of some physical features. Finally, step back and look for contour artifacts. These generally appear as sharp bends or long parallel lines, but they may take other forms. My most embarrassing map contained an errant contour line in the shape of a male sexual organ. Thank God the project manager, Ted, spotted it before the meeting with the client. That's what reviews are supposed to do!

Documentation

It's always handy in a review to say that all the documentation was not included. But let's be realistic. Even an average statistical analysis can generate a couple of inches of paper. A good statistician will provide what's relevant to the final results. If you're not going to look at it probably no one else will either. Again, waste your time on other things. On the other hand, if you really need some information that was omitted, you can't be faulted for making the comment.

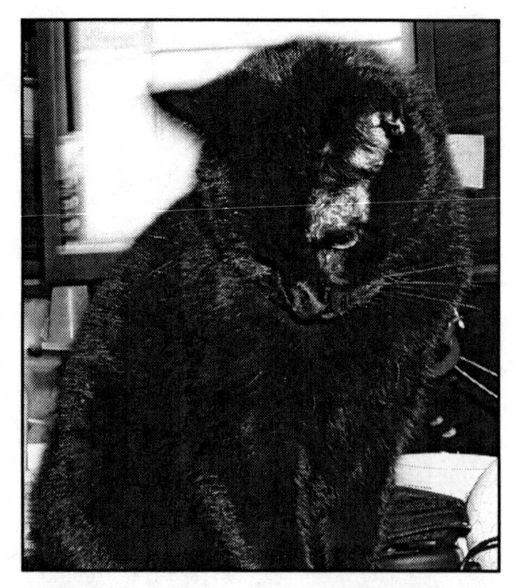

I'm sorry. I ate your documentation.

---

4 For more information on this topic, see M. Monmonier, 1996. *How to Lie with Maps* (2nd ed.). Chicago: University of Chicago Press.

You've Got Nothing

Well, I suppose now is the time for me to say something profound… Nothing comes to mind.

Jack O'Neill, in "The Serpent's Lair [2.1]" on *Stargate SG-1*

If, after reading the report cover-to-cover, you can't find anything to comment on, you can sit back and relax. If you're the suspicious sort, though, there are two more things you can try.

First, take a look through Table 24. This is a summary of the ten problems I've seen most often in statistical analyses. If none of these problems seem to fit, go to Plan B. Say you are concerned that the samples may not fairly represent the population being analyzed. Expressing concern over the representativeness of a sample is like questioning whether a nuclear power plant is safe. No matter how much you try, there is no absolute certainty. Even experienced statisticians will gasp at the implications of a comment concerning the sample not being representative of the population. That one problem could undermine everything they've done. Here's what to look for in a response. If the statistician explains the measures that were used to ensure representativeness, prevent bias, and minimize extraneous variation, the sample is probably all right. If the statistician mumbles about not being able to tell if the sample is representative and talks only about the numbers and not about the population, there may be a problem and the statistician might even know it. If the statistician ignores the comment or tries to dismiss it with a stream of unintelligible jargon and meaningless generalities, there is a problem and the statistician probably knows it. If he or she won't look you in the eyes, you've definitely got something.

Now It's Up to You

Concern for man and his fate must always form the chief interest of all technical endeavors. Never forget this in the midst of your diagrams and equations.

Albert Einstein, U.S. physicist

So that's my quick-and-dirty guide to critiquing statistical analyses. Sure there's a lot more to it, but you should be able to find something in these tips that you could apply to almost any statistical report you have to review. At a minimum, you should be able to provide at least some constructive feedback that will benefit both the writer and the report. Maybe you'll even be able to prevent a catastrophe. If nothing else, you'll have earned your day's pay, and if you critique constructively, the respect of the report writer as well.

Table 24. Top Ten Flaws in Data Analyses

Where's the Beef?	In a way, the worst flaw a data analysis can have is no analysis at all. Instead, you get data lists, sorts and queries, and maybe some simple descriptive statistics but nothing that addresses objectives, answers questions, or tells a story. If that's all you want, that's fine. But a report is not a data analysis. Reports provide information; analyses provide knowledge. Look beyond the tables for models, findings, conclusions, and recommendations. If they're not there, you didn't get an analysis.
Phantom Populations	If there were to be a fatal flaw in an analysis, it would probably involve how well the samples represent the population. Sometimes data analysts don't give enough thought to the populations they want to analyze. They use samples to make inferences to a population that doesn't exist. Populations must be based on some identifiable commonalities that would meaningfully affect some characteristic. A group of anomalies would not be a population. Opinion polls sometimes suffer from phantom populations. Say you surveyed people wearing red shirts. Could you then generalize to everyone who wears red shirts? Make sure the population being analyzed is more than an illusion. For example, Canadian researchers found one such phantom population when they tried to create a control group of men who had not been exposed to pornography.[3]
Wow, Sham Samples	Sometimes the population is real and well defined, but the samples don't represent it adequately. This is a common criticism of opinion polls, especially election polls. It was the reason cited for why exit polls during the presidential election of 2004 indicated that John Kerry won many precincts that ballot counts later awarded to George Bush. Medical and sociological studies may have sham samples because it is often difficult to select subjects to match some target demographic. Likewise, environmental studies can suffer from inconsistencies between soil types or aquifers. To identify sham samples, look for three things: (1) a clear definition of a real population, (2) a description of how samples were selected so that they represent the population, and (3) information about any changes that occurred during sampling, such as subjects being dropped or samples moved.
Enough Is Enough	The number of samples always seems to be an issue. For too few samples, question confidence and power. For too many samples, question meaningfulness. Usually analysts are ready for this question but beware if they cite the old familiar fable about using 30 samples (see Chapter 14 "Samples and Potato Chips"). It may indicate their understanding of statistics is not as formidable as you supposed. Also, if they appear to be using a reasonable number of samples but then break out categories for further analysis, make sure each category has an appropriate number of samples for the analysis they are doing.
Indulging Variance	Most people don't appreciate variance. They don't even know it's there. If their candidate for office is up by two percentage points in a poll, they figure the election is in the bag. Even professionals like scientists, engineers, and doctors don't want to deal with it. They ignore it whenever they can and just address the average or most common case. Business people talk about variances all the time, only they mean differences rather than statistical dispersion. Baseball players thrive on variance. Where else can you have two failures out of every three chances and still be considered a star? Data analysts have to understand variance and address it at every step of a project. Look for how variance will be controlled in study plans. Look for variance to be reported with results. And most importantly, look for some assessment of how uncertainty affects any decisions made from the analysis.
Madness to the Methods	NASA uses checklists to ensure that every astronaut does things the same way. See if the analysis you are reviewing takes the same care. If they use multiple data collection points, is there a standard protocol or script? Be especially concerned if the data are collected over multiple years. Better and cheaper methods and equipment are continuously being developed. Be sure they are compatible. Look for homoscedasticity in the data. Don't let Frankendata corrupt the analysis.
Torrents of Tests	If a study conducts a statistical test, false positives and false negatives can be controlled, or at least, evaluated. But if there are many tests, you can bet there will be false results just because of Mother Nature's wicked sense of humor. In groundwater testing, for example, there may be a test for every combination of well, analytes, and sampling rounds, resulting in literally hundreds of tests. There are strategies for dealing with this type of situation, such as hierarchical testing and the use of special tests (look for the term ***Bonferroni***). Be careful of bad decisions based on a small proportion of the tests being (apparently) significant.

Significant Insignificance and Insignificant Significance	Here's where you have to use your gut feel if you're not a statistician. If a test is statistically significant and you don't believe it should be, ask about the confidence level and whether the size of the difference is meaningful. Just as correlation doesn't necessarily imply causation, significance doesn't necessarily imply importance. If something is not statistically significant and you believe it should be, ask about the power of the test and the size of the difference the test should have detected. Be sure the study looked at violations of assumptions. Also, look for what's ***not*** there. Sometimes studies do not report nonsignificant results. Such results could be exactly what you're looking for.
Extrapolation Intoxication	Make sure the data spans the parts of the variable scales about which you want to make predictions. If a study collects test data at ambient indoor temperature, beware of predictions made under freezing conditions. Likewise, be careful of tests on rabbits that are extrapolated to humans, maps showing information beyond the limits observed, surveys of one demographic extrapolated to another, and the like. The best example of appropriate statistical extrapolation is time-series analysis. You have to extrapolate to predict the future. The issue is how far into the future is reasonable, which will depend on the degree of autocorrelation, the stability of the data, and the model.
Misdirected Models	Models are great tools for helping you understand your data. Statistical models are based on data but deterministic models rely on theories, or at least the theories believed by the researcher. These models are no better than the theories on which they are based. Misdirected models involve researchers creating models based on biased or mistaken theories, and then using the model to explain data or observed phenomena in a way that fits the researchers preconceived notions. This flaw is more common in areas that tend to be more observational than experimental, such as economics.

Uncertainty isn't just something you have to deal with in a statistical model. Each of us has to deal with our own personal uncertainties. If you're doing a statistical analysis for the first time, you are probably uncertain about your ability to do the analysis. You may be uncertain about where to start. All that uncertainty is perfectly normal.

Statistics are like presidential advisors. If a president has no preconceived notion about an issue, he will listen to the opinions of his advisors and then make a decision. If he has an intuitive feeling about an issue, he will listen to the opinions of his advisors and then send the advisors that don't agree with him off on a tour of New Jersey's landfills. Data analysis should be viewed the same way. Sometimes a particular statistic or statistical test doesn't agree with other analyses, so you conclude it's a mathematical aberration and send it to the landfill.

You also have to consider when it might be time to get opinions from other advisors. Just as the president might bring in consultants or other experts, you don't have to limit yourself to statistical and graphical analysis techniques. There may be opportunities to use data mining, differential equations, linear programming, fractals, physical modeling, visualization, and deterministic modeling, not to mention the scores of methods of probability and statistical analysis.

Sometimes a particular statistical analysis makes intuitive sense. The results are what you expected. You get the same answers when you analyze the data in different ways. The results are consistent with findings made independently by data analysts. No problem. But, sometimes a statistical analysis doesn't yield the result you expected. You may be able to rationalize some meaning, but is that explanation valid? Was there a flaw somewhere in your data collection or analysis? Do you bury the results, start from scratch, or hold a press conference?

Good data with a good analysis will produce good conclusions. Good data and a bad analysis produce misleading conclusions or doctoral dissertations. Bad data and a good analysis lead to hair-pulling frustration often culminating in a statistical consultant's paycheck. Bad data and a bad analysis produce political campaign ads.

So a big part of successful data analysis is cultivating the ability to know when to trust the statistics or other analysis method and when to trust your own intuition. When are anecdotes or outliers meaningful clues to answers and when are they tempting distractions? When do you need to consider whether your data may be flawed somehow? The ability to make those kinds of decisions is something you can learn only with practice, practice, practice.

The real world is a challenging place. There will never be enough data to prove a point beyond question. Even if there were, there will never be enough time and money to analyze them. But that doesn't mean you have to rely on anecdotes and celebrity testimonials. You can use probability and statistics, mathematics, graphics, or another data analysis technique to convert information into knowledge and knowledge into wisdom. One way or another, you're the one responsible for the results. So grab a cup of coffee and start getting that experience.[5]

5 Vini. Vidi. Caffeini. (I came. I saw. I drank coffee.)

25 The TerraByte Zone

This Is the Dawning of the Age of Analysis

You're traveling through another dataset, a dataset not only of means and variances but of models and hypothesis testing; a journey into a wondrous land whose limits are those of statistics—next stop, the TerraByte Zone.

The academic world of statistics stresses understanding theory, types of analyses, calculations and interpretations. That's a lot! In the world of profit-driven business and government regulations, though, there's even more to consider. How you propose to do the work, how you support the data generation, how you interact with others, how you package the results, and what you recommend your client should do are all part of the big picture. Here is an opportunity to practice all the skills that a statistician needs to have beyond number crunching.

Things to Think About

This section presents hypothetical situations in which someone might be asked to do (or review) a statistical analysis. The best way to use these exercises is to discuss them in groups, perhaps as class projects or lunchtime discussions at work. The scenarios are very brief descriptions, devoid of many essential facts. There's a reason for that.[1] You should customize the scenarios to stimulate your own discussions. Change the person who is the subject of the scenario. Add data or other information so that the scenario can be developed into a work plan. There will be no unique correct answers. A student will answer differently from a seasoned professional. An individual trained in statistics will answer differently from an individual trained in sociology or biology. What's important is that you visualize how you might go about approaching these scenarios. Here are some things to think about:

1 It was easier to write that way.

- What is the problem or question that needs to be addressed by the analysis? How important would the analysis be? Should the work even be done? Is there a better way to answer the question or solve the problem than statistics?
- Who would do the work, you or some hired help? What help, (e.g., people, information, tools) might you need to obtain?
- Are there valid data available or would they have to be generated? If you need to generate data, how will you control bias and variability? Could publicly available information be used to augment the data?
- What data analysis techniques would need to be used? What software and special expertise would you need? Are there technical assumptions or caveats that should be considered?
- Could the analysis be kept small (e.g., relatively unsophisticated descriptive statistics and graphs), completed in steps (e.g., initially at a small scale like a pilot study), or would the study need to be thorough and technically defensible?
- How long might it take to complete the analysis? Where might the schedule for getting the work done be problematical? How much funding would be needed? Where might it come from? Might the source of funding introduce any unintentional bias or apparent conflict of interest?
- Is there likely to be media attention or legal proceedings associated with the results? Are there any potential ethical dilemmas or political complications? Are you competing with someone else for the work? Might the results produce some undesirable outcome? What other risks might there be in the work?

Oh, and don't forget all the statistical specifications and decisions you have to address. You'll find discussions of these considerations in all the other statistics books. Look at a few. Visit some relevant websites. Talk to people who do data analysis as part of their livelihood. It'll all help.

Analyzing data is like taking a long road trip. Most of the trip has nothing to do with your destination but you have to go through it to get there. If you're not proficient in data analysis, it can be like the last bridge, tunnel, or traffic jam you have to get by, white-knuckled and sweating, before you reach your destination. If you're a statistician, it's more like the last rest stop where you can relieve your pent up anxiety before you cruise home. Whether your analysis will be an aggravating traffic jam or a tranquil rest stop will depend on your confidence. Confidence comes from practice.

With that in mind, here are a few road trips you might consider taking.

Hypothetical Situations

1. You are Seymour Krelborn, operator of a small nursery that supplies local florists. You have developed a unique cross between a *Dionaea* and a *Pinguicula*, which you named Audrey. You are convinced Audrey will be a big seller and maybe even open national or international markets for your business. Unfortunately, Audrey

plants aren't very robust and propagation is difficult. You want to improve the viability of the plant so that you can put it into commercial production. You're thinking of changing the fertilizer and other soil amendments you use and want to run tests to find the best nutrients for Audrey. What would you do?

2. You are Donna Burke, a member of the local school board. You are leading negotiations with the teachers' union over salaries. The union has prepared a report comparing the average salaries of teachers in the fifty-eight counties in the state, which indicates that your district is far below the state average. A national taxpayers' group prepared a contrasting report indicating that the district spends more on salaries per student than 75 percent of the districts in the state. What would you do?

3. You are Rosa Stallman, manager of the returns department of a large home improvements store. The store manager, Donald, has asked you to analyze product returns for the past year. He wants you to recommend changes that should be made in the store's sales practices to minimize the number of returns as well as develop a plan to speed up and lower the cost of your department restocking the returned products. You have heard from other employees that three other department heads have been given comparable assignments. You suspect that the assignment is a test to identify who Donald will select to be his new assistant. What would you do?

4. You are Harry Bailey, a graduate student in the English Department of a large university. One of your advisors wants you to conduct a stylometric analysis of F. Scott Fitzgerald's novels. You have no idea what stylostatistics is. What do you do?

5. You are Jeffrey Bezillionaire, founder of a business involving sales through an Internet website. The firm that programmed your website linked a database to the site to record information about visitors, including their Internet location, the pages viewed on your site, purchases, and the sites they visited before and after your site. You now have a decade of data and want to do something with the information. What would do you do?

6. You are Harris Telemacher. You live near the freeway at the intersection of two busy streets controlled by four-way stop signs. You are disturbed at the number of cars that only slow down at the signs but do not stop. You want to make a case to the township to replace the stop signs with a traffic light. What would you do?

7. You are Dean Vernon, long-time head of Phaber College. Several student organizations have complained that there are differences in the grades assigned by some professors on the basis of sex and fraternity membership. A member of the Sociology Department analyzed historical grade data and concluded that the students' complaint has merit. If you can't defuse the situation quickly, law suits are inevitable. A former colleague in the Political Science Department has offered to reanalyze the data and says he is confident that he can obtain a finding more friendly to the college. What would you do?

8. You are Leo Getz, the accountant for a medical practice consisting of ten doctors. The doctors are dissatisfied with the profits of the practice and want you to conduct an analysis of the doctors and cases to identify any patterns that could improve the efficiency of the practice. What would you do?

9. You are George Bailey, manager of a branch of a chain of twenty banks. The number of tellers you are allowed to employ is based on a formula based on the number of accounts. However, you want to make the case to the bank directors that you need to hire more tellers because you service many accounts with frequent transactions making the wait times of your customers longer than at other branches. What would you do?

10. You are Dwayne Schneider, manager of an office building that houses twenty businesses, ranging in size from six to sixty employees. You are concerned that some tenants may leave at the end of their leases because of the availability of alternative space, so you want to conduct a survey of the tenants to identify what aspects of the building and the building services you provide (e.g., cleaning, maintenance) need to be improved. What would you do?

11. You are Charlie Eppes, an independent consultant specializing in data management and analysis. The president of a financial management firm is raising funds to build a memorial to soldiers killed in the Iraq War. He wants you to take the firm's client database and use the information to predict which potential donors would be most likely to give to the charity. What would you do?

12. You are Bunsen Honeydew, a member of the research department of a chemical manufacturer. Your supervisor has assigned you to do a review of a statistical analysis of toxicity data on the plant's wastewater that the company's lawyer wants. The report was prepared by an economics professor at a local college for the state environmental agency. What would you do?

13. You are C. P. Ray, an accountant for a medium sized travel agency where part of your job is to process expense reports from the field salesman. One day, you overhear two agents discussing their expense reporting and suspect that some of their claims may not be valid. What would you do?

14. You are Carl Kolchak, an investigative reporter doing a story on burgeoning crime rates in the city. Your evidence is anecdotal but compelling. Just as you are going to send your story to the editor, the Police Department releases their compilation of crime statistics that shows incidences of most crimes to have decreased over the past year. What would you do?

15. You are Charlie Terrabyte, the lone statistician in a management consulting firm. It's 11:00 AM on Friday when you get a call from a young business analyst in a branch office. He took it upon himself to do a complex statistical analysis for one of the firm's largest clients, but his boss told him to have you review it before sending it to the client. It's due on Monday. What would you do?

16. You are Aurora Beth Allen, a college student in a sociology class. Your semester

assignment is to conduct an actual survey of students involving some aspect of campus life. Because you arrived late for class, you ended up in a group consisting of a jock, a nerd, a miscreant, a snob, and a stoner. The group decides to study preferences in coffee consumption by comparing the preferences of one hundred students for coffee from three local coffee shops. The group wants to control for the demographics of the students and the chemistry of the coffee, and is considering double-blinding the study. What would you do?

17. You are Tess McGill, the administrative assistant to the CEO of a small corporation. An important part of your job is scheduling the CEO's meeting and other activities. You've noticed that regular meetings on some topics and meetings with certain individuals either tend to run long and delay subsequent activities or finish early leaving the CEO impatient for the next activity to begin. Because the CEO's time is so valuable, you want to see if you can fine tune your scheduling to accommodate these tendencies. What would you do?

Lap. Lap. Yuk, why does he drink this stuff? Lap. Lap. Lap. It's awful. Lap. Lap. Lap. Lap.

18. You are Scott Montgomery, a retired engineer and homeowner. Over the past twenty years, you have replaced windows, added insulation, and completed other energy conservation projects. Now you want to know which projects significantly lowered your consumption of electricity. You have all your old utility bills, which list the monthly energy consumption and the average temperature for the month. What would you do?

19. You are Mike Flaherty, an advisor to the mayor of a large city who is running for reelection. You recently conducted a poll for the campaign in which 60 percent of the respondents were in favor of building a new sports stadium. Your opponent, however, also conducted a poll indicating that 60 percent of the respondents were against building the stadium. The two candidates will debate in two weeks and the topic is sure to come up. What would you do?

20. You are Reverend Jim, the new pastor in a parish with six priests, four support staff, and scores of volunteers. In reviewing purchase records with the church treasurer, you notice that the parish purchases five times as much sacramental wine as the similarly-sized parish you were with previously. Church records include fifty years of attendance and fifteen years of bank records of all purchases. What would you do?

21. You are Vincent Forte, plant manager of a metals finishing shop. Toxic chemicals have been identified in residential wells down gradient of your facility. There are also several other facilities in the area, however, that could be the source of the pollution. What would you do?

22. You are David Fisher, the owner of a family business. You currently have advertisements of various sizes in fifteen print outlets, including newspapers, phone books, and trade journals. You plan to add a website so you want to reduce the

cost of your print advertising, probably by eliminating some of the print outlets. At the same time, you are also considering increasing the size of some of the more effective ads. What would you do?

23. You are Larry Daley, director of security at a large electronics store. The store is scheduled to undergo renovation in two years and you have been asked to provide input on what products should be kept in theft-proof displays. Because the displays are much more expensive than regular displays, their number will be limited. You have "shrinkage" reports (counts of products missing from inventory) for the past five years. What would you do?

24. You are Arthur Himbry, Principal of a large suburban high school. One of your responsibilities is security and you have been inundated with demands by some parents to be more proactive in the wake of Columbine and other shootings in schools. You have compiled five years of reports on illegal and otherwise prohibited items that have been confiscated from students, ranging from weapons and drugs to cell phones and gum. Each report can be supplemented with information on the student's grades, home setting (demographics and socioeconomics), and prior disciplinary actions. You want to develop a way to predict which students would be most likely to be caught with contraband. What would you do?

25. You are Jack Wells, the director of the state's fish and game agency. A local environmental group held a press conference to announce that they tested a dozen fish caught in the state's largest lake and found dangerous concentrations of a pollutant in three of the samples. The governor wants you to reassure the public that the fish are safe to eat, especially because tourism is an important part of the state's economy. What would you do?

26. You are Lois Gibbs, an average resident of a blue-collar community in a small tourist town. Over the past two years, you've heard from at least a dozen neighbors about someone in their families being diagnosed with a rare form of brain cancer. You asked local health officials to investigate the occurrences but they dismissed your concerns as coincidental. What would you do?

27. You are Thomas Banacek, an accountant working as an independent consultant. A mutual acquaintance introduces you to the director of product safety for an appliance manufacturer. The manufacturer recently had a number of reports of fires started by their top-of-the-line crock pot. The Product Design Department hired a business management consultant to do a statistical analysis of the reports, and concluded that the reports were coincidences. The director of product safety offers you the job of reviewing the report for validity. What would you do?

28. You are a Winnie Cooper, a student intern working for the summer at a small nonprofit organization dedicated to consumer protection. The local public transit system is in the process of requesting additional funding from the state and is touting improvements in its traditionally poor on-time performance as evidence

of the effectiveness of the new management. Your organization has obtained public records for the past five years on the system's riders and on-time performance. Because you plan to major in statistics, the organization has asked you to analyze the data to see if the transit system's claims are true. What would you do?

29. You are Carey Drew, a personnel officer in the Human Resources Department of a large manufacturer. Your department is responsible for hiring and training new employees in the manufacturer's processes. You took notice at the last company picnic that a sizable proportion of the employees in attendance were mature, perhaps at least fifty years old. You are concerned that a large percent of the staff will reach retirement age at the same time, making effective hiring and training problematical. What would you do?

30. You made a bet with your friend Bob that your favorite weather forecaster, Ororo Munroe, is more accurate than his favorite forecaster, David Drake. You want to design a study that you and Bob can carry out to determine who the better weather person is. What would you do?

Statistical Self-Amusement

It is by stats alone I set my data in motion. It is by the use of software that thoughts acquire speed, relationships penetrate the brain, the brain sires understanding. It is by stats alone I set my data in motion.

The Mentat Mantra for Statisticians

We are all awash in statistics. Every day, we see the probability of precipitation, the results of opinion polls, changes in the stock market, your grades in school, or the batting average of the baseball team you follow. So it's surprising that many people believe that data analysis is something you do only in school or at work. This is a ridiculous misconception.

Just as the evolution of computer hardware and software fed the growth of statistics and other breeds of data analysis in business, more recent advancements are cultivating the proliferation of personal data collection and analysis.

Four things changed. First, electronic sensors got smaller and better. Second, people started carrying powerful computing devices, typically disguised as mobile phones. Third, social media made it seem normal to share everything. And fourth, we began to get an inkling of the rise of a global superintelligence known as the cloud.

Gary Wolf[2]

2 From "The Data-Driven Life" by Gary Wolf in *The New York Times Magazine* on April 26, 2010. The article is available online at: www.nytimes.com/2010/05/02/magazine/02self-measurement-t.html?scp=9&sq=weird%20self&st=cse#. A version of this article appeared in *The New York Times Sunday Magazine* on May 2, 2010 on page MM38.

No skill improves without practice but with all the data exhaust you're emitting from your communications and activities, you have many opportunities to analyze your own data-driven life. Practice your data analysis skills at home but don't make it feel like homework. Start with something you love to do, like your favorite hobby or interest. Design a study to answer some question that is interesting to you. Collect the data and then do your analysis and see what happens. Here are a few ideas for how to do that.

Personal Behaviors

Ever wonder where the time goes? Time is a major component of many data analyses, so what better place to start than an analysis of your own time. Keep a timesheet of what you do each day for at least a month. For example, categorize how you spend your day into work/school, commuting, chores, errands, sleep, and personal time. Before you start, write down how you think you allot your time to each category. Then calculate the percentages from your data. How close are your predictions to the actual percentages? Do the percentages change much from day to day? Do they vary by day of the week?

There may also be some specific activities you might want to collect data on, like how much you smoke, drink, do drugs, look at porn, gamble, curse, and watch reality TV. Keep this data in a hidden directory on your computer.

Consumption

Have you ever been on a diet and kept a food diary? You can expand this concept to create a dataset. Convert the types and amounts of foods you eat in a day into estimates of calories. Get a pedometer to estimate your exercise. Record your weight. Then after a year or so, see if you can see any correlations between your weights, the foods and calories you eat, your exercise, the season, and anything else you record. You might find the information quite valuable.

Don't make me hungry. You wouldn't like me when I'm hungry.

You may already be keeping track of your car's mileage and fuel costs. It's a good way to see the effects of driving styles, seasons, maintenance, and other factors on your miles per gallon. If you keep your household financial data on software, like Quicken, you can do many analyses and graphs of your spending patterns. For example, do you spend more on lattes than laundry?

If you have a cell phone, put your usage records in a spreadsheet. Figure out the minimum, maximum, and average amount of time you spend on the phone in a day. Who do you talk to the most often and for the longest duration? What is your most connected time of day and day of the week? Save these records so your family can sue Nokia in thirty years after you die from brain cancer.

Other good sources of data you can analyze are your utility bills. Some utility companies will report your past year of electricity, gas, oil, and water consumption, as well as some supporting information such as average temperature. In fact, they may have many years of your energy usage data that they can retrieve for you. You can use the data to test the effects of seasons, vacations, holidays, energy conservation measures, and more significant lifestyle changes, like the kids finally moving out.

Screen Time

Do you relax by watching TV, surfing the Internet, playing video games, or all three? Keep a log of how much time you spend in front of a view screen. You might record date, day of the week, the weather, hours watching TV, hours surfing the Internet, hours playing video games, hours sleeping, and so on, every day for a month. What is the average proportion of your day that you spend looking at a view screen? Does it vary by day of the week or by weather? At the end of a month, revise your data collection to look at other ways you spend your time? Does the act of collecting the data influence how you spend your free time?

Hobbies

Everybody has hobbies and interests that they enjoy, so why not use your favorite pastimes as opportunities to design statistical studies and collect data you can practice analyzing. Here are a few ideas for what you might do.

- **Blogging**—If you write blogs, keep track of how long it took you to write each article, when (time, day, date) you posted it, and how many views or comments you got. After you accumulate enough history, see if you can see any patterns. Is there an optimal day and time to post? Is your readership increasing over time? Are your articles changing in length or the time it takes you to write them?
- **Hunting and Fishing**—Record where you hunt or fish, what bait or other aids you use, the time, the weather, and what results you had. Likewise with treasure hunting, record where you search, what detector settings you use, the time, the weather, and what results you had. Are there any notable patterns?
- **Gardening**—Keep a diary (or better, a spreadsheet) of how much time you spend in your garden, what you do, and the weather. What proportions of your time do you spend planting, weeding, maintaining, and harvesting? How do the percentages change with the date and the weather? If you plant seeds, do you get similar germination rates for the same plant from different suppliers?
- **Reading**—Keep a log of what you read and when you read it. How much of your reading is for enjoyment versus work? What are your reading preferences? Format (books, ebooks, magazines)? Genre (e.g., nonfiction, science fiction, religion, mystery, romance)? Are there differences in how fast you read different genre or formats?
- **Music**—Build a database of music—music you like and music you don't like. Include variables like genre, length, artist, year released, theme of lyrics, instruments, time, key, and so on. Set up a rating scale for each song as the dependent

variables and see if you can find patterns that explain why you like or dislike the music that you do. Extend your findings to artists you haven't listened to before. You may even find something unexpected, like Prisencolinensinainciusol. All right?

Kids and Other Pets

If you have a youngster in the family, start early recording height (length) and weight. Don't just make marks on a doorframe; set up a spreadsheet to organize your data. Do this daily for a few months. How much variation in height and weight occurs from day to day? Is the variation natural or attributable to how you measure the variables? Graph the data over time. Are there changes in growth rates? How do the height and weight compare to standards for the age and species? How much can you change the frequency of data collection without losing the resolution you need to see changes?

Medical Conditions

If you have any chronic medical condition, start collecting relevant data. For example, you can collect a variety of medical parameters using equipment you can find at most drugstores. For example, you might record your weight, heart rate, blood glucose, blood pressure, and temperature. Be sure to note the date and time of each measurement. You can also record qualitative variables like what and when you ate, how you feel, what exercise you did, and so on. Put the data in a graph and show your doctor on your next visit. He or she may be impressed enough to prescribe you some medical marijuana.

Sports

No matter how you like sports—professional, amateur, personal, or fantasy—you'll always be served a side dish of statistics. Relish the experience by analyzing data in ways no one else has. Google "sabermetrics" to see what I mean. Figure out what baseball player is paid the most per hit. What basketball player scores the most points per minute played? Is there a relationship between height and the number of catches a receiver makes? You can find data for almost every sport imaginable on the Internet. Go to it.

Politics

Don't get me going on politics. Suffice it to say that you could spend a lifetime and not analyze all the data that is currently available for free from government websites. If you come up with anything good, write a blog about it. Most political blogs are fanatical fluff made of anti-data. Annihilate them with a real analysis.

Coda

Hey little buddies, soft and silky night walkers.
Dangerous species, tiptoe menace long grass stalkers
On my bed, no butter melting in your jaws.
Bonding monster, lethal weapon wearing claws.
Let's go out and hunt by numbers.

from *Hunt by Numbers* by Ian Anderson

So it's been over a hundred thousand words since Professor Zahlmeister's statistics class and what a long, strange journey it's been. By now, I'm sure you've figured out that there's much more to data analysis than just calculating a few averages and graphing. It's not easy. Even so, there are more pollsters than ever and baseball announcers still talk endlessly about statistics between pitches.

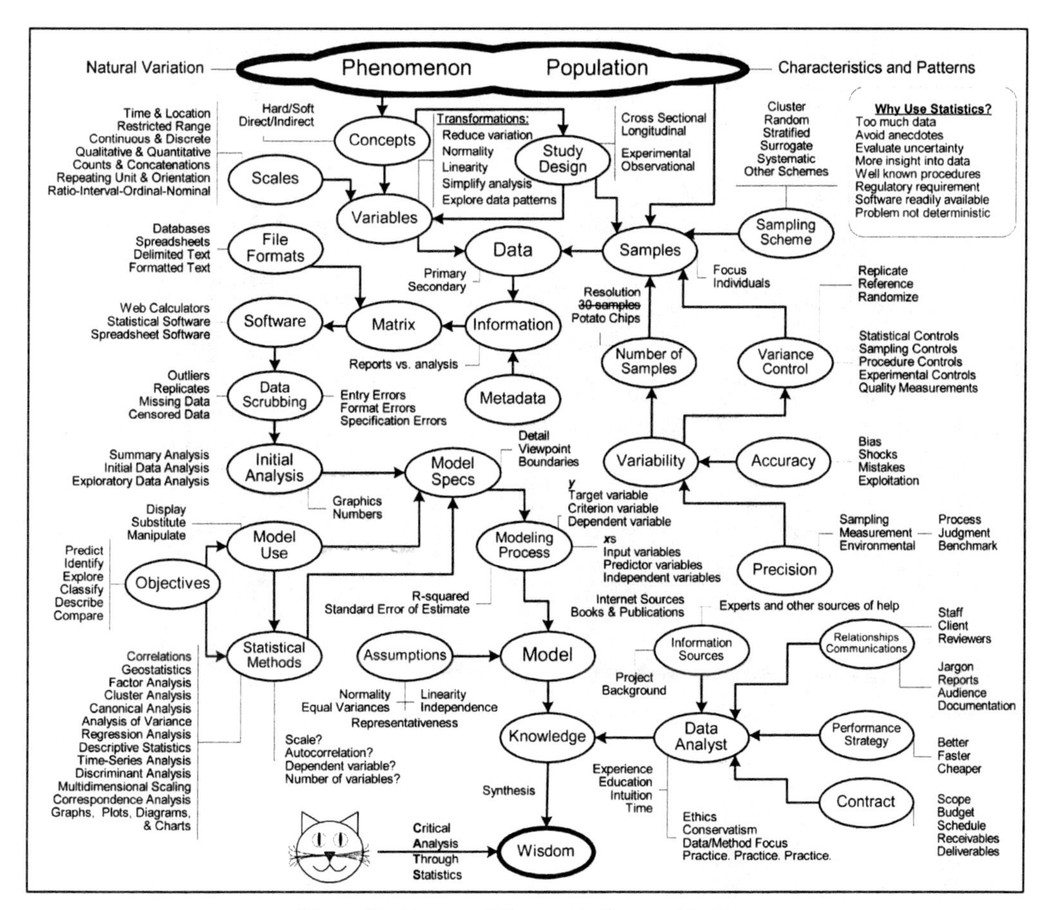

Figure 47. Concepts Discussed in Stats with Cats.

Probably most of the things you've read about in this book were never mentioned in your Statistics 101 class. You'll have to know about these things, though, if you want to analyze your own data at home and at work. It looks formidable at the beginning, and it is. Consider Figure 47, a map of the concepts discussed in *Stats with Cats*.

You use statistics because you can; you have the knowledge and software is readily available. You use statistics because you need to, to analyze uncertainty, especially when there are too many data to just make a graph. You use statistics when you have to, such as when the problem can't be solved any other way or when regulations mandate their use.

As a data analyst, you have to know many things, not just about statistics and the project background, but also about the project's contract, scope, schedule, budget, and deliverables. You have to communicate effectively, both in speech and in writing, and establish good working relationships with project stakeholders. You have to decide on a performance strategy, ensure you get paid, and never compromise your ethics. Finally, you have to have the expertise and time to do the work, and above all, you have to practice, practice, practice.

Data analysis begins when you want to investigate some phenomenon that occurs in a definable population. You collect samples of the population using an appropriate sampling scheme and other measures to control variance and avoid biases so that you will meet your targets for precision and accuracy. You may need to collect more (or less) than thirty samples to meet the resolution you need for the analysis. You measure variables relevant to the phenomenon on appropriate scales. These measurements are the data, which along with the metadata, form the information you structure in a file format your software can recognize as a matrix. Your objectives and aims for model use, together with the scales and natures of your variables, enable you to select appropriate statistical methods. You scrub the information and do an initial analysis, which together with the objectives and methods, lead to your model specifications. Using the specs, you go through the steps of the modeling process to develop, and calibrate a model and evaluate possible violations of assumptions. From the model, you build on your knowledge of the phenomenon, which eventually, you can synthesize into the wisdom you need to make informed decisions.

Those three paragraphs describe the contents of this book in 350 words, without the cats of course. See what a big difference they make?

Over time, as you analyze different datasets, you'll become more comfortable with the process. You'll learn shortcuts to doing things. You'll develop an instinct for things that will work and things that won't. You'll even be able to impress your friends and co-workers with all the new jargon you've learned. You might also learn a bit about yourself. Are you more of a right-brained, intuitive, visual, big-picture, inductive thinker or are you more of a left-brained analytical, verbal, detail-oriented, deductive thinker. Understanding your own preferred thought processes will help you find your best paths in life as well as data analysis.

Being able to analyze data is the asset that sets the knowledge-wielding experts apart from the arm-waving storytellers. Don't wait for your boss or teacher to send you out into the many unmarked routes of the databahn. Journey to the land of data analysis at your own speed along paths you're comfortable with. Don't just endure a data analysis project. Make the journey as fulfilling as the arrival. Make data analysis your passion.

Any questions?

Glossary

Aitchison's Adjustment—Method of adjusting means of censored datasets

Algorithm—Step-by-step procedure for performing some operation, often by computer.

Alphanumeric—Consisting of letters and numbers.

Alternative Hypothesis—The hypothesis that is adopted when the null hypothesis is rejected on the basis of a statistical test.

ANCOVA—Analysis of Covariance, a statistical technique for removing the effects of one or more continuous variables to help detect differences between averages of a continuous dependent variable in groups defined by one or more qualitative independent variables.

ANOVA—Analysis of Variance, a statistical technique for detecting differences between averages of a continuous dependent variable in groups defined by one or more qualitative independent variables. There are many variations of the basic ANOVA method including, one-way, two-way, and n-way designs, fixed-effects, random effects, and mixed effects designs, factorial designs, and repeated measures designs.

ARIMA—Autoregressive, Integrated, Moving Averages model, a statistical technique for forecasting time-series. Also referred to as Box-Jenkins modeling or time-domain analysis.

Array—Matrix.

Association Rules—A method to explore relationships between qualitative variables. Results often take the form "if you did this, you might also want to do these things, too."

Attribute—Characteristic of an individual sample. Another term used for variable.

Autocorrelation—Synonymous with serial correlation if the data points are collected at a constant time intervals.

Autoregressive—Prediction of a variable from other measurements on the same subject, usually over time.

Bayesian Statistics—Methods based on Bayes Law that seek to refine probability estimates by successively incorporating new information.

Benchmark—Reference point

Benford's Law—The probability that the first digit in a set of data values being a low number is much higher than the probability that the digit is a higher number.

The distribution of first digits in a dataset that follows Benford's Law is logarithmic. Benford's law, also called the first-digit law, is often used in fraud detection.

Bias—Systematic deviation, whether intentional or not.

Binomial Regression—Like logistic regression except that values of the dependent variable are either 0 or 1.

Bivariate—Two variables, usually referring to scatter plots.

Blanks—Samples collected to assess variance in studies involving chemical analyses.

Blocking—Process of grouping samples in ANOVA to try to control extraneous variability.

Box-Whisker Diagram—a statistical graphic consisting of a rectangle (the box) with lines extending from two opposing ends (the whiskers). Box plots depict several different statistical measures of central tendency and spread of a dataset. The box usually represents the interquartile range, which is the center 50 percent of the dataset. The small square in the center of the box represents the median. Sometimes a line is used instead of a square. The whiskers represent the upper and lower ends of the data distribution ending with the minimum and maximum. There are quite a few variations of box plots. For example, instead of the whiskers representing the total data range, they can represent an expected range or statistical interval and then outliers are shown as circles or stars at the ends of the whiskers. Instead of the median, the mean can be used.

Bubble Diagram—A two-variable scatter plot in which a third variable is represented by the size of the data points.

C&RT—Classification and Regression Trees, a nonparametric, nonlinear method of predicting continuous dependent variables or classifying categorical dependent variables using a hierarchical, binary decision structure.

Canonical Analysis—A statistical method for examining relationships between multiple dependent variables and multiple independent variables (or any two sets of variables).

Cats—Critical Analysis Through Statistics. Also refers to the Rulers of the Universe who visit Earth disguised as small furry creatures with whiskers and claws.

Cell—The area of a spreadsheet created by the intersection of a row and a column.

Censoring—Data that is hidden because of limitations of a scale, measurement device, or other aspect of the data collection process.

Census—Collection of data from all members of a population.

CHAID—Chi-squared Automatic Interaction Detector, similar to C&RT except that the classification decisions are not binary.

Chemometrics—Analysis of data from studies involving chemistry.

Chernoff Faces—A multivariable icon graphic in which variable values are represented by facial characteristics of a cartoon figure.

Chi-squared test of Normality—A statistical comparison of the difference between a histogram of the data and a histogram of the theoretical distribution. It is sensitive to the classes used for the histograms and requires at lease five samples in each class.

Cluster Analysis—A statistical technique for sorting samples into groups based on continuous-scale variables.

Codes—Values, usually numeric, used to represent the levels of a variable measured on an ordinal or nominal scale.

Coefficient—A number, usually a constant, which is multiplied by a variable, such as in a regression equation.

Cohen's Adjustment—Method of adjusting means of censored datasets.

Comparison—Same as a statistical test.

Confidence—Probability of not making a Type I (false positive) error.

Contours—Lines of equal value used on maps and some diagrams.

Coordinates—Values that define an object's position in space or time.

Correlation Coefficients—Measures of the strength of a data relationship. There are many types of correlation coefficients. The Pearson Product correlation coefficient is used when two variables are measured on a continuous scale. There are several variations of the Pearson Product correlation coefficient. The multiple correlation coefficient indicates the strength of the relationship between a dependent variable and two or more independent variables. The partial correlation coefficient indicates the strength of the relationship between a dependent variable and one or more independent variables with the effects of other independent variables held constant. The adjusted (or shrunken) correlation coefficient indicates the strength of a relationship between variables after correcting for the number of variables and the number of data points. There are also correlation coefficients for variables measured on noncontinuous scales. The Spearman rho, Kendall tau, polyserial, and polychoric correlations are used when at least one of the variables is measured on an ordinal scale. The point-biserial, biserial, rank biserial, phi, and tetrachoric coefficients are used when at least one of the variables is measured on a binary scale.

Correspondence Analysis—A statistical method for analyzing multi-way frequency tables analogous to Factor Analysis.

Covariance—A measure of how much two variables vary relative to each other.

Covariates—Continuous-scale variables used to control extraneous variance, usually in ANCOVA.

CPM—Critical Path Method is an algorithm for scheduling a set of project activities (and resources) so that the activities that control the schedule, called critical path activities, can be managed more effectively. CPM is an important tool in Six Sigma and project management.

Critical Path—The activities in a project that must be completed on time for the entire project to be completed on time. Critical path activities have zero float.

Cumulative probability plot—a plot of the proportion of data in a dataset on the vertical axis versus the values of the variable on the horizontal axis. Probability plots show a comparison of the sample data to a Normal distribution (which plots as a straight line).

Data Mining—Use of statistics, mathematics, and computer algorithms to recognize patterns in data, usually for the purpose of prediction. Data mining usually involves large amounts of data and is sometimes conducted in real-time.

Data Scrubbing—Removing nonrepresentative or inappropriate data from a dataset prior to analysis. Sometimes referred to as data cleansing.

Database, Relational —A data management tool that stores data in tables formed by rows and columns, akin to a spreadsheet, only there may be many tables that are related to each other by some key variable. Relational databases can store and sort data more efficiently than spreadsheets.

Deduction—Using general rules to make suppositions about individual cases.

Delimiters—Commas, tabs, spaces, semicolons, and other typographical items used to separate data in delimited file formats.

Demographics—Characteristics of individual samples used to ensure representativeness of a population or to stratify the sample.

Deviations—Differences between observed values and values predicted from a statistical model. Also called residuals, errors, noise, dispersion, spread, and differences.

df—Degrees of freedom, the number of independent pieces of information that must be known in order to estimate the value of a parameter.

Discriminant Analysis—Method of analysis for models having a qualitative dependent variable.

Distribution Model—A mathematical equation of a curve that is used to represent the frequencies of items in a population. The most commonly used distribution model is the Normal distribution (a.k.a. bell curve, Gaussian model). Other distribution models include: (for continuous data) t, F, Chi^2, Lognormal, Weibull, Rayleigh, triangular, rectangular, logistic, Laplace, gamma, Cauchy, exponential, and beta; and (for discrete data) binomial, multinomial, Poisson, geometric, and Bernoulli.

DOE—Design of Experiments, an area of statistics involving optimization of ANOVA designs.

Econometrics—Statistics involving data on economics.

EDA—Exploratory Data Analysis.

EFT—Earliest Start Time. The earliest date a project activity can begin if all prior activities are completed on time.

Errors—Differences between observed values and values predicted from a statistical model. Also called residuals, deviations, noise, dispersion, scatter, spread, and differences.

Estimates—All statistics that put a value on a population parameter on the basis of a sample of the population are estimates.

Extrapolation—Prediction of values outside the span of the data used to create the prediction model.

Factor Analysis—A statistical data reduction technique in which the original variables are recombined into a smaller number of factors, which can then be used as new, more efficient variables in other statistical procedures.

Float—The amount of time that a project activity can be delayed without causing a delay in the project. Also called slack.

Forecasting—Prediction of values from a time-series model.

Frankendata—Data collected by different researchers, at different times, for different purposes. Data hash.

Geostatistics—Approach to analyzing location-dependent data, involving variogramming and kriging.

GIS—Geographic Information System, used in the analysis and presentation of location dependent data.

Gridding—Interpolation.

Heteroscedasticity—Unequal variances between data groups.

Hierarchical Samples—Samples that can be combined into larger units or split into smaller units for analysis. For example, a study of educational performance could focus on individual students, classes, schools, school districts, counties, states, or nations.

Histograms—vertical bar charts in which the horizontal axis represents data ranges (called bins) and the vertical axis depicts the number of samples in each data range. A histogram for a Normally distributed dataset has the largest number of samples at the center of the dataset with progressively fewer samples toward the tails of the distribution.

Homoscedasticity—Equal variances between data groups.

Icons—Small graphics that depict relative values of two or more variables, including star charts, ray charts, spark lines, and Chernoff faces. Icons are often plotted on larger graphs to depict additional data dimensions.

ID—Identification, identity, or identify.

IDA—Initial Data Analysis.

Imputation—The replacement of a missing data point with a surrogate, such as a mean value, so that the sample can be included in an analysis.

Independence—Statistical independence refers to an assumption made by most statistical modeling techniques. The assumption is that the residuals (errors) from the model are independent (uncorrelated with) each other. This is about the same as assuming that the data values themselves are independent of each other. When data are correlated, such as with spatial and time-series data, special techniques have to be used to account for the correlation.

Individual—A word used to refer to an individual sample (e.g., observation, subject, record, case, individual, or what the entity is, such as patient, respondent, or student).

Induction—Using individual cases to reach general conclusions.

Inference—The process of using samples that are representative of a population to make general statements about the population. Statistical inference usually involves testing or interval estimation.

Information—Data with its associated metadata.

Interpolation—Prediction of values within the span of the data used to create the prediction model. Interpolation can be either inexact (sample values are assumed to have some variability associated with them) or exact (sample values are assumed to have no variance). There are a variety of algorithms for interpolation, including: nearest neighbor, inverse distance, spline, and moving average.

Intervention Analysis—A method of analyzing time series for a shock or intervention that may have changed the nature of the series.

Kolmogorov-Smirnov (K-S) test—a statistical test of Normality. The K-S test compares the maximum vertical distance between a probability plot of the data and a probability plot of the Normal distribution or other theoretical distribution being considered. The test tends to overemphasize deviations near the center of the distribution and the population distribution parameters must be known for the test to be exact. The Anderson-Darling test is a modification of the K-S test in which more emphasis is placed on deviations in the tails of the distribution. The Lillifors test is another modification of the K-S test which is used when the mean and standard deviation of the hypothesized Normal distribution are not known (i.e., they are estimated from the sample data).

Kriging—Interpolation method similar to inverse-distance interpolation weighted for the correlation structure of the data, used in geostatistics.

Kurtosis—A measure of the relative proportion of area in the tails of a distribution relative to the center of the distribution. A negative kurtosis indicates that there is a greater frequency of values in the center of the sample distribution and the data tails are short compared to a Normal distribution Such sample distributions are said to be platykurtic. A positive kurtosis indicates that there is a greater frequency of values in the tails of the sample distribution with a relatively tall, thin peak at the center. This form of sample distribution is said to be leptokurtic. Some older references calculate kurtosis so that a Normal distribution has a value of three instead of zero.

Lags—As used in time series analysis, a first-order lag is the value of a variable before the current value. A second-order lag is the value of a variable two time periods before the current value. In general, an n-order lag is the value of a variable n time periods before the current value. The duration between measurements of the time-series variable must be constant in order to compute lags.

Levels—The different possible designations for values of a variable measured on an ordinal or nominal scale.

LFT—Latest Start Time. The latest date a project activity can begin without delaying the entire project.

Linear Regression—A statistical method for fitting a trend line through a set of data such that the variance is a minimum. Simple regression involves one dependent variable and one independent variable. Multiple regression involves one dependent variable and more than one independent variable. Stepwise regression involves the sequential selection of variables in a multiple regression model such that the variables accounting for the most variance are selected first.

Linearity—Refers to the assumption that a statistical model can be expressed as the sum of constant coefficients multiplied by variable values. If the coefficients are not constant, the model is considered to be non-linear.

LOCF—Last Observation Carried Forward. A method of replacing missing values in a time series with the previous non-missing value.

Logistic Regression—A regression technique used when the range of the dependent variable is 0 to 1. Logistic regression, or logit regression, is most often used for models involving the prediction of a probability. The logistic trend line is an S-curve in which the tails of the trend flatten as they approach 0 and 1.

MANOVA—Multivariate Analysis of Variance, a statistical technique for detecting differences between averages of more than one continuous dependent variables in groups defined by one or more qualitative variables.

MAR—Missing At Random, means that there is a pattern to the occurrence of the missing values related to some other data attribute in the model.

Matrix—Mathematical array of numbers consisting of columns and rows.

Maximum—The highest data value.

MCAR—Missing Completely At Random, means that there is no pattern to the occurrence of missing values.

MDS—Multidimensional Scaling is a data-reduction, variable rearrangement technique like Factor Analysis, the difference being that MDS isn't restricted to using the correlation/covariance matrix as its measure of similarity.

Mean—An estimator of the center of a distribution of data measured on a continuous scale. A mean is usually an arithmetic average but may also be calculated from logarithms (geometric mean), reciprocals (harmonic mean), or quartiles (trimean).

Measure—An attribute of an individual sample. A variable.

Median—An estimator of the center of a distribution of data.

Messy Data—Data with outliers, censored data, errors, and missing entries that have to be scrubbed before analysis.

Metadata—Data about how a data value was generated.

Minimum—The lowest data value.

MNAR—Missing Not At Random, also called Non-Ignorable missing data, means that the pattern of the missing values in a variable is related to the non-missing values in the same variable.

Mode—The value that appears most commonly in a dataset.

Model—A representation of some phenomenon, typically used in place of the phenomenon to manipulate a process or display a result. Models can be physical, written as descriptive text, drawn, or consist of mathematical equations or computer programming.

Monte Carlo Simulation—An approach to evaluating models using datasets randomly selected from a known theoretical distribution.

Multicollinearity—High correlations between predictor variables, which can result in unstable results for the correlated predictors in a regression model.

Multivariable—More than one variable.

Multivariate—More than one dependent variable.

N or n—N is the number of individuals in a population; n is the number of individuals in a sample.

Names—Many statistical procedures are named for the person or persons who developed the procedure. If you hear a statistician mention one of these names, you can guess the topic from this list:

Normality testing—Kolmogorov–Smirnov, Lilliefors, Anderson–Darling, Shapiro–Francia, or Shapiro–Wilk.

Homogeneity-of-variances testing—Levene.

Nonparametric testing—Kruskal-Wallis, Wilcoxon, McNemar, Kendall, or Mann-Whitney.

Post hoc testing in an ANOVA—Scheffé, Tukey, Bonferroni, or Dunnett.

If you're adventurous, try going in the other direction. Throw out a name associated with the topic you're discussing. For example, if you're discussing Normality testing, just say Lillifors. You may give a condescending statistician a heart flutter. It's like the episode of *M*A*S*H* when Trapper and Hawkeye taught Radar to say, "Ah Bach," to sound more educated.

Neural Networks—A class of nonlinear statistical modeling methods that identify, segment, and model patterns in data and then combine the models into a single network of models used for prediction.

Non-linear Regression—Statistical methods for fitting nonlinear trends, such as curves, waves, and steps, through a set of data such that the variance is a minimum.

Nonparametric Statistics—Descriptive and inferential statistics and procedures that do not use theoretical population frequency distribution models.

Normality—Refers to the property of datasets that have a frequency distribution that mimics a Normal distribution. The Normal distribution model refers to a specific mathematical equation. In this book, the term "Normal" is capitalized whether it is referring to the distribution model, assumptions about the distribution model, or tests of assumptions about the distribution model. The word "normal" is also used in this book as a common noun meaning typical, a mathematical term meaning standardized, and a chemistry term referring to a unit of concentration. Research has shown that these differences in meaning cause copy editors to suffer brain spasms.

Nugget—The spatial variance of co-located samples.

Null Hypothesis—The hypothesis in a statistical test that is presumed to be true. The test is then conducted to see if the hypothesis can be rejected. Usually, the null hypothesis involves there being no difference or no change in some values.

Objectives—The objective of a statistical analysis might be characterization, identification, classification, detection, prediction, or explanation.

Observation—A word used to refer to an individual sample (e.g., subject, record, case, individual, or what the entity is, such as patient, respondent, or student).

Orientation Statistics—Statistics for cyclic data represented on a circle or sphere.

Outlier—A data point that is far greater or less than other data points in a dataset. Outliers may provide information on special conditions that affect a phenomenon or occur for no known reason.

Overfitting—The inclusion of many variables in a model, especially relative to the number of samples and the simplicity of the data pattern. Overfit models usually perform poorly on new samples from the same population.

Parameter—A number that defines some characteristic of a population or a model. For example, the mean and variance of a Normally distributed population are the population's parameters.

Path Analysis—A statistical technique for evaluating causal relationships among variables.

PCA—Principal Components Analysis. A special case of factor analysis in which the original variables are recombined into a smaller number of uncorrelated factors, which are then used in regression analysis to evaluate the relative importance of the factors to the dependent variable.

Pilot Study—A small-scale study conducted before large-scale research to assess the practicality, accuracy, and precision of the data generation process and to improve the statistical design of the research.

Population—A collection of individuals or items that have some common set of characteristics. Samples of the individuals or items must be representative of the entire population.

Power Analysis—An evaluation of the validity and reasonableness of statistical tests involving number of samples, effect size (resolution) and the probabilities if Type I and Type II errors. Confidence is equal to (1 - probability of Type I error). Power is equal to (1 - probability of Type II error).

Prediction—The estimation of new, unmeasured values of a dependent variable in a statistical model. If the prediction involves extrapolation into the future, the term *forecast* is used instead of prediction.

QA/QC—Quality Assurance/Quality Control. Quality control refers to actions taken to obtain data of a consistent level of quality. Quality assurance refers to actions taken to ensure that all necessary quality control measures exist and are used consistently.

Qualitative—Scales consisting of any combinations of numbers and letters that indicate classes or groups.

Quantitative—Scales consisting entirely of numbers representing a natural progression.

Randomization—The act of randomly assigning subjects to treatments in experimental studies. Sometimes also used to refer to the random selection of samples from a population.

Range—The maximum minus the minimum. Also refers to the distance at which spatially correlated variables become independent (uncorrelated).

Reciprocal—The reciprocal of a value, x, is 1/x.

Recoding—Changing the codes used to represent levels of a variable so they are more appropriate for an analysis.

Regionalized variable—A spatially correlated variable.

Repeatability—The ability the benchmark and process portions of a measurement system to produce consistent results. Repeatability does not consider variability attributable to the person making the measurement.

Replication—Replication is used in a variety of ways to assess or control variability. Samples or variables can be replicated to test for consistency. Most often, replication refers to the practice of repeating an entire study, particularly in many of the sciences, to verify previously determined results.

Representativeness—The property of a sample related to how similar the sample is to its parent population on the characteristics being studied.

Reproducibility—The ability of the measurement system and the people making the measurements to produce consistent results. By comparing reproducibility to repeatability, the effects of the judgments made by the people making the measurements can be assessed.

Resampling statistics—A variety of methods for using subsets of a dataset to refine or verify a model or statistic. Bootstrapping involves taking many random subsamples of a dataset to create more robust estimates of distribution statistics. Jackknifing involves recomputing statistics for a dataset by systematically leaving out one data value at a time and then using the jackknifed statistics to evaluate the variability and bias of the statistic. Cross validation involves leaving a subset of the data out of an analysis and using it to verify the model created with the remaining data.

Rescaling—Changing the scale of a variable by adding or removing some information.

Residuals—Differences between observed values and values predicted from a statistical model.

Ridge Regression—A form of linear regression in which a small amount of bias is added during the modeling calculations resulting in regression coefficients that are slightly biased but much less variable. Ridge regression is used to address the problem of multicollinearity in the independent variables.

R-square or R^2—Coefficient of Determination. Interpreted as the proportion of variance in the dependent variable that is explained by the independent variables.

Sabermetrics—Analysis of data from baseball.

Sample—A portion of a population. May refer to a single individual sample or a collection of many individual samples.

Sampling Design—A strategy for collecting samples in a manner that will best represent the population. Commonly used sampling designs include: stratified, random, systematic, surrogate, and cluster

SAS—Statistical software, formerly Statistical Analysis System.

Scale—Scales describe the relationships between numbers used to measure some attribute of an object. Scales can be qualitative or quantitative. They can be continuous or discrete. They can be classified as ratio, interval, ordinal, or nominal. They can represent counts, orientations, times and locations. These distinctions are important because they influence how the measurements can be analyzed.

Semivariogramming—Analyzing the relationship between spatial variance and the distance between samples. Same as variogramming.

Serial Correlation—The correlation between data points with the previously listed data points, termed a *lag*. Serial correlations can be calculated for any number of lags, although usually only a few are important. If the data points are collected at a constant time interval, the term *autocorrelation* is more typically used.

Shapiro-Wilk test—A statistical comparison of a sample distribution to a Normal distribution. The vale of the Shapiro-Wilk test is like a correlation coefficient between a probability plot of the data and a probability plot of the Normal distribution. It is considered by many statisticians to be the best test of Normality in common use. The Shapiro-Frankia test is a modification of the Shapiro-Wilk test that is used when there are more than fifty samples.

Shocks—Uncontrollable short-duration conditions or events that can influence a single data point, a cluster of data points, or even most of the data in a dataset.

Significance—Statistical significance means that the magnitude of a number is large relative to the variability in the measurement of the number such that the number couldn't have occurred by chance. A statistically significant result may or may not be meaningful. Conversely, a meaningful result may or not be statistically significant. Results that are significant-but-meaningless or insignificant-but-meaningful occur when the statistical test is not designed correctly, such as there being too few or too many samples.

Sill—The maximum spatial variance in a location dependent variable.

Six Sigma—Six Sigma refers to a broad range of activities and tools used to address the objectives of corporate leaders and project managers, such as customer satisfaction, worker productivity, fewer defects, reduced waste, increased profits, more market share, and so on. The tools of Six Sigma include traditional management tools, like people skills, planning, and critical-path scheduling but then also incorporate data and analysis. Six Sigma relies heavily on hypothesis testing, process control charts, correlation and regression, ANOVA, cause-and-effect diagrams, pareto charts, decision matrices, flowcharts, failure modes and effects analysis, check sheets, histograms, box-and-whisker diagrams, surveys, affinity diagrams, benchmarking, brainstorming, relations diagrams, tree diagrams, matrix diagrams, process decision program charts (PDPC), and many other analytical tools.

Skewness—The symmetry of a frequency distribution along the axis of the variable's values.

SOP—Standard operating procedure.

Sparklines—A type of data icon. Sparklines are small lines used to express changes in several variables.

SPC—Statistical process control, statistical techniques involving control charts for detecting patterns in manufacturing data.

Spectral Analysis—A mathematical technique for separating patterns in time series data into sine and cosine functions of different frequencies and amplitudes, to try to identify important cycles. Also referred to as frequency domain analysis.

Spline—Splines are mathematical algorithms for interpolation that produce smooth surfaces while minimizing residuals at data points. Sometimes referred as piecewise polynomials.

SPSS—Statistical software, formerly Statistical Package for the Social Sciences

Statistic—A number, usually carrying information about some sample, population, or phenomenon.

Stochastic—Random.

Structural Equation Modeling—A mode of multivariate analysis with a wide variety of applications including causal modeling, constrained regression modeling, and confirmatory and higher-order factor analysis.

Study Design—The perspective of a study in terms of the role of the experimenter and the function of time. In observational studies, the experimenter observes naturally occurring conditions and events. In experimental studies, the experimenter controls the conditions and events. In cross-sectional studies, data are collected at the same time (or at least over a finite duration). In longitudinal studies, data are collected over a significant period of time.

Stylostatistics—The statistical analysis of style, such as in literature and art.

Subject—A word used to refer to an individual sample (e.g., observation, record, case, individual, or what the entity is, such as patient, respondent, or student).

Survey—A measurement device consisting of questions, which is used for collecting data from human subjects about their opinions and perceptions.

Tail—The extreme ends of a distribution where the distribution approaches the x-axis.

TerraByte—Charlie Kufs.

Time-Series Analysis—Analysis of time-dependent data. There are many methods for analyzing time-series data including: repeated measures ANOVA, intervention analysis, ARIMA, seasonal decompositions, time-series regression, and smoothing algorithms.

Transformation—The application of a mathematical function to data for a variable to give the variable more desirable properties (e.g., Normality, linearity).

Treatments—The conditions or actions tested in an experimental study

Trimean—The sum of the 25th percentile plus twice the 50th percentile (i.e., the median) plus the 75th percentile, divided by four.

t–Test—The same as a z-test except that the probability of a Type I error is calculated from the t-distribution instead of the Normal distribution. The t-distribution has more of its area in the tails than a Normal distribution so it is often used to compensate for a small number of samples.

Type I Error—False positive, the probability of rejecting a null hypothesis when it is true.

Type II Error—False negative, the probability of not rejecting a null hypothesis when it is false.

Uncertainty—Variability. Inexactness. Error.

Unimodal—Refers to frequency distributions that have only one "high" point. The Normal distribution is a unimodal distribution. Frequency distributions with two peaks are called bimodal.

Units—The benchmark that describes how the measurements relate to the phenomenon being measured. This is different from scale, which is how the numbers relate to each other.

Validation—The process of ensuring that the data were generated in accordance with quality assurance specifications.

Variability—Uncertainty. Inexactness. Error.

Variable—The columns of a data matrix that contain the pieces of information you collect from or about each of your samples.

Variance—A measure of dispersion. Also, a difference from a benchmark in a data report.

Variogram—Graph of the spatial variance on the y-axis versus the distance between samples on the x-axis, used in geostatistics.

Variogramming—Analyzing the relationship between spatial variance and the distance between samples. Same as semivariogramming.

Verification—The process of ensuring that each value in the dataset is identical to the value that was originally generated.

Weighting—The process of applying a multiplier to the data value for a sample to give the value either more or less influence in a subsequent calculation.

Winsorizing—The process of adjusting for extreme values by replacing the same number of extreme values on each side of the distribution with the next "non-extreme" value.

Violations of Assumptions—There are four basic assumptions in statistical modeling—linearity, independence of errors, Normality of errors, and homogeneity of variances.

» Violations of the linearity assumption are usually substantial, but they are relatively easy to detect and correct.

» Violations of the independence assumption are much more difficult to detect and correct. The calculated probability that a population and a fixed value (or two populations) are different will be underestimated if the correlation of the errors is negative, or overestimated if the correlation of the errors is positive. The magnitude of the effect is related to the degree of the correlation.

» Violations of the Normality assumption may involve symmetry (measured by skewness) or form (measured by kurtosis). If the distribution of errors in the model is asymmetrical (i.e., skewed), there will be little effect on two-sided tests. However, for one-sided tests of the truncated (short-tailed) side of the distribution, both the level of significance and the power can be seriously affected. If the distribution is leptokurtic (i.e., has more values in the tails of the distribution than a Normal distribution), the calculated level of significance will be slightly more than the true level of significance, and the calculated power will be less than the true power, especially for small sample sizes. If the distribution is platykurtic (i.e., has more values in the center of the distribution than a Normal distribution), the calculated level of significance will be slightly less than the true level of significance and the calculated power will be greater than the true power, especially for small sample sizes.

» Violations of the homogeneity-of-variances assumption will be small if the ratio of the variances is near 1 and the sample sizes are about the same for all values of the independent variables. However, as differences in both the variances and the numbers of samples become large, the effects can also be great

z-Test—A statistical comparison between a difference in a variable (usually either between two means, or between a mean and a data value, or between a mean and a constant) relative to the variability of the variable. The null hypothesis is usually that the difference is zero. Large differences (relative to the variability) are considered evidence that the null hypotheses should be rejected. The probability that a difference of a similar magnitude could have occurred by chance, a Type I error, is calculated from the Normal distribution.

Index

A

B

C

D

E

F

G

H

I

J

K

L

M

N

O

P

Q

R

S

T

U

V

W

Z

CPSIA information can be obtained at www.ICGtesting.com
Printed in the USA
BVOW01s0137151214

379123BV00005B/109/P

9 781604 944723